Economic Analyses at EPA

Economic Analyses at EPA

Assessing Regulatory Impact

edited by
Richard D. Morgenstern

Resources for the Future
Washington, DC

Printed in the United States of America

Published by Resources for the Future
1616 P Street, NW, Washington, DC 20036–1400

Library of Congress Cataloging-in-Publication Data

Economic analyses at EPA : assessing regulatory impact / edited by Richard D. Morgenstern.
p. cm.
Includes bibliographical references.
ISBN 0–915707–83–7 (alk. paper)

1. Environmental policy—Economic aspects—United States—Case studies. 2. Environmental protection—Economic aspects—United States—Case studies. 3. United States. Environmental Protection Agency. I. Morgenstern, Richard D.
GE180.E36 1997
363.7'056'0973—dc21 97–14606
CIP

The paper in this book meets the guidelines for permanence and durability of the Committee on Production Guidelines for Book Longevity of the Council on Library Resources.

This book was typeset in Palatino by Betsy Kulamer and its cover was designed by Joanne Zamore of Zamore Design.

RESOURCES FOR THE FUTURE (RFF) is an independent nonprofit organization engaged in research and public education on natural resources and environmental issues. Its mission is to create and disseminate knowledge that helps people make better decisions about the conservation and use of their natural resources and the environment. RFF takes responsibility for the selection of subjects for study and for the appointment of fellows, as well as for their freedom of inquiry. RFF neither lobbies nor takes positions on current policy issues.

Because the work of RFF focuses on how people make use of scarce resources, its primary research discipline is economics. Supplementary research disciplines include ecology, engineering, operations research, and geography, as well as many other social sciences. Staff members pursue a wide variety of interests, including the environmental effects of transportation, environmental protection and economic development, Superfund, forest economics, recycling, environmental equity, the costs and benefits of pollution control, energy, law and economics, and quantitative risk assessment.

Acting on the conviction that good research and policy analysis must be put into service to be truly useful, RFF communicates its findings to government and industry officials, public interest advocacy groups, nonprofit organizations, academic researchers, and the press. It produces a range of publications and sponsors conferences, seminars, workshops, and briefings. Staff members write articles for journals, magazines, and newspapers, provide expert testimony, and serve on public and private advisory committees. The views they express are in all cases their own and do not represent positions held by RFF, its officers, or trustees.

Established in 1952, RFF derives its operating budget in approximately equal amounts from three sources: investment income from a reserve fund; government grants; and contributions from corporations, foundations, and individuals. (Corporate support cannot be earmarked for specific research projects.) Some 45 percent of RFF's total funding is unrestricted, which provides crucial support for its foundational research and outreach and educational operations. RFF is a publicly funded organization under Section 501(c)(3) of the Internal Revenue Code, and all contributions to its work are tax deductible.

Contents

Foreword

Requirements that federal regulatory agencies such as the U.S. Environmental Protection Agency subject their proposed rules to benefit-cost analysis extend back to the 1970s. Interest at Resources for the Future in the application of benefit-cost analysis to environmental regulatory problems goes back at least a decade more. For it was in the mid-1960s that Allen Kneese and his colleagues at RFF began extending benefit-cost principles that had theretofore been used for the evaluation of federal water resource projects—dams, irrigation projects, and stream channelization, for instance—to the analysis of programs that would improve water *quality*. Soon their work expandedto include air quality regulations, drinking water programs, and other environmental issues that are familiar parts of the regulatory landscape today.

For this reason, we are very pleased to be publishing this book edited by Richard D. Morgenstern. It represents by far the most careful review to date of the role that benefit-cost analysis and other analytical techniques have played in EPA's decisionmaking over the last two decades. Moreover, it is hard to imagine a better person than Morgenstern to have identified and commissioned the preparation of the case studies, convened a conference at which they were reviewed, and distilled important insights into the conduct, uses, and abuses of economic analysis. Not only is he an accomplished economist who has conducted original research on the types of issues discussed below, but, for more than a decade, he also has been a senior policy official working within EPA to improve the quality of the analysis done in support of major rulemakings. He has, in other words, seen it from both sides.

Though it should never be the only factor on which to base a regulatory decision, economic analysis in its many forms should always be one of the factors. It is our hope that this book will be both informative about the past role of analysis and suggestive as to ways that it can play a larger, more constructive role in future decisions.

Paul R. Portney
President
Resources for the Future

Preface

The idea for this volume arose several years ago when students in my course at American University wanted to know how economic analysis was actually used in environmental decisionmaking. More recently, the interest of policymakers in the role of economics in regulatory decisions has been rekindled. My hope is that this volume informs both audiences. Clearly economics is neither inherently evil nor the sole method on which decisions should be made. Rather, it is an important tool for the regulatory process. This volume argues that when economic analysis is performed, it should be done well, and transparently. Stakeholders should be involved from the outset.

Many people made this book possible. My biggest debt of gratitude, of course, is to the individual chapter authors. They tirelessly plowed through their files and personal recollections to develop comprehensive case studies of regulatory analyses they were previously involved in, either as analysts at the U.S. Environmental Protection Agency or, in some instances, as consultants. An equally important debt is owed to the several dozen peer reviewers who provided detailed comments on the cases and actively participated in a conference involving authors and reviewers held at Resources for the Future in June 1996. They are listed in the appendix. Elizabeth Farber, who initially interned at RFF while completing her graduate work in policy studies at the Johns Hopkins University, played an invaluable role in providing the case study authors and me with insightful critiques of drafts and assisting in revising chapters and sharpening presentations throughout. I assume full responsibility for the selection of the case studies and for any errors of interpretation.

Resources for the Future provided an ideal environment in which to develop a project of this sort. Knowledgeable and supportive colleagues have helped in a variety of ways. Several served as peer reviewers of individual chapters and provided useful comments. As colleague and director of RFF's Center for Risk Management, Terry Davies has been a constant source of support and insight in ways large and small. His encyclopedic knowledge of regulatory and policy issues has helped expand the breadth of this volume considerably. As a colleague and president of RFF, Paul Portney has provided both intellectual insights and practical support on this project. His own scholarship in the field has been influential to many of us. Sue Lewis tirelessly and graciously provided support in managing the conference of authors and reviewers and in organizing key aspects of the project. Eric Wurzbacher cheerfully served as editor of the volume and helped integrate the many disparate parts. Outside of RFF, two individuals deserve special mention. Vic Kimm, most recently at the University of Southern California, has been enormously helpful to several case study authors and to me in thinking through problems and incorporating the practical side of regulatory decisionmaking. Marc Landy of Boston College, my coauthor of Chapter 16, provided a range of useful ideas, particularly in the formative stages of the project.

I am grateful to many friends and colleagues at the Environmental Protection Agency—my intellectual home for more than a decade. The agency, which has one of the hardest jobs in Washington and one of the most dedicated staffs, provided both the opportunity and the support to carry out this project. I also want to thank the Carnegie Corporation of New York for financial support to RFF specifically for this effort.

My wife, Devra Lee Davis, has played an important role throughout the development of this work. Partly in connection with a course we are jointly teaching, she has helped clarify my thinking about the appropriate role of economics in environmental decisionmaking. She has challenged me to distinguish what we know from what we believe. I am truly blessed to have her as a partner and colleague. Our children, Aaron and Lea, have shown exceptional tolerance and support for my preoccupation with this project and have prodded me to make it relevant to their generation.

Finally, I dedicate this volume to the memory of my recently departed father, Nathan Ralph Morgenstern, LL.B. A Russian Jewish immigrant to this country, he taught me the value of hard work and, particularly, the importance of piercing the veil of complex issues to get to the heart of the matter.

Richard D. Morgenstern
Resources for the Future

Economic Analyses at EPA

1

Introduction to Economic Analyses at EPA

Richard D. Morgenstern

Throughout history, the messenger of bad news is often attacked by those who do not like the message. Recently, economic analysis has come under attack from several quarters as an inherently flawed environmental management tool. On the one hand, environmentalists criticize those economic studies that reveal high costs associated with proposed actions. On the other hand, industry challenges economic assessments that indicate high benefits. Thus, in this conflict-ridden world, economic analysis of environmental policies inevitably comes under attack as the vehicle for conveying information considered bad news by somebody. In point of fact, economic analysis is inherently neither good nor evil. Rather, it can serve as an analytical tool for estimating real world trade-offs and for identifying cost-effective approaches to environmental management.

In order to determine what constitutes well-conducted economic analyses, this book reviews the execution and use of economic analyses in support of twelve major regulations issued over the past decade by the U.S. Environmental Protection Agency (EPA) under six different statutes. They include both highly sophisticated analyses as well as simpler ones. The authors are all economic analysts who were involved in the individual rulemakings, either as EPA employees or as consultants.

The premise of this book is that stigmatizing economic analysis *in general* for the flaws of *specific* economic analyses is misguided. Here, economic *analysis* is viewed as a potentially powerful set of methods that can be systematically applied to the study of environmental issues in order to achieve better policies. Economic *analyses* are understood as specific applications of economic methodologies that may or may not be done well.

This volume is an attempt to learn as much as possible about the real-world practice of economic analysis as applied to environmental issues.

- How and to what extent have economic analyses contributed to better environmental rules?
- Have they helped reduce the costs of environmental regulations? Have they helped develop more stringent rules?
- Have they delayed regulatory actions?
- Do economic analyses generally consider a broad array of options?
- Do barriers to quality analyses exist? If so, what are they and how can they be overcome?
- Can analytical or institutional procedures be changed to facilitate more informed environmental decisions and, if so, in what ways?
- All said, what can we learn from our experience with economic analysis to improve environmental decisionmaking in the future?

The estimated benefits of just three of the rules examined in this volume exceed the estimated costs of all twelve rules combined. Yet, unless one simply seeks to praise or damn the regulatory system, the conclusion that the benefits do or do not exceed the costs for any group of rules is not terribly helpful. The real goal of this book is to elucidate the processes that will improve our abililty to develop sound environmental policies. As former EPA administrator William Reilly has noted, "we can probably afford to spend three percent of the nation's wealth on environmental protection, but we can't afford to spend it in the wrong places (Stevens 1991)."

Following this introductory chapter, Chapter 2 provides background on the legal, institutional, and cultural context applicable to the conduct and use of economic analysis at EPA. Chapter 3 presents the basic rationale for economic analysis as applied to environmental issues, considers some of the controversies surrounding that rationale, and reviews key elements of the analyses. Here, a theme of this book—that economic analysis can play an important role in environmental decisionmaking—is developed in some detail. Chapters 4 through 15 constitute the core of the volume with cases covering air and water pollution, drinking water, municipal wastes, toxics, and pesticides. Most focus on human health, a few on ecological concerns, and one solely on aesthetic considerations. Finally, Chapter 16 assesses the contribution that economic analysis has made to the environmental rules under study and, within the confines of the existing laws, makes recommendations to improve the process.

Given the diverse array of issues considered in this volume, it is difficult to draw any general conclusions. Yet, one of the clear findings is that, despite their limitations, the group of economic analyses studied here made significant contributions to the improvement of the agency's regulations. According to the case study authors, the economic analyses helped reduce the costs of all twelve of the rules studied and, at the same time they helped increase the benefits of five of the rules. The value of these rule improvements likely dwarfs the costs of conducting the economic analyses.

Another finding of this volume is that, even though the economic analyses clearly helped improve the quality of the rules, in many instances the economic analyses played only a minor role in actual decisionmaking. While many factors, including politics, could explain this situation, three possible explanations for the limited role of economic analyses are examined:

- The underlying scientific and risk information was so uncertain that it provided an insufficient basis on which to conduct an economic analysis.
- The economic analysis itself was technically flawed in one or more critical ways.
- The economic analysis was not designed to address a sufficiently rich array of policy options and was thus rendered irrelevant to the actual policy and regulatory decisions.

In fact, all three explanations have currency, although the first and, particularly, the third explanation seem to be the most important. In some instances the economic analysis served little purpose as the underlying risk assessment on which it was based was so weak. In other instances, the policy options explored were so narrow and the range of outcomes so limited that the resulting economic analyses were of limited relevance to the policy process.

Consistent with recent environmental legislation, such as the Safe Drinking Water Act Amendments of 1996, and in an attempt to bring economic analyses into the mainstream of EPA rulemakings, it is proposed that a more open, stakeholder process be developed to conduct economic analyses. Toward that end, economic studies should be begun earlier and, throughout the process, studies should involve public input regarding both their design and conduct. The 1978 revisions to the regulations implementing the National Environmental Policy Act, which created a public process (referred to as a "scoping" process), offers one model for opening economic analyses to greater public involvement.

Fundamentally, there is no escaping economic analysis. If it is not done explicitly, with a careful consideration of alternative options, it will occur implicitly. In that case, decisionmaking will be driven by public fears, special interest lobbying, and bureaucratic preferences. The resulting policies are not likely to reflect the best interests of our citizenry. A key conclusion of this volume is that within the existing legal framework a series of administrative and procedural changes could help move economic analysis of environmental rulemakings more into the bright light of public view and the mainstream of agency decisionmaking.

REFERENCES

Stevens, William K. 1991. What Really Threatens the Environment? *New York Times* January 29.

2

The Legal and Institutional Setting for Economic Analysis at EPA

Richard D. Morgenstern

Americans have complicated and sometimes paradoxical attitudes toward the environment. More than ninety percent identify themselves as pro-environment and, in response to surveys, routinely express strong support for clean air, clean water, and generally stringent environmental goals. Yet surveys also find that individuals often resist specific policies, particularly when these policies entail direct financial or inconvenience costs. For instance, stringent automobile inspection and maintenance programs are opposed by almost as many people as favor clean air, even though the former help ensure the latter.

Not surprisingly, the laws and institutions that shape environmental policies in the United States—cutting across all branches of government—reflect some of these same paradoxes. Congress enacts the laws and, through the oversight process, influences their implementation. The courts interpret ambiguous legislative language and, sometimes, force agencies to meet specific statutory or court-ordered deadlines. Implementation responsibilities rest exclusively with the executive branch although, increasingly, these responsibilities are being delegated to the states. The U.S. Environmental Protection Agency (EPA) is primarily responsible for administering the major environmental statutes, with some duties also carried out by other departments of government. (Some environmental laws are not implemented by EPA. For instance, the Endangered Species Act is administered by the Fish and Wildlife Service, and various federal wetlands policies are administered by a number of agencies, including the Department of Agriculture and the Army Corps of Engineers).

This chapter focuses on the uneven and sometimes inconsistent manner in which economic considerations enter into environmental decisionmaking. The focus is on the laws, the executive orders issued by Presidents over the past quarter century, and the culture of EPA, the principal implementing institution. A short section of this chapter reviews the recent substantive and procedural provisions enacted in 1995 and 1996.

THE LAWS

Unlike most other agencies (such as the National Highway Traffic Safety Administration), EPA does not administer a single, organic statute. Instead, the EPA administrator implements nine major laws and more than a dozen minor statutes. Among the major statutes, six of them form the basis for the rules examined in this volume. Three of the laws—the Clean Air Act (CAA), Clean Water Act (CWA) and Safe Drinking Water Act (SDWA)—are based on the environmental medium in which pollution occurs; the Resource Conservation and Recovery Act (RCRA) focuses primarily on a single medium (land) but deals with other matters as well; the Federal Insecticide, Fungicide and Rodenticide Act (FIFRA) deals with a particular set of products; and the Toxic Substances Control Act (TSCA) deals with chemicals in general. While all these laws share a common theme of "protection of human health and the environment," they differ in many respects, including the way economic considerations enter into the design and implementation of environmental policies.

Many of the environmental statutes set goals that, if interpreted literally, would virtually eliminate pollution or any harms therefrom. The Clean Air Act, for instance, directs the EPA administrator to set primary ambient air quality standards that, "...allowing an adequate margin of safety...protect the public health." [Clean Air Act, Section 109 (b) (1)]. Economic considerations are not mentioned in the section dealing with setting these standards. The Clean Water Act states that "... it is the national goal that the discharge of pollutants into navigable waters... and...the discharge of toxic pollutants in toxic amounts be eliminated" [Clean Water Act, Section 101 (a)]. Yet, stated so broadly, these goals beg the question: how much protection is required? Since environmental protection almost always requires some effort, how much effort is enough? How much is too much? How much degradation can be allowed until the environment is no longer considered "protected"? Interestingly, it has been argued that this statutory commitment to perfection is an impediment to the functioning of the agency.[1]

Most statutes leave it up to EPA to determine what specific requirements or limitations should be placed on the conduct of regulated entities. Typically, statutory language is general in nature, granting broad discretion to agency decisionmakers. These decisionmakers, in turn, are empowered to write rules and standards, to issue permits, and to develop and oversee requirements to help achieve environmental goals. Current methods of addressing environmental problems include:

- Ambient media standards, used as benchmarks for subsequent, more narrowly defined requirements (for instance, ambient air or water quality standards)
- Emission/effluent standards, which focus on pollution at the point of release into the environment and can themselves be defined in a variety of ways including:
    - limits on total amounts released
    - limits on concentrations discharged to effluent streams
    - percentage reduction from uncontrolled levels
    - rate of emission per unit of output (such as CWA effluent limits)
- Controls on the sale and use of products that have environmental effects when used or disposed of (for instance, mobile source regulations under Clean Air Act, controls on materials affecting stratospheric ozone, and pesticide regulation under FIFRA)
- Controls on contaminants in products directly consumed (such as the SDWA)
- Controls on operations of activities that manage or use hazardous materials (RCRA requirements for management, disposal, transportation of hazardous waste)
- Targets for remediation of past releases (such as RCRA corrective action)
- Controls to prevent degradation of targeted areas or resources (such as prevention of significant deterioration requirements under the Clean Air Act or nondegradation requirements under the Clean Water Act)
- Requirements for public reporting of information (such as the SWDA and EPCRA)
- Protection of workers from exposure during employment (such as FIFRA worker protection standards)
- Decision processes, procedures and certification training requirements for private abatement of hazards (such as the TSCA lead abatement program)

Much as the implementing mechanisms vary across statutes, so do the requirements for considering costs when a regulation is being developed. Sometimes the agency is granted broad discretionary authority to consider the economic impacts of its decisions. At times it is specifically required to consider economic factors. At other times it is explicitly prohibited from considering costs. Several court decisions have held that benefit-cost studies cannot be considered by agencies unless expressly authorized by statute.

Table 1 reviews the oportunities to use economic analysis under the six major statutes considered in this volume. TSCA and FIFRA contain explicit mandates for economic analysis. Under the statutes, EPA is required to balance costs and benefits in the screening and regulation of chemicals and pesticides. Apart from TSCA and FIFRA, several statutory provisions require that "costs must be reasonable." However, a 1989 court decision does not interpret this to require a formal benefit-cost analysis or even a precise calculation of costs [F.2d 177, 226 (5th Cir. 1989), modified 885 F.2d 253 (5th Cir. 1989)]. Several other provisions mandate that costs be balanced with benefits but fall short of the formal procedures used in a benefit-cost analysis. The courts generally interpret such statutes to require the agency to consider benefits in light of costs and to avoid setting limits with costs wholly disproportionate to the benefits [*Association of Pacific Fisheries v. EPA*, 615 F.2d 794, 805 (9th Cir. 1980)]. In a 1989 Clean Water Act case, the court held that the cost-balancing test precluded EPA from giving costs primary importance [*Chemical Mfrs. Association v. EPA*, 870F.2d 177, 201 (5th Cir. 1989), *modified* 885 F.2d 253 (5th Cir. 1989), *American Iron and Steel Institute v. EPA*, 526 F.2d 1027, 1051 (3rd Cir. 1977)]. A number of provisions require the agency to consider costs, among other factors, when developing regulations, typically without specifying what weight cost should be given (Downing 1995, 56).

Unlike most major statutes governing environmental regulation, the Regulatory Flexibility Act (RFA) specifically requires agencies to determine whether a regulation has significant economic impact, at least on small businesses and other small entities. When such a finding is made, agencies must identify alternative regulatory approaches for such entities that still meet the statutory objectives, albeit in a less burdensome manner. The RFA has been hailed by the Small Business Administration (SBA) as "small business' most significant mechanism for influencing Federal regulations (SBA 1996, 1)." Yet, implementation of this statute is generally recognized as spotty. Perhaps this can be traced to the fact that, at least prior to 1996, the RFA was not subject to judicial review.

Finally, it is worth noting the Administrative Procedures Act (APA), which establishes the ground rules for public involvement in the regulatory development process. Key APA provisions require that specific procedures are to be followed in proposing and promulgating regulations and that potentially impacted parties have an opportunity to comment on proposed actions. Like EPA's major implementing statutes (and unlike the RFA), failure to adhere to the APA is judicially reviewable. Not surprisingly, procedural issues are given great weight in EPA decisionmaking.

Overall, there is no simple formula by which cost considerations enter into environmental statutes. Key provisions of the various statutes differ markedly, although cost is rarely the pivotal factor determining

Table 1. Analysis Allowable under the Environmental Statutes.

	Benefit-related factors			*Cost-related factors*			
	Pollution reduction	*Health*	*Welfare*	*Technical feasibility*	*Affordability*	*Cost-effectiveness*	*Benefit/ cost*
Clean Air Act (CAA)							
NAAQS /primary		X					
NAAQS /secondary			X				?
Hazardous air pollution		a	a	b	b	b	b
Automobile engines	c	c	c	c	c	c	c
Fuel standards	c	c	c			c	c
New source standards	X			X	X	X	X
Clean Water Act (CWA)							
Effluent guidelines, industrial sources	X	X	X	X	X	X	?
Safe Drinking Water Act (SDWA)							
Maximum contaminant levels		X	X	X	X	X	X
Toxic Substances Control Act (TSCA)		X	X	X	X	X	X
Resource Conservation and Recovery Act (RCRA)		X	X	X	?	?	?
Federal Insecticide, Fungicide and Rodenticide Act (FIFRA)		X	X		X	X	X

?: Uncertain if allowable under this statute.
a: Only marginally relevant in the initial MACT (maximum available control technology) phase; principally relevant for residual risk phase.
b: Affordability, etc. are relevant only within a narrow framework for MACT determinations.
c : Statute contains many specific directives limiting considerations of costs, health, and welfare.

Sources and notes: This table builds off one created by Hahn (1994, 331). The only modifications made to Hahn's table incorporate the 1996 Amendments to the Safe Drinking Water Act. (These recent changes do not apply to the cases in this volume.) Hahn's table in turn was derived from Fraas (1991) and Blake (1991).

program design or the stringency of emission limits. Unless precluded by statute, administrative discretion is accorded a significant weight in establishing most environmental standards. Former EPA Deputy Administrator Alvin Alm was on target when he likened the legislation implemented by EPA to an archeological dig. "Each layer," he wrote, "represents a set of political and technical judgments that do not bear any relationship to other layers (U.S. EPA 1990)."

EXECUTIVE ORDERS: EXPLICIT CALLS FOR ECONOMICS

In the late 1960s and early 1970s a variety of regulatory policies were established in the areas of health, safety, and the environment. These were soon followed by presidential attempts to exert greater executive-branch influence over the regulatory process. Central to this executive branch interest, economic analysis was often seen as a way of influencing the substantive content of federal rules. During the 1970s, under the labels "Quality of Life Reviews" and "Inflation Alerts," Presidents Nixon, Ford and Carter all issued executive orders calling for limited economic analyses of major rules. Further, they sought interagency review of major rules and the accompanying economic analyses as a means of assuring that broad public policy concerns, as opposed to the more compartmentalized issues of the individual agencies, were considered. In effect, the Office of Management and Budget (OMB), representing the executive office of the president, became, in the words of economist Charles Schultze, "the lobby for economic efficiency (Schultze 1977)."

Within a month of taking office, President Ronald Reagan issued Executive Order 12291, which substantially strengthened the requirements for both economic analysis and executive branch review. Described by some scholars as possibly the foremost development in adminstrative law of the 1980s, E.O. 12291 (46 *Federal Register* 13193; 3 CFR, February 17, 1981) required regulatory agencies to prepare *regulatory impact analyses* (RIAs) on all major regulations and to submit them to OMB for review before taking regulatory action (Pildes and Sunstein 1995, 3). In effect, by requiring agencies to consider the gains from regulation on an equivalent footing with the costs, it attempted to change the "yardstick" used to evaluate environmental regulations for which such considerations are not specifically precluded by law. Most importantly, E.O. 12291 required that "the potential benefits outweigh the costs," and that "of all the alternative approaches to the given regulatory objective, the proposed action will maximize net benefits to society."

Mandating such an economic yardstick, of course, was not entirely consistent with the spirit of most environmental statutes. As noted, these statutes have, for the most part, focused on noneconomic criteria such as protection of human health "with an adequate margin of safety" or on specific environmental goals like "fishable and swimmable" waters rather than economic efficiency. Further, by requiring that agencies obtain prior approval from OMB before placing either proposals or final rules in the *Federal Register*, E.O. 12291 gave OMB a clear mechanism by which to enforce the provisions of the Order.

President Bill Clinton issued E.O. 12866 in September 1993. It superceded the Reagan executive order and replaced the stipulation that benefits "outweigh" costs with "...a reasoned determination that the benefits of the intended regulation justify its costs [E.O. 12866 1(a), 3 CFR at 638–39 (1995)]". By it, agencies are to "...include both quantifiable measures (to the fullest extent that these can be usefully estimated) and qualitative measures of costs and benefits that are difficult to quantify," and to "select those approaches that maximize net benefits (including potential economic, environmental, public health and safety, and other advantages; distributive impacts; and equity) unless a statute requires another regulatory approach." This formulation endorses benefit-cost analysis as a tool for helping choose among alternative regulatory (and nonregulatory) options while not requiring that benefits quantitatively "exceed" costs. Under President Clinton's executive order, agencies are required to make the implications of various policies explicit but are not forced to adhere to any rigid decisionmaking formula.

Legal scholars have praised this executive order as "a dramatic step:"

> First,...it maintains the basic process...that major regulations be submitted to OMB for general review and oversight.... [It] also maintains the... emphasis on cost-benefit analysis as the basic foundation of decision. President Clinton thus rejected the view that an assessment of costs and benefits is an unhelpful or unduly sectarian conception of the basis of regulation.... Executive Order 12866 includes a set of innovations.... [It] addresses unnecessary conflicts between agencies and OMB, and the appearance (or perhaps the reality) of factional influence (Pildes and Sunstein 1995, 6–7).

Perhaps the most significant change of E.O. 12866 is that it effectively distinguishes between two aspects of benefit-cost analysis: an "accounting framework" for tracking and exploring social decisions versus an "optimizing tool" for attaining maximum social welfare (Lave 1996, 129–30). By dropping the decision rule developed in E.O. 12291 that benefits must "outweigh" costs in favor of the more flexible term, E.O. 12866 clearly endorses the notion of an accounting framework for exploring social decisions. The Clinton rule places much greater emphasis on non-quantifiable benefits including "...public health, and safety, and other advantages; distributive impacts; and equity." By embracing social welfare considerations that may not be easily quantified, such as public health and distribution impacts, E.O. 12866 effectively rejects the idea of using solely quantified benefit-cost analysis as an optimizing tool in favor of a more general approach, relying less on quantitative analysis.

As the Reagan administration did in 1982, Clinton's OMB issued guidelines in early 1996 that lay out the key steps agencies should take in performing economic analyses pursuant to E.O. 12866. The specifics of these guidelines are discussed in the next chapter.

EPA'S CULTURE AND INCENTIVES FOR POLICY ANALYSES

Unlike OMB, which operates on what might be thought of as traditional utilitarian principles, EPA is a mission-oriented agency with the overall goal of protecting human health and the environment through administration of a complex set of statutes. Even apart from any legal restrictions, EPA's broad scope often works against a consistent use of economics in agency decisionmaking. The agency culture is probably best described as a legal culture, buttressed, in large part, by scientific considerations and, to a far lesser extent, by economic factors. The culture is also significantly shaped by the large number of both congressional and court-ordered deadlines. Most outsiders do not realize the burdens these deadlines place on the agency's ability to function and, particularly, on its ability to use state of the art science and economics in rulemaking. Former Administrator William Ruckelshaus has suggested that EPA suffers from "battered agency syndrome...not sufficiently empowered by Congress to set and pursue meaningful priorities, deluged in paper and lawsuits, and pulled on a dozen different vectors by an ill-assorted and antiquated set of statutes (Ruckelshaus 1995, 3)."

The founding of EPA in 1970 reflected a sea change in thinking about the environment. Broadly stated, it valued esthetics and biology above economic efficiency and commerce. An EPA pamphlet reflects the spirit of that time:

> The subtle metaphor of a "web of life," in which all creatures depended upon one another for their mutual perpetuation, gained currency. Hence, the powerful reaction to Rachel Carson's 1962 classic *Silent Spring* (U.S. EPA 1992, 6).

The notion of "Spaceship Earth" that R. Buckminster Fuller and other "wholistic" thinkers and social critics devised was powerfully reinforced in the public's mind by the first photographs of the "whole Earth." Taken by the Apollo astronauts and widely displayed at the first Earth Day celebration in 1970, these images provided a haunting reminder of the planet's fragility.

In setting up EPA, Congress established an agency that differed from "old line" regulatory agencies like the Federal Deposit Insurance Corporation, the Securities and Exchange Commission, and others in at least three significant ways (Eads and Fix 1984, 12–15; Portney 1990, 7–25). First, many older agencies were set up to control a perceived failure of the market, typically related to monopoly power, fraud in advertising, or unsound financial practices. In contrast, EPA was founded to deal with what economists refer to as externalities and/or the problems associated with highly imperfect information regarding the nature and consequences of environmental discharges. Second, EPA differs from the older agencies in the breadth of its mandate. Whereas most of the older agencies focus on a single industry, EPA is expected to cover a vast territory including virtually every economic sector. Third, at the older regulatory agencies recent legislation and administrative actions have generally curtailed intervention in the markets they regulate—witness the demise of the Civil Aeronautics Board in 1985. In contrast, Congress has repeatedly expanded the scope of responsibility and authority of EPA, at least up until the early 1990s. All of these differences are significant because they impose special problems for effectively managing the agency.

In creating this new agency with its broad mandate, personnel were recruited from the Departments of Interior, Agriculture, and Health, Education and Welfare. As Thomas McGarity describes it, "...President Nixon assembled a loose amalgam of bureaucrats with widely varying institutional backgrounds from several existing regulatory programs and called it the EPA (McGarity 1991, 57)." The new agency also attracted a number of activists from the early environmental movement. The importance of these activists in shaping the agency culture is widely disputed. Some scholars believe they were a dominant influence, others do not.[2] McGarity, for example, argues that "While some of the new hires at EPA joined the agency out of strong ideological desire to protect the environment, most of the original employees that came from existing departments were anything but environmental zealots (McGarity 1991, 57)."[3]

While it is difficult to convey the essence of the early agency staff, former EPA Administrator Ruckelshaus has noted that "I've never worked anywhere where I could find (the great interest, excitement, challenge or fulfillment) to quite the extent as at EPA.... At EPA, you work for a cause that is beyond self-interest and larger than the goals people normally pursue (U.S. EPA 1993, 36)." In a similar vein, former EPA Administrator William Reilly has referred to "the quality of the people who work at EPA, their zeal, their commitment, the fact that for them it's not just a job, they really believe in what they're doing (U.S. EPA 1995, 79)."

The Educational Backgrounds of EPA Employees

The EPA workforce is one of the most highly educated, technically sophisticated, and decidedly interdisciplinary in government. More than two-thirds of the staff hold a college degree; more than half of those college graduates also hold one or more advanced degrees. Table 2 shows a distribution of EPA employees with graduate degrees, according to twenty-seven separate academic disciplines, as recorded in the EPA personnel records as of November 1996. The most popular disciplines, in descending order, are law (18%), biological or life sciences (16%), engineering (16%), physical science (14%), business management and administrative services (7%), public administration including public policy (7%), social science other than economics (5%), health sciences (3%), conservation and natural resources (3%), and economics (2%).

Despite their relative minority status, the agency does employ a considerable number of economists. In fact, there are probably more economists working on environmental issues employed at the EPA than at any other single institution in the world. Perhaps the most relevant question, however, is what influence economists and economic reasoning have on the EPA culture and, specifically, on regulatory decisionmaking.

It is certainly true that being trained in economics does not necessarily make you a strong advocate of economic reasoning. Conversely, you don't have to be an economist to think like one. Nonetheless, as a proxy for their influence, it is useful to consider the number of economists working where individual regulations are written, or in senior management positions within the agency. Table 3 indicates that while more than half of the economists are in the program offices, they constitute only about 3–5% of staff with graduate training in those offices. In contrast, about one-third of the agency's economists are in the Office of Policy, Planning and Evaluation (OPPE) that, historically, has assisted in rule development, but has never had responsibility for issuing rules. In OPPE, about one in seven individuals with a graduate degree specializes in economics. In terms of senior management positions, EPA had 255 members of the Senior Executive Service (SES) as of November 1996. Of the 196 SES members with a graduate degree, almost one-third held a law degree,

Table 2. Profile of EPA Employees with Graduate Degrees.

Discipline	*Doctorate*	*Master's or J.D.*	*Total number*	*Total percent**
Law	50	1,201	1,251	18.1
Biological science/life science	500	624	1124	16.2
Engineering	104	992	1,096	15.8
Physical science	323	613	936	13.5
Business management and administrative services	6	463	469	6.8
Public administration	9	442	451	6.5
Social sciences and history/not including economics	52	294	346	5.0
Health professional and related sciences	31	157	188	2.7
Conservation/renewable natural resources	23	154	177	2.6
Economics	31	85	116	1.7
Agriculture	34	76	110	1.6
Architecture	2	101	103	1.5
Education	5	91	96	1.4
Multi/interdisciplinary studies	11	76	87	1.3
Psychology	25	28	53	0.8
Computer/information science	4	45	49	0.7
English language and literature	5	40	45	0.7
Communications	2	34	36	0.5
Library science	0	30	30	0.4
Philosophy and religion	21	9	30	0.4
Liberal arts and sciences, general studies and humanities	0	23	23	0.3
Foreign language/literature	2	11	13	0.2
Visual and performing arts	0	10	10	0.1
Home economics	5	4	9	0.1
Ethnic/cultural studies	1	7	8	0.1
Theological studies/religious vocations	2	6	8	0.1
Other	25	86	111	1.6
Total**	1,248	5,674	6,922	100*

* As percent of total employees at EPA with graduate degrees

** Totals may not add due to double-counting of employees with more than one graduate degree.

Source: EPA Personnel Office, November 1996.

twenty percent held a science degree, and four individuals held their graduate degree in economics.

Not surprisingly, different disciplines rely on distinct approaches to problem definitions and solutions. These differences have been cogently characterized:

> Lawyers, having read hundreds of cases in law school, learn there are two sides to every argument. Since defendant and

Table 3. EPA Employees with Graduate Degrees in Economics.

EPA office	*Doctorate*	*Master's*	*Total graduate degrees*	*SES** graduate degrees*
Program offices				
OAR	4 / 68	24 / 457	28 / 525	0 / 13
OPPTS	11 / 239	10 / 370	21 / 609	1 / 19
OSWER	1 / 30	11 / 262	12 / 292	0 / 7
OW	1 / 43	10 / 226	11 / 269	0 / 12
Multimedia offices				
OPPE	13 / 41	21 / 152	34 / 193	3 / 6
OARM	0 / 21	2 / 259	2 / 280	0 / 14
OE	0 / 21	1 / 334	1 / 355	0 / 20
ORD	1 / 531	2 / 441	3 / 972	0 / 29
Regional offices	0 / 238	1 / 2665	1 / 3024	0 / 44
All offices***	31 / 1,250	82 / 5,491	113 / 6,931	4 / 193

*Notes:*The figures in the four rightmost columns are numbers of degrees in economics in comparison to the total number of degrees for each column's category

*Doctorates of law included **SES: EPA's Senior Executive Service ***Includes offices without any economists.

OAR—Office for Air and Radiation; OPPTS—Office of Prevention, Pesticides, and Toxics Substances; OSWER—Office of Solid Waste and Emergency Response; OW—Office of Water; OPPE—Office of Policy, Planning and Evaluation; OARM—Office of Administration and Resources Management; OE—Office of Enforcement (now OECA, Office of Enforcement and Compliance Assurance); ORD—Office of Research and Development.

Source: EPA Personnel Office, November 1996.

plaintiff alike present cogent theories and precedents, lawyers learn that disagreements cannot be resolved by appealing to shared ideas. Hence, fair procedures (such as bargaining) may be the only way to resolve conflicts.

Engineers, on the other hand, are trained to solve problems, not resolve them. The formulae and rules on which they base their calculations are often arbitrary....They come to believe that there are right answers to problems and that those can be arrived at by manipulating data according to a unique "best practice."

Unlike either lawyers or engineers, economists are trained to view all variables as continuous. Regardless of whether price, production, or consumption is at issue, choices are not "yes" or "no" but matters of amount or degree. Thus, economists instinctively see all issues as arenas for trade-offs, and outcomes that produce "a little of this and a little of that" are often judged desirable. (Landy, Roberts, and Thomas 1994, 10)

Simply put, the aversion to "yes" and "no" choices, and the mentality of seeking trade-offs rather than bright lines rests at the core of much economic analysis. Yet, this outlook world view contrasts with that of the professions that dominate the EPA culture. And the fact remains that despite their absolute numbers, as a proportion of the EPA workforce, there are relatively few economists employed at EPA, particularly in program offices and in the ranks of senior managment. One study on the attitudes and beliefs of EPA employees concludes that "educational background, rather than office or job responsibilities of EPA officials, appears to play the dominant role in shaping perspectives about the utility of risk assessment and benefit-cost analysis, the valuation of life, and the distribution of risk (Rycroft, Regan, and Dietz 1989, 419)."

Recent Agency History: Economics Under Seige?

By the late 1980's *Time* magazine proclaimed that "we're all environmentalists now." Concurrent with this "mainstreaming of the environment," and perhaps, in part, because of it, naive notions that economic analysis itself is invariably bad for the environment have found ready adherents. In fact, a series of interviews conducted for this book reveals an unmistakable view within the agency that, within recent years, economics has had to fight harder for its place at the table.

No single factor explains the de-emphasis of economics, but two competing explanations are generally offered. First of all, noneconomists tend to fault the discipline of economics itself, claiming that it is no longer relevant (if it ever was) to environmental issues. In place of neo-classical economics they cite the emerging field of "ecological economics" or the "no-cost" views of environmental protection often associated with Professor Michael Porter of the Harvard Business School. (For further discussion of this perspective, see Porter and van der Linde 1995. For a counter view, see Palmer, Oates, and Portney 1995.)

The idea that government can develop policies that systematically enable us to have a cleaner environment without sacrificing anything else seems improbable to most economists. Economists approach their discipline from the perspective that in a world of limited resources, it is important to try to figure out how to make the most of such resources. Since environmental problems generally derive from "market failures," some form of government intervention is often warranted. Yet, economists emphasize the need for credible estimates of potential damage to human health or the environment as a basis for action.

Secondly, and in contrast to the first view, economists generally see the decline in their influence as part of a larger trend in public affairs to focus on "good news" and to avoid tough choices. They point to the downgrading of economists and of policy offices all around government—not just at EPA. As Phil Lee, a former assistant secretary of health in both the Johnson and Clinton administrations, put it, "nowadays, decisionmakers consult their PR people and interest groups more often than the experts."[4]

Managerial Considerations of Economic Analysis

Albert Nichols observes, in Chapter 4, that economic analysis at EPA "is likely to have more influence when it identifies previously unrecognized opportunities for cost-effective health/environmental gains, than when it concludes that a proposed action fails to yield benefits commensurate with costs." Yet, even when they do make a case for strong regulation, economic analyses can create headaches for agency management by pointing out some contrary information or by making transparent certain costs that, even if clearly outweighed by benefits, are still seen as burdens to some group in the society. Such "on the one hand and on the other hand" sorts of statements provide ammunition for the policy's opponents to use in court or in the media, particularly when taken out of context.

Such barriers are not entirely insurmountable. Before the Congressional Budget Office (CBO) came into existence, few Congress watchers would have predicted that an analytic office within the Congress would acquire the prestige and influence that now attaches to it. It succeeded because it developed a reputation for analytic excellence and fairness and it actively worked to build constituencies appreciative of its product. Unlike EPA, of course, CBO does not carry the burdens of a regulatory agency. Nonetheless, the CBO example suggests that it is possible to develop at least some constituencies for economic efficiency.

Finally, as is well known, EPA managers are very often under significant political pressures from both Congress and the Courts. The famed "Prune Book," which lists the toughest jobs in government, quotes a former EPA administrator who referred to the pressures of his job: "...(it's) like beating a train across a grade crossing—if you make it, it's a great rush. If you don't, you're dead (Trattner 1988, 250)."

A recent report of the National Academy of Public Administration (NAPA), addressing some of the pressures and problems faced at EPA, placed much of the blame on Congress:

> The EPA lacks focus, in part, because Congress has passed more than a dozen environmental statutes that drive the agency in a dozen directions, discouraging rational priority-setting or a coherent approach to environmental management. EPA is sometimes ineffective because, in part, Congress has set impossible

deadlines and unrealistic expectations, given the agency's budget. The EPA can be inefficient, in part, because Congress has attempted to micro-manage the agency through prescriptive legislation, earmarked appropriations, and direct pressure (NAPA 1995, 8).

Again, Ruckelshaus' comment is on point: "The people who run EPA are not so much executives as prisoners of the stringent legislative mandates and court decisions that have been laid down... for the past quarter century (Ruckelshaus 1995, 4)."

Notwithstanding the view that some of the disinterest in economic efficiency within EPA may stem from congressional actions that "...discourag[e] rational priority setting or a coherent approach to environmental management (NAPA 1995, 8)," the responsibility for the disaffection regarding economic analysis is, undoubtedly, much broader. Certainly it is fair to say that economists themselves bear some of the blame, as do those who have undermined the overall legitimacy of economic considerations and the ability to carry out competent and timely analyses. The next section of this chapter addresses some changing congressional winds that may presage new directions for the agency.

NEW LEGISLATIVE DEVELOPMENTS

The 1970s, when most environmental statutes were first enacted, is considered the "environmental decade." In contrast, the 1990s have been a time of reevaluation, when the complexity of environmental management has proved especially challenging. All the individual regulations examined in this volume were issued under statutory authorities that were in place prior to 1995. Both the 103d and 104th Congresses have considered risk assessment and benefit-cost isssues in some depth. Although none of the more sweeping proposals were adopted,[5] enactement of both the Safe Drinking Water Act (SDWA) Amendments of 1996 and the Small Business Regulatory Enforcement Fairness Act of 1996 (SBREFA) can be read as clear signs of congressional concerns about the absence of economic considerations in environmental management.

In what is undoubtedly the most far-reaching requirement for economic analysis in any environmental law, the SWDA Amendments mandate the use of benefit-cost analysis for all major drinking water rules. They specifically require the use marginal (as opposed to average) analysis, consideration of risk-risk trade-offs, and the analysis of a rule's impacts on the broader population, not just its effect on so-called maximally exposed individuals. The statute also calls for explicit consideration of nonquantifiable effects, uncertainties, and the degree and nature of the risk being controlled.[6]

It is no coincidence that the drinking water statute is the first substantive law to include such explicit use of economic analysis. Concerns about drinking water are widely shared in our society. The drinking water industry consists primarily of urban and rural water systems, many of which are municipally operated. In contrast to other pollution issues, there are few large industrial polluters of drinking water, and treatment costs are generally passed along directly to customers.

SBREFA is a procedural statute that modifies the requirements for economic impact analysis under the Regulatory Flexibility Act (RFA). SBREFA responds to claims from the small business community that some agencies have failed to comply with RFA. It mandates more rigorous regulatory flexibility analyses and, in cases involving either EPA or the Occupational Safety and Health Adminstration, establishes a small entity stakeholder process including the Small Business Administration. SBREFA also empowers Congress to review and potentially disapprove all individual regulations whether or not they have significant small business impacts.

These provisions present a small but significant change in thinking about the environment. In addition to mandating additional analyses and congressional review, both the SDWA Amendments and SBREFA specifically allow judicial review of any agency's conduct and use of economic analysis carried out pursuant to those statutes. Whether the courts or Congress will prove capable of implementing these mandates in a consistent and balanced manner remains to be seen. In theory, these new provisions are intended to introduce greater rigor into the conduct and use of economic analysis. In practice, neither Congress nor the courts are likely to muster the necessary expertise or resources required to undertake such reviews. Thus, it is questionable whether these solutions will prove fully successful. With these changes, however, Congress is clearly expressing concerns about the need for greater consideration of economic factors in environmental management.

CONCLUSION

The U.S. public has struggled with contradictory desires to protect the environment and avoid the inevitable burdens associated with doing so. As a consequence, the laws and institutions in the United States have evolved somewhat inconsistently. Various statutes forbid, inhibit, tolerate, allow, invite, or require the use of economic analysis in environmental decisionmaking. Most environmental laws place limits on the consideration of economic factors but do not preclude them entirely. For almost three decades, various executive orders had required that economic analyses be conducted. Later chapters of this book show that these Presidential directives have so far failed to produce consistently high quality and relevant economic assessments. The EPA culture has, at best, tolerated the goal of economic efficiency. Recent congressional actions such as the amended Safe Drinking Water Act have expanded the role of economic considerations. Whether this will ultimately provide a more reasoned approach to agency decisionmaking remains to be determined.

ENDNOTES

[1]For example, William D. Ruckelshaus, former administrator of EPA has stated that "the nation was committed to a sort of pie in the sky at some future date.... Each time a new generation of clean technologies came into use, the response from EPA had to be: 'That's great—now do some more," whether that 'more' made any sense as an environmental priority or not." (From a speech at the Environmental Law Institute, October 18, 1995, p. 7.)

[2]For an example of those who believe it was dominant, see Harris and Miklis (1989, 231).

[3]See also Marcus (1980, 35–43). In a similar vein, Mark Landy, Marc Roberts, and Stephen Thomas argue: "The agency...was initially staffed primarily by bureaucrats transferred from other federal departments. Mostly scientists and engineers, they had long labored in the bowels of large departments....(T)hey brought with them the concepts, attitudes, and skills that had served their former agencies....By both training and conviction, they were not prepared to shift from being aid and advice givers to aggressive violation hunters (Landy, Roberts, and Thomas 1994, 34)."

[4]Interview, October 23, 1996.

[5]Both the Johnston Amendment, introduced by former Senator J. Bennett Johnston (D-La.) in 1993 and then again in 1994, as well as provisions of the 104th Congress' legislative "Contract with America," introduced by the newly empowered Republican majority in early 1995, would have made benefit-cost analysis a more central element—some might argue *the* central element—in decisionmaking on major environmental rules. By some interpretations, the analysis itself would have been subject to judicial review, thereby establishing the courts as the arbiters of quality. Other provisions of the "Contract" would have increased procedural hurdles, thereby raising significantly the cost and complexity of rulemaking.

[6]Specifically, the statute calls for EPA to consider the following: the quantifiable and unquantifiable benefits of the health risk reduction associated with the contaminants being controlled; the quantifiable and unquantifiable benefits of the health risk reductions of the co-occuring contaminants likely to also be controlled; the quantifiable and unquantifiable costs of compliance; the incremental costs and

benefits of each level being considered; the effects of the contaminant on the general population and any vulnerable subpopulations; any increased health risks that may be introduced as a result of compliance and other factors, like the quality of the data, the uncertainties in the analyses, and the degree and nature of the risk being controlled. (Safe Drinking Water Act Amendments of 1996, section 103)

REFERENCES

Blake, F. 1991. The Politics of the Environment: Does Washington Know Best? *American Enterprise* 6(1).

Downing, Donna Marie. 1995. *Cost Benefit Analysis and the 104th Congress: Regulatory Reform, "Reg-icide" or Business as Usual?* Master's thesis. Washington, D.C.: George Washington University, National Law Center.

Eads, George C. and Michael Fix. 1984. *Relief or Reform? Reagan's Regulatory Dilemma.* Washington, D.C.: Urban Institute Press.

Fraas, A. 1991. The Role of Economic Analysis in Shaping Environmental Policy. *Law and Contemporary Problems* 113(54).

Hahn, Robert W. 1994. United States Environmental Policy: Past, Present and Future. *Natural Resources Journal* 34(1).

Harris, Richard A. and Signey M. Miklis. 1989. *The Politics of Regulatory Change: A Tale of Two Agencies.* New York: Oxford University Press.

Landy, Mark K., Marc J. Roberts, and Stephen R. Thomas. 1994. *The Environmental Protection Agency: Asking the Wrong Questions from Nixon to Clinton.* New York: Oxford University Press.

Lave, Lester. 1996. Benefit-Cost Analysis: Do the Benefits Exceed the Costs? In Robert W. Hahn (ed.). *Risks, Costs, and Lives Saved.* Washington, D.C.: American Enterprise Institute.

Marcus, Alfred A. 1980. *Promise and Performance: Choosing and Implementing an Environmental Policy.* Greenwood Press.

McGarity, Thomas O. 1991. The Internal Structure of EPA Rulemaking. *Law and Contemporary Problems* 54(4).

NAPA (National Academy of Public Administration). 1995. *Setting Priorities, Getting Results: A New Direction for the Environmental Protection Agency.* Washington, D.C.: NAPA.

Palmer, Karen, Wallace E. Oates, and Paul R. Portney. 1995. Tightening Environmental Standards: The Benefit-Cost or the No-Cost Paradigm. *Journal of Economic Perspectives* 9(4):119–32.

Pildes, Richard H. and Cass R. Sunstein. 1995. Reinventing the Regulatory State. *The University of Chicago Law Review* 62(1).

Porter, Michael E. and Claas van der Linde. 1995. Towards a New Conception of the Environment-Competitiveness Relationship. *Journal of Economic Perspectives* 9(4): 97–118.

Portney, Paul R. 1990. EPA and the Evolution of Federal Regulation. In Paul R. Portney (ed.) *Public Policies for Environmental Protection*. Washington, D.C.: Resources for the Future.

Ruckleshaus, William D. 1995. Speech at the Environmental Law Institute, October 18.

Rycroft, Robert W., James L. Regan, and Thomas Dietz. 1989. Incorporating Risk Assessment and Benefit-Cost Analysis into Environmental Management. *Risk Analysis* 8(3).

Schultze, Charles L. 1977. *The Public Use of Private Interest*. Washington, D.C.: The Brookings Institution.

SBA (Small Business Administration) 1996. Highlights of Small Business Regulatory Enforcement Fairness Act of 1996. Fact sheet, March 29. Washington, D.C.: SBA. Office of Chief Counsel for Advocacy.

Trattner, John H. 1988. *The Prune Book: The 100 Toughest Management and Policy-Making Jobs in Washington*. Lanham, Maryland: Madison Books.

U.S. EPA (Environmental Protection Agency). 1990. *EPA Journal* 13 (September/October). Washington, D.C.: U.S. EPA.

———. 1992. *The Guardian: Origins of the EPA*. EPA Historical Publication 1. Spring. Washington, D.C.: U.S. EPA.

———. 1993. *William D. Ruckelshaus Oral History Interview*. Interview 1. January. (202-K-92-0003). Washington, D.C.: U.S. EPA.

———. 1995. *William K. Reilly Oral History Interview*. Interview 4. September. Washington, D.C.: U.S. EPA.

3

Conducting an Economic Analysis

Rationale, Issues, and Requirements

Richard D. Morgenstern

To most policy professionals, the usefulness of economic analysis in decisionmaking is clear: it can identify trade-offs among alternatives, reveal less costly ways to achieve specific goals, and build support for policies that serve the public interest. Economics attempts to systematically integrate information from multiple disciplines to assess broad implications of policy choices. Economists see their discipline as a tool for helping make the most out of society's limited resources.

Despite the growing support for economic analysis in many public policy areas, such as health care and education, there is considerable hostility to its use in environmental decisionmaking. Economists have been lambasted as knowing "the price of everything but the value of nothing." Some consider economics to be too narrow a discipline to deal with the moral and ethical issues associated with environmental policy. Others charge the discipline with making excessive claims about its ability to measure the broad array of health and environmental benefits, the so-called "disease of false precision."

Notwithstanding such criticism, the U.S. EPA has been conducting various types of economic analyses of regulations since its founding in 1970. In the early years, the focus was exclusively on cost-effectiveness analysis, as a means of identifying the least expensive way to achieve specific environmental objectives. During the Carter administration benefit-cost analysis began to be used on a limited basis, in both EPA and the traditional economic agencies (CEA, OMB), to help clarify the suitability of alternative environmental objectives. Since the issuance of Executive Orders (E.O.) 12291 and 12866, EPA has conducted more than one hundred regulatory impact analyses (RIAs), also known as *economic analyses* (EAs). Formally, E.O. 12291 called for the conduct of *regulatory impact analyses*. E.O. 12866, which mandates almost identical studies, labels them *economic analyses*. Reflecting current practice, these two terms are used interchangeably throughout this volume. With a few exceptions, RIAs or EAs are mandated on major rules, defined as those with expected annual costs or benefits in excess of $100 million. Cost-effectiveness analysis is common to all RIAs; this involves comparing the costs of different approaches for achieving the same goal. Many of them use benefit-cost techniques as well; this involves estimating benefits and costs of any proposed regulatory action.

The purpose of this chapter is twofold. First, it seeks to lay out the broad rationale for economic analyses of environmental regulations—principally, benefit-cost analysis—as well as some of the pertinent critiques. Why should public agencies invest the resources to conduct these analyses? What can they expect to gain from such activities? And what are the reasons to be wary of the results? The second objective of this chapter is to acquaint the reader with the specific requirements for conducting economic analysis of a proposed regulation in a federal agency. Under E.O. 12291 and, more recently, under E.O. 12866, OMB has issued specific guidance regarding the conduct of these analyses. Issues such as baseline development, consideration of uncertainties, discounting, and other matters are addressed in these Guidance documents and are discussed in the second part of this chapter.

THE CASE FOR CONDUCTING ECONOMIC ANALYSES OF ENVIRONMENTAL REGULATIONS

Economic analysis of rules can help improve the allocation of society's resources while at the same time engendering an understanding of who benefits and who pays for any given regulatory action. Additionally, properly conducted economic analysis encourages transparency and accountability in the decisionmaking process, and provides a framework for consistent data collection and identification of gaps in knowledge. In theory, EAs allow for the aggregation of many dissimilar effects (such as those on health, visibility, and crops) into one measure of net benefits expressed in a single currency.

The literature on environmental economics has expanded rapidly over the past twenty years. There is now broad agreement in the economics profession on the theoretical bases for estimating the benefits and costs of alternative courses of action. (See, for instance, Oates and Cropper 1992.) This is not to suggest that all the methodological issues and data problems related to benefit and cost estimation have been resolved. Certainly the premise of the various executive orders is that a great deal can be gained by applying the concepts of modern welfare economics, including the notion of willingness-to-pay, to regulatory issues. Not knowing how to do everything is no excuse for failing to apply the considerable analytic and practical knowledge in this area that has been developed in recent decades.

A distinguished group of economists, including Nobel laureate Kenneth Arrow, has identified three major functions of economic analysis in the regulation of health, safety, and the environment: arraying information about the benefits and costs of proposed regulations; revealing potentially cost-effective alternatives; and showing how benefits and costs are distributed, for example, geographically, temporally, and among income and racial groups. However, even those who advocate the use of benefit-cost analysis also recognize some limits:

> ...[In] many cases, benefit-cost analysis cannot be used to prove that the economic benefits of a decision will exceed or fall short of the costs....[But] benefit-cost...[analysis] can provide illuminating evidence for a decision, even if precision cannot be achieved because of limitations on time, resources, or the availability of information (Arrow and others 1996, 5).

In contrast, some economists take the case further and argue, for example, that benefit-cost analysis can be used to identify a policy option that is truly "best" for society in the sense that it maximizes overall social well-being. Most economists who see these exceptional powers of their discipline tend to be theorists, but some policy officials have also taken that view. For example, James C. Miller, Director of the Office of Management and Budget during the Reagan administration, has written: "[E.O. 12291] ... is no more than common sense....Without some check, like that provided by President Reagan's executive order, regulatory decision makers might be tempted to pander to such moralism (that disdains rational analysis)—or to other special interests (Miller 1989, 91)."

The mainstream view generally does not agree that benefit-cost analysis is capable of identifying the truly "best" option. Arrow and others (1996, 7), for example, argue that "Agencies should not be bound by a strict benefit-cost test, but should be required to consider available benefit cost analyses." They note that there "may be...[other] factors... that agencies want to weigh in their decisions, such as equity within and across generations... [or they] may want to place greater weight on particular characteristics of a decision, such as irreversible consequences." They also recommend, however, that "... for regulations whose expected costs far exceed expected benefits, agency heads should be required to present a clear explanation justifying the reasons for their decisions."

ISSUES IN THE USE OF ECONOMIC ANALYSIS IN ENVIRONMENTAL DECISIONMAKING

Economic analysis is seen by many as an aid to improving the allocation of society's resources while encouraging transparency and accountability in the decisionmaking process. Yet, it is clear that this view is not universally held. Even within the economics community some theorists argue that the fundamental basis of modern welfare economics is flawed, largely because it is infeasible, and certainly uncommon, to compensate the "losers" from any given policy. (A recent nontechnical review of that literature can be found in Lave 1996.) According to these theorists economic analysis is thus limited to the search for Pareto-efficient gains; that is, the (very limited) set of alternatives that improve overall well-being but make no individual worse off. Noneconomists typically articulate three areas of concern: moral and ethical issues, subjectivity, and equity.

Moral and Ethical Issues

Neoclassical economics is based on the preferences of individuals, as measured in terms of their willingness to pay for gains and willingness to be compensated for losses. Where willingness to pay or to be compensated cannot be directly measured by market transactions, as is the case with many environmental problems, economists often resort to verbal indicators of preference (for instance, contingent valuation).

Critics of economic analysis argue that the paradigm of individual preferences does not provide a broad-enough basis on which to make environmental decisions. Environmental issues, it is suggested, involve *public* as well as private preferences. Public preferences are said to involve opinions and beliefs about what *ought to be* and are thus more than the summation of individual preferences. Critics of economics suggest that public preferences are the basis of many social norms including those governing environmental issues. It has also been argued that nature has intrinsic or inherent value independent of humans. Based on survey results, one group of researchers concluded that Americans' environmental values derive from two sources apart from the traditional anthropocentric or utilitarian bases: religion, whether traditional Judeo-Christian religious teaching or a more abstract feeling of spiritual concern for the Earth; and biocentric (living-thing-centered) values, which grant nature itself intrinsic rights, particularly the rights of species to continue to exist (Kempton, Booster, and Hartley 1995).

Some religious and ethically-oriented groups completely reject any application of economic analysis by arguing that the environment should not be treated like a commodity. (A clear and readable introduction to the ethical and philosophical issues in benefit-cost analysis can be found in Sagoff 1993 and Kopp 1993.) These groups believe that people have a moral responsibility to show deference and respect for the environment, independent of the utility of doing so. In fact, from Native Americans to Henry David Thoreau and John Muir, this is an old and familiar theme in American culture. More recently, Pope John Paul II declared that "...the ecological crisis is a moral issue (John Paul II 1990)." Philosophers like Mark Sagoff argue that environmental policymaking is irreducibly about values, quintessentially "political," and therefore not amenable to welfare-maximizing "economic" calculations (Sagoff 1982).

It is certainly true that the right to a safe and healthy environment and the rights of nonhuman species can clash with the preferences of some individuals. Yet rights are typically nonnegotiable. When issues are framed in terms of conflicting rights, it becomes virtually impossible to conduct meaningful policy deliberations. Even the most ardent moralists would recognize that, as a practical matter, decisions about which pollutants to control, and how much control to impose, are not readily resolvable on moral grounds. What to do? Philosopher Thomas M. Scanlon suggests that:

> Satisfaction of people's "manifest" preferences is not an adequate index of well-being because there are conceivable circumstances in which these preferences might be satisfied even though the individuals' true interests were far from being served.... I ... [advocate] ... an index of well-being not in terms of preference satisfaction but in terms of the availability of goods and conditions deemed important for a good life (Scanlon 1991).

In rejecting the framework of individual preference satisfaction, Scanlon implies that individuals cannot be relied on to seek their own collective interests and that specific indicators must be used to help define these social interests. Economic analysis offers a variety of indicators for determining what may best serve the public interest with respect to the environment. At a minimum, economists would argue for articulating the efficiency trade-offs when one moves away from the individual preference framework.

The Subjectivity Issue

It is sometimes claimed that because of the primitive level of our knowledge about the basic physical effects associated with pollution, not to mention some of the issues associated with the valuation of benefits and costs, the economic analysis of environmental regulation is too readily manipulated. Donald Hornstein, for example, argues that "the technical process of making risk-based decisions is more accessible to those special interests that can consistently afford to deploy technically competent scientists, economists, attorneys, and public relations firms to represent their interests...(Hornstein 1994, 151)."

Critics are certainly correct to observe that where you stand depends on where you sit when it comes to estimating. In situations of high uncertainty about how to quantify benefits and costs, the results of economic analyses may vary across interest groups. Often, industry estimates of costs are greater and estimates of benefits are less than environmentalists' calculations.

The critics are wrong, however, to presume that since some analyses are biased, all analyses will be biased. As Herman Leonard and Richard Zeckhauser point out:

> Any technique employed in the political process may be distorted to suit parochial ends and particular interest groups. Cost-benefit analysis can be an advocacy weapon, and it can be used by knaves. But it does not create knaves, and to some extent it may even police their behavior. The critical question is whether it is more or less subject to manipulation than alternative decision processes. Our claim is that its ultimate grounding in analytic disciplines affords some protection (Leonard and Zeckhauser 1986, 31).

In a similar vein, Lester Lave argues, "Partial information can mislead, and analysts can, even inadvertently, become advocates. It is not enough to provide a caveat that such analysis may be under (over) estimating costs...or benefits. We must investigate the extent of possible bias (Lave 1996, 130)." As discussed in Chapter 16, increased public scrutiny throughout the conduct of economic analysis, via both peer and public review, can serve as an added protection against biased analyses.

The Equity Issue

A third area of concern with benefit-cost analysis involves the potentially negative distributional or equity implications of efficiency-oriented policies. As Richard Zeckhauser has argued:

> The basic dynamic of political life may be at variance with the efficiency orientation which underlies the discipline of benefit-cost analysis. The political process recoils from the "let the chips fall where they may" nature of traditional efficiency maximization. In particular, it attempts conscientiously to redistribute

> resources so that cost impositions are reduced, even if this can only be achieved at the expense of substantially greater reduction in the sought-after benefits (e.g., environmental benefits, in the case of environmental policy) (Zeckhauser 1981, 218).

The merits of this argument are widely recognized in policy circles. By the early 1990s, for example, the environmental justice movement had raised concerns that both the burdens of pollution as well as the costs of reducing pollution may be falling disproportionately on the poor and minority communities. While the literature in this field is still scant, one of the best researched issues involves lead poisoning of children, the avoidance of which is a major benefit category for two of the case studies in this volume. Lead poisoning disproportionately affects low-income children and, even after adjusting for income, is still greater among blacks than among whites. (See, for instance, U.S. EPA 1992.)

No one questions that such inequities have to be redressed. Economic analysis can provide valuable tools for determining effective ways of assessing the magnitude of the problem and for developing solutions. Despite its focus on traditional efficiency concerns, mainstream economics encourages consideration of these effects. The Clinton administration executive order, E.O. 12866, has emphasized the importance of equity issues in the consideration of environmental issues, including the conduct of benefit-cost analysis.

The import of all three criticisms is somewhat muted in the consideration of marginal policy changes. And they lose virtually all their punch when economics is defined as an information or accounting framework rather than as a means of optimizing social well-being. As Peter Railton notes, "once benefit-cost analysis is understood as a process meant to yield information rather than to make decisions...one need not take sides on controversies on the nature of justice (Railton 1990, 62)."

GUIDELINES FOR CONDUCTING ECONOMIC ANALYSES OF MAJOR REGULATIONS

In January 1996, OMB issued guidelines reflecting the Clinton administration's thinking on the best practices to be followed in conducting an economic analysis of proposed regulations. While these guidelines reflect the philosophical approach of E.O. 12866, they are strikingly similar to those issued by the Reagan administration in 1982 for conducting economic analyses pursuant to E.O. 12291.

This section presents some of the principal issues involved in conducting an economic analysis of proposed federal regulations. The emphasis is on benefit-cost analysis as opposed to cost-effectiveness analysis. However, cost-effectiveness analysis can be interpreted as a special case of benefit-cost analysis where the regulatory goals are fixed and the analysis attempts to highlight the least-cost means of achieving them. Accordingly, most of the discussion in this chapter about desirable practices in conducting benefit-cost analysis generally apply as well to cost-effectiveness analysis. Most of the references are to E.O. 12866, although the major substantive differences between the guidelines issued pursuant to E.O. 12291 and E.O. 12866 are noted.

Conducting an economic analysis of a proposed environmental regulation involves addressing a number of key questions:

- Is the regulation addressing a significant problem that requires fixing?
- Is federal regulation—as opposed to a nonregulatory or perhaps a non-federal solution—the preferred approach?
- If federal regulation is appropriate, do the benefits of the proposed action justify the cost? At a minimum, is the preferred regulatory option the least-cost way of achieving a specific goal?

Often, answering these questions involves complex analytic issues. In fact, one of the first environmental management decisions requires determining how much detailed information and analysis is necessary for making any given regulatory or policy decision. Sometimes a simple economic analysis, the "back-of-the-envelope"approach, is sufficient. In other cases, an in-depth effort is required. Since benefit-cost analysis is costly to perform; it ought not be applied to every problem or applied lightly.

Statement of Need for Proposed Action

The first step in conducting an economic analysis of an environmental rule is to determine if there really is a problem that requires fixing. To the economist that means a regulation should address a market failure and improve upon the operation of the economy. One of the key reasons for market failure is that prices fail to capture so-called external costs (or benefits) imposed on others, such as pollution. Formally, an externality exists when an activity by one individual harms or causes a loss of well-being to another individual and the loss of well-being is uncompensated.

The economic definition of pollution is dependent on some physical effect associated with discharges into the environment and a human perception or response to that physical effect. While the physical effect is typically biological or chemical, the human reaction—the loss of well-being—shows up as an expression of human distress, concern, or unpleasantness. In the extreme case, it shows up as death. Both a physical insult (human health or environmental) and a willingness to pay to correct the situation are essential for pollution to be considered a market failure. Although a rarity, polluters may compensate victims for damages. In that case the effect is said to be internalized and it is no longer considered a market failure. Thus, the physical presence of pollution does not necessarily mean that "economic" pollution exists. To be considered "economic" pollution, and thus a market failure, the physical discharges must cause some reduction of human well-being and be uncompensated.

Another type of market failure addressed in the guidelines that is relevant to environmental issues concerns the problems of inadequate and/or asymmetric information. It is well known that inadequate information in the marketplace or large disparities in the amount or quality of information among different market participants can lead to inefficient resource allocation. The 1996 Guidance contains expanded treatment of this issue, reflecting recent research in the field. The Guidance cautions that while these information problems may form the basis of government regulation or other intervention, the agency should determine what information is actually available in the market before assuming the information is adequate. For example, the Guidance cites third-party provision of information, insurance markets, and warranties as important sources of information that may obviate the need for a regulatory or other government intervention.

A further point with regard to the statement of need for the proposed action concerns the importance of the particular market failure. Since the potential for uncompensated losses in human welfare caused by others is virtually unlimited, the Guidance cautions that the market failure should be "significant." There are also questions of whether the market failure can, in fact, be ameliorated by a regulatory solution and whether the "unintentional harmful effects" of the regulation outweigh the original benefits. Recent research, for example, in the areas of pesticide regulation, as well as lead recycling, suggests that the unintentional harmful effects of some regulations may be more important that previously thought. (For an interesting set of essays on the issue of unintended consequences of regulatory action, see Graham and Weiner 1995.)

Examination of Alternative Approaches

After clarifying the need for the proposed action, the regulatory agency should consider a broad range of alternative approaches, while at the same time avoiding the trap of "paralysis by analysis."

E.O. 12866 calls for a level of analysis broadly commensurate with the potential value of information. Requirements for full assessments apply only to "significant" regulatory actions, or those expected to have impacts on the economy in excess of $100 million per annum. Within that limit, moreover, it is still possible to scale the depth of analysis according to the specific situation.

The Guidance instructs the agency to consider at least eight issues:

- *Performance-oriented standards*—These are generally preferred to engineering or design standards because they provide flexibility to achieve the regulatory objective in a more cost-effective way.
- *Differential standards*—These are not generally preferred because they have the potential to load on the most productive sectors costs that are disproportionate to the damages they create.
- *Stringency*—Since marginal costs generally increase with stringency, whereas marginal benefits decrease, it is important to consider alternative levels of stringency.
- *Compliance dates*—Since the costs of a regulation may vary substantially with different compliance dates, a regulation that provides sufficient lead time is likely to achieve its goals at a much lower overall cost than one that is effective immediately although, as the Guidance notes, the benefits could also be lower.
- *Compliance assurance*—When alternative monitoring and reporting methods vary in their costs and benefits, alternatives should be considered. For example, in some circumstances random monitoring will be more cost-effective than continuous monitoring.
- *Informational remedies*—A regulatory measure to improve the availability of information, particularly about the concealed characteristics of products, is often preferred to mandatory standards or bans since it gives consumers a greater choice.
- *Market-oriented appproaches*—These include fees, subsidies, penalties, marketable permits or offsets, changes in liabilities, property rights or required bonds, and insurance or warranties, all of which are generally more cost-effective than so-called command and control approaches, and should be explored.
- *Specific statutory requirements*—When a statute establishes a specific regulatory requirement and the agency has discretion to adopt a more stringent standard, the agency is advised to examine the benefits and costs of the specific requirement as well as the more stringent alternative and present information that justifies the more stringent alternative if that is what the agency proposes (OMB 1996).

Analysis of Benefits and Costs

Calculations of benefits and costs, whenever possible, should use measures of consumer and producer surplus as measures of economic well-being. An extensive theoretical and practical literature is available on how to carry this out, but the basic idea is that benefits and costs should reflect the valuation that individuals place on the effects in question. This section lays out some of the key analytic and methodological issues likely to be encountered in the analysis of alternative regulatory options.

The Baseline. The baseline refers to the health, environmental and economic conditions that occur in the absence of a contemplated policy intervention. Selecting an appropriate baseline may seem straightforward: the baseline should present the best assessment of the "way the world would look absent the regulation." Yet, determining this is often a very difficult task. What else is going on in the market other than responses to the proposed rule? What level of compliance with current regulations should be assumed? It may be especially difficult to decide what should be in the baseline when there are multiple regulatory actions under consideration, or when new rules are being promulgated before existing rules are fully implemented.

An example may clarify this issue. Consider the landfill rule presented in Chapter 9. One of the key issues in this case was determining what controls the states would likely issue on their own. When development of the federal regulation began, only a few states had developed their own rules. By the time the rule was actually promulgated, six years later, many states had adopted rigorous requirements. Failure to recognize these new state requirements in the economic analysis led to an overstatement of both costs and benefits of the federal rule. In general, when more than one baseline appears reasonable or the baseline is highly uncertain, alternative baselines should be examined as a form of sensitivity analysis.

Discounting. One of the most vexing difficulties faced by environmental analysts is comparing benefits and costs occurring in different time periods. Because investments are expected to be productive, resources on hand today are generally considered to be more valuable than resources available at a later date. By the same token, benefits (both environmental as well as standard consumption such as food and travel) accruing in the future are generally considered to be less valuable than those accruing today. As Peter Wentz has argued:

> At a 5% discount rate...one life today is equivalent today to two lives in 14.4 years, four lives in 28.8 years, eight lives in 43.2 years, sixteen lives in 57.6 years, and so forth....In 489.6 years, for example, doubling of value has taken place thirty four times (when the discount rate is 5%) making one life today worth more than sixteen billion lives. So from the perspective...(of benefit-cost analy-

> sis), if the human population is twelve or fifteen billion 500 years from now, one life beginning today is worth considerably more than the present worth of all human beings who will exist in 500 years. In other words, a policy that saved one life today would be justified even if it caused the extinction of our species, so long as the harmful effects were delayed for 500 years (Wentz 1988, 224).

At least when considering conventional pollution problems where the costs and benefits are experienced within a single generation, the standard approach is to deflate or discount future benefits and costs, thereby placing them on a comparable basis with current benefits and costs. On these points there is broad agreement. However, selecting the appropriate discount rate remains a controversial matter.

For many years OMB recommended that all regulations be evaluated on the basis of a 10% annual rate of discount, expressed in real terms. Many observers believed this was excessively high, treating corporate investments as the true opportunity cost of capital. In 1990, OMB revised its discounting guidance to reflect the rates of return commonly obtained on relatively low yielding forms of capital, such as housing, as well as the higher rates of returns typically yielded by corporate capital. The average rate on this mixed set of investments is currently estimated to be about 7% in real terms.

Reflecting the notion that the opportunity cost of capital and the willingness to pay to bear risk should be the bases for discounting, current economic thinking favors what is known as the shadow price of capital model (Lind 1990).[1] The shadow price reflects the "opportunity cost" of capital. The basic premise of this approach is that economic welfare is ultimately determined by consumption. Investment affects human well-being only to the extent that it affects current and future consumption. Thus, government outlays or mandated private expenditures must be converted to a stream of effects on consumption before being discounted. With all regulatory impacts converted to consumption equivalents, one can discount streams of benefits and costs at a rate that reflects intertemporal consumption trade-offs. While it can be difficult to make these calculations, this approach can yield considerably lower discount rates than the standard opportunity cost of capital approach. Current estimates of the shadow price of capital can be as low as a 3% real rate for the United States.

Despite the theoretical appeal of the shadow price of capital approach, there is little practical experience in applying it to regulatory issues. Further, it can be quite sensitive to means of calculating displaced consumption associated with reduced investment. As noted, the 7% OMB rate reflects the overall average return to capital in the economy. This rate is easier to apply than the shadow price of capital approach and may be a reasonable approximation when regulations primarily affect domestic investment rates over the short and medium time frame. At a minimum, agencies are encouraged to conduct sensitivity analysis to gauge how benefits and costs change with alternative discount rate approaches.

Special problems may arise when discounting over very long periods of time. Various alternatives have been proposed to address the issues raised by Wentz at the beginning of this discussion on discounting. They generally involve using a lower discount rate (such as the long-term rate of economic growth). This lower rate reflects, in part, the presumed greater ability of future generations to shoulder economic burdens. This topic remains an active area of inquiry. One prominent researcher has even acknowledged the possibility of a negative discount rate (Weitzman 1994).[2]

The cases in this volume generally use the standard opportunity cost of capital approach, including sensitivity analysis of alternative discount rates. Although it preceded issuance of the new Guidance, the case on CFCs, which involves intergenerational issues, uses an approach that is generally consistent with the intergenerational issues raised in the new Guidance, which takes into account intergenerational effects.

Treatment of Risk and Uncertainty. Whatever we do we are likely to encounter risks. Yet, risks, benefits, and costs are rarely known precisely. Thus, most economic analysis must rely upon statistical probability distributions for the values of key parameters to be used in the analysis. Key issues in both the scientific and economic treatment of risk and uncertainty involve the quality and reliability of data, models, assumptions, and scientific inferences. Research and analysis of these issues has evolved considerably since the 1982 OMB guidance was issued. The 1996 Guidance represents a significant expansion over the earlier document. Fundamentally, the new Guidance recommends incorporating the uncertainties directly into the analysis, consistent with principles of full disclosure and transparency. Especially where there are a limited number of studies with divergent results, the Guidance recommends use of a range of plausible scenarios, together with any information that can help in providing a qualitative judgment of which scenarios are more scientifically plausible. The economic analyses of individual regulations reviewed in this volume reflect the evolving thinking on these issues. Although many of the cases do consider risk and uncertainty, the treatment of these issues is generally less extensive than might be implied by the 1996 Guidance. EPA itself issued guidance on risk characterization in 1992, and this was reaffirmed in 1995. The following discussion addresses some of the key issues.

One of the key dilemmas for agencies conducting economic analyses of health related issues is that different methodological approaches are commonly used in the different academic disciplines. Economists typically rely on a graded continuum with a dose-response description of risk. Reflecting this view, the 1996 Guidance calls for presenting data and assumptions in a manner that "...permits qualitative evaluation of...incremental effects (OMB 1996, 17)." In contrast, many health scientists typically utilize a risk-assessment aproach, attempting to establish a single target level of "safe" or acceptably low level of risk (such as a maximum allowable level of a contaminant in an environmental medium or food product). Determining such a "safe" contamination level involves a combination of scientific analysis, scientific opinion, and value judgments. Such a "safety" oriented approach may seem particularly appealing when the data and/or the science for understanding dose-response relationships are poor. Yet, the establishment of a single target level of concentration or exposure necessarily precludes economic analysis of alternative standards, including a comparison of the target level to the status quo. For risk assessments to provide information that can be used in economic analyses, it is important to enlarge the capacity to express how risks vary with changes in concentration or exposure—whether this is done through basic research on molecular structure, development of biologic markers, epidemiological studies, or other means.

Another key methodological difference between the approaches of economists and many health scientists concerns the calculation of individual versus population risk estimates. Individual risk is the risk posed to a hypothetical individual (often a "worst case," or a "maximally exposed individual"). In contrast, population risk is the risk posed to the entire population, which may be estimated by multiplying a unit risk from a change in exposure for each population segment by the size of that population subgroup.

The economic paradigm is often concerned with aggregate well-being and thus tends to focus on the entire population. Yet, even if one were performing a cost-effectiveness analysis, where abatement costs were being compared to the risk level of the maximally exposed individual, the use of individual risk estimates could produce paradoxical results. For example, consider two abatement strategies that cost the same amount. One involves exposing a small number of people to a high risk, while the other strategy exposes many more people to a smaller individual risk. A cost-effectiveness analysis based on individual risk could lead to the first strategy being adopted, though the latter strategy might generate greater total economic benefits. The Guidance recommends that: "A full characterization of risks should include findings for the entire affected population and relevant subpopulations (OMB 1996, 17)."

To reflect the various uncertainties, the risk assessment paradigm often adds safety factors into the assumptions used to estimate risks. This approach, combined with a focus on individual risks, tends to increase overall estimates of risk. In contrast, the economic paradigm attempts to distinguish between the distribution of risks within the population and preferences for risk aversion. With this paradigm, the analyst attempts to describe the distribution of risks in the population and leaves it to the decisionmaker to decide which strategies yield an acceptable degree of protection. The risk assessor's use of "worst case" estimates provides one point on the risk distribution. While worst case analysis is not precluded, the economic approach seeks to develop a broader set of options by requiring information on the shape of the dose response curve above and below the "safe" level. The Guidance specifically calls for the use of statistical techniques to combine uncertainties about separate factors into an overall probability distribution for a risk.

Nonmonetized Benefits and Costs. Almost anything can be quantified if the requirement that the results be convincing and methodologically sound is abandoned. Requiring the quantification of all effects is an invitation to shoddy analysis and would likely provide no more information, and possibly more sources of dispute and wasted effort, than a careful attempt to qualitatively consider the effects in question.

Several possible approaches to the treatment of nonquantitative effects have been used in the literature. One approach involves estimating the net benefits for the quantifiable elements and asking how large the nonquantifiable elements would have to be to alter the conclusion of the analysis. If the nonquantifiable factors consisted of only benefits (as opposed to costs) and other net (quantified) benefits were previously determined to be positive, this information would add little value to the decision. If the quantified net benefits were negative, the nonquantifiable benefits would have to be large enough to reverse the outcome of the analysis. The decisionmaker could then make an explicit judgment about whether the nonquantifiable factors were likely to be greater than this amount, often an easier judgment than one about the actual size of the nonquantifiable benefits. This type of approach could be performed in either a quantitative way or as a matter of judgment.Of course, neither of these approaches is practical if both costs and benefits have significant unquantifiable elements. In that case, special care should be taken to describe the nonquantifiable elements as precisely as possible, to best provide information to the agency decisionmakers.

E.O. 12866 emphasizes that nonquantifiable elements should not automatically be given less weight than quantified elements. It may be useful to understand why the particular elements have not been quantified. For example, some elements may not have been quantified because researchers and funders have accorded other areas of study a higher priority. Other areas may have received attention in the research community but so far defied quantification, although ample qualitative information exists to be concerned. Some elements may have simply been overlooked. Depending on the situation, the agency may want to make some judgment about the importance of the nonquantified effects.[3]

Distributional Effects and Equity. Consistent with the emphasis given in E.O. 12866, the new Guidance contains an expanded treatment of distributional and equity issues. Distributional effects are defined as the "net effects of a regulatory alternative across the population and economy, divided up in various ways (for instance, by income groups, race, sex, industrial sector). Benefits and costs of a regulation may also be distributed unevenly over time, perhaps spanning several generations (OMB 1996, 23)."

Where distributional issues are thought to be important, the new Guidance encourages quantitative treatment, to the extent possible, including the magnitude, likelihood, and incidence of effects on particular subgroups. Notwithstanding that information needed for subgroup benefit estimates may be more difficult to obtain than estimates for the overall population, agencies are advised "...to be alert for situations in which regulatory alternatives result in significant changes in treatment of outcomes for different groups (OMB 1996, 24)."

The rise of the environmental justice movement in the early 1990s has heightened sensitivity to situations where the burdens or risks of particular policies fall disproportionately on ethnic or racial minorities. Thus, the 1996 Guidance notes:

> ...there are no generally accepted principles for determining when one distribution of net benefits is more equitable than another. Thus the economic analysis should be careful to describe distributional effects without judging their fairness. These descriptions should be broad, focusing on large groups with small effects per capita as well as on small groups experiencing large effects per capita. Equity issues not related to the distribution of policy effects should be noted when important and described quantitatively to the extent feasible (OMB 1996, 24).

Benefits Estimation. Economic theory suggests that environmental investments should be undertaken until the marginal (as opposed to the average) benefits justify the marginal (as opposed to the average) costs. Then and only then can one be certain that society's scarce resources are used in their most efficient way.

Most of the controversy that surrounds the use of economic analysis for environmental decisionmaking stems not from doubts about the usefulness of knowing about the benefits or costs of proposed policies, but from the difficulty of defining and measuring the true marginal benefits and marginal costs of environmental regulation. As the Guidance acknowledges, the extent to which different concentrations of pollution actually affect human health or the well-being of ecosystems is, in many instances, poorly understood and difficult to measure. Even when a so-called dose-response function can be estimated, the economists' challenge is to determine society's willingness to pay to avoid particular health or environmental damages.

The Guidance recommends that an attempt be made to monetize incremental benefits to the maximum extent feasible:

> Any benefits that cannot be monetized, such as an increase in the rate of introducing more productive new technology or a decrease in the risk of extinction of endangered species, should also be presented and explained (OMB 1996, 24).

For valuing the marginal benefits of goods directly traded in markets, market prices are generally the best estimate of the willingness to pay. For example, if air pollution damages crops, one of the benefits of controlling that pollutant would be the value of the crop saved as a result of the controls. However, if the price of that crop is regulated in some way, then the regulated price may not represent the best measure of the society's willingness to pay. In that case, the Guidance recommends using other techniques to determine the willingness to pay for marginal changes.

For valuing benefits of goods or services that are only indirectly traded in markets, for example, health and safety risks, explicit measures of willingness to pay are needed. These measures generally apply statistical techniques to distill from observable market transactions the portion of willingness to pay that can be attributed to the benefit in question. For example, the property value approach for measuring benefits—that was used (in a limited way) in the municipal landfill rule—is based on the assumption that the value of different pieces of land depends in part on their relative environmental amenities. Given that different locations have varied environmental attributes, such variations will result in differences in property values. With the use of appropriate statistical techniques, this approach attempts both to identify how much of a property value differential is due to a particular environmental difference between properties as well as to infer how much people are willing to pay for an improvement in the environmental quality they face. Other examples include estimates of the value of environmental amenities derived from travel-cost studies, and statistical studies of occupational-risk premiums in wage rates.

The value of fatality risk reduction figures prominently in assessment of environmental benefits. In the case of air pollution, the reduced risk of death often accounts for the largest single component of the dollar value of environmental benefits. The term "statistical life" refers to the sum of risk reductions expected in a population. The value of changes in fatality risk is sometimes expressed in terms of the value of statistical life. As the Guidance notes:

> These terms are confusing at best....It should be made clear that these terms refer to the willingness to pay for reductions in risks of premature death (scaled by the reduction in risk being valued). That is, such estimates refer only to the value of relatively small changes in the risk of death. They have no application to an identifiable individual (OMB 1996, 29).

In separate guidance issued by EPA, a range of approximately $2 to $10 million per statistical life has been recommended for use in conducting economic analyses.[4]

Another way of expressing reductions in fatality risks is in terms of the value of statistical life years extended. This approach allows distinctions in risk-reduction measures based on their effects on longevity. Older people tend to be more vulnerable to certain pollutants, for example, fine particles, whereas toxics may affect all age segments of the population. While the new Guidance encourages agencies to conduct sensitivity analyses involving the value of statististical life years, it stops short of fully endorsing the use of this approach:

> While there are theoretical advantages to using a value of statistical life-year extended approach, current research does not provide a definitive way of developing estimates...that are sensitive to such factors as current age, latency of effect, life years remaining, and social valuation of different risk reductions (OMB 1996, 31).

Quite a few of the cases in this volume develop estimates of "statistical lives saved" and apply a range of dollar values to these estimates. To date, the value of life years extended has not been widely used in economic analyses, but the concept has a certain appeal and there is growing interest in applying it.

Considerable research has been devoted to estimating willingness to pay for reduced risk of many types of illness, although chronic illnesses have been less well studied. Considerable research has also been devoted to estimating the value of lost recreational opportunities, including swimming, boating, and fishing. Similarly, the value of reduced visibility associated with air pollution has been extensively studied.

Economics research is probably on the weakest footing when it comes to developing willingness-to-pay estimates for the preservation or protection of resources that individuals do not directly use now or that have no current commercial value. Yet these so-called non-use values reflect common and widespread public sentiments. To date, few market-oriented approaches for valuing these benefits have been developed and economists have been forced to use survey techniques, the best known of which is contingent valuation. Contingent valuation methods are a way of estimating nonuse values by conducting a survey and asking what the respondent would be willing to pay to avoid some hypothetical incident or damage to a resource. The canonical example is a survey asking residents of Maine what they would be willing to pay to avoid reduction of visibility in the Grand Canyon or an oil spill in a body of water they have never visited.

The basic criticism of such efforts to elicit nonuse values is that the way in which surveys are conducted has an inordinate influence on the estimates, and that respondents often have neither the incentive nor the information to give meaningful responses. Critics have also noted that responses sometimes fail tests of logical consistency. Counterarguments are that, under strict guidelines, surveys can produce meaningful response and that some information on nonuse values is better than none. At a minimum, this argues for great care in the design and execution of the surveys.

In this area of valuing ecological benefits, the neoclassical economic model has come under particular attack. There is a real debate about whose values we should be attempting to measure. Should we really be concerned with the typical consumer's valuation of these resources or do we want the views of a minority, even an elitist social group? Notwithstanding the pitched battles between economists and ecologists, one of the cases in this volume—the Great Lakes Water Quality Guidance—makes extensive use of contingent valuation studies to estimate ecological benefits.

Cost Estimation. The estimation of cost can be as difficult an undertaking as the estimation of benefits—a fact not often appreciated by even the most knowledgeable practitioners of benefit-cost analysis. Cost means a foregone opportunity, that is, something that is given up as a result of a particular policy. The banning of chlorofluorocarbons (CFCs) like freon in the late 1980s brought forth a variety of new chemicals to replace them. Where would one look for the costs of this policy? As Kopp, Krupnick, and Toman (1996) suggest, the costs could be found in any (or all) of the following areas:

- The diverted resources necessary to develop the substitute products.
- The value of services of specialized capital (such as machines that can only be used to make freon) and technology necessary to manufacture the banned products.
- Increased costs of production for the replacement product vis-à-vis the banned product.
- Differences in the retail price of the replacement product.
- Decline in the quality of the replacement product.

This enumeration of cost categories further emphasizes that the total compliance cost of regulation consists of more than the typically reported measures of compliance cost. None of the case studies reported in this volume, including the one involving CFCs, was able to estimate this full set of costs.

Some have argued it is possible for regulation to have positive offsetting effects as well, by stimulating productivity increases or promoting innovation. Although there have been a number of case studies on this subject, there has not been widespread empirical or theoretical support for this view. (See Porter and van der Linde 1995; Palmer, Oates, and Portney 1992.) The Guidance advises that some attempt should be made to take into account potential future changes in technology and productivity. Such changes could include innovation spurred by regulation, as well as the potential that regulation could slow technical progress or deflect it into regulatory response from other areas of the economy.

A further cost issue concerns the secondary effects of a proposed action, as its impact ripples through the whole economy. In a study of thirty-six U.S. production sectors, Michael Hazilla and Raymond Kopp (1990) found that, while pollution control investments were required in only thirteen of the sectors, costs increased, and output and labor productivity decreased, in all thirty-six sectors as a result of the Clean Air Act and the Clean Water Act. To understand their results, consider the impact of these environmental laws on the finance, insurance, and real estate sectors. These sectors were not required to invest in pollution abatement equipment and did not incur higher operating costs as a direct consequence of the Clean Air and Clean Water Acts. They would thus appear not to bear any costs under a standard approach to cost estimation. However, using a large, complex economic model of the U.S. economy, Hazilla and Kopp found that by 1981 the cost of production in the finance, insurance, and real estate sectors rose by about 2% as a result of the indirect impacts of air and water regulations, specifically, as the result of higher factor (input) prices. None of the case studies reported in this volume consider indirect costs of the type studied by Hazilla and Kopp. And there is not a strong a priori basis to believe that regulations focused on single sectors of the economy will involve significant secondary impacts. However, regulations affecting highly integrated sectors of the economy, for example, those producing widely used intermediate products like energy, may require this type of analysis.

A particularly controversial issue in economic analysis is the treatment of impacts on employment and local economic activity. Various analyses have purported to calculate how many jobs regulation "creates"(ICF 1992). Such studies typically treat the supply of labor as unlimited, so that increases in employment as a consequence of regulation have no opportunity cost in terms of the diversion of productive activity from other parts of the economy. Economists are generally skeptical about the job-creating (or job-destroying) potential of regulation, though there can be circumstances when regulation increases the utilization of unemployed or underemployed workers. In the absence of evidence that such conditions prevail, employment effects may be noted but should not counted as net benefits of regulation.

CONCLUSION

Economic analysis can assist environmental decisionmaking in at least three principal ways:

- by systematically arraying the favorable and unfavorable impacts of alternative policy choices and thereby helping determine how the benefits of specific environmental policy alternatives compare to their costs,
- by revealing whether particular environmental benefits are readily achievable at lower costs, and
- by showing how benefits and costs are distributed; for example, geographically, temporally, and among income and racial groups.

The OMB Guidance represents a catalog of "best practices" to be observed when conducting an economic analysis. One of the key issues concerns the need for the proposed action: is there a market failure? As regards the analysis of benefits and costs, important issues arise in establishing the baseline, discounting future benefits and costs, the treatment of risk and uncertainty, nonmonetized benefits and costs, distributional effects and equity, and the valuation of certain types of benefits, particularly premature death. Cost estimation is also fraught with potential problems, although these have generally received less attention in the literature.

No Guidance document can truly dictate the specifics or even the scope of an economic analysis. At best it can lay out some general principles and approaches to be followed. Ultimately, good analysis demands a commitment to follow systematic scientific principles in framing issues and conducting the studies. That commitment should be reinforced by a rigorous and open review process. The new OMB guidelines lay out a set of requirements that, while flexible enough to be tailored to individual circumstances, provide a norm against which assessments can be judged. Even where statutory language limits the scope for consideration of economic factors in rulemaking, economic studies can grease the wheels of democracy by informing policymakers, legislators, and the general public about the consequences of regulation; helping build support for regulation that serves the public interest; and highlighting opportunities for regulatory or legislative improvements.

ENDNOTES

[1]However, in a paper delivered at a November 1996 conference cosponsored by the Energy Modeling Forum and Resources for the Future, Lind himself has voiced concern about this approach (Lind 1996).

[2]See also IPCC (1996). See also Toman (1994) for a discussion of an alternative "safe minimum standard" and when to use it.

[3]A recent example of using this latter approach involves the Clinton administration's proposed changes to the Clean Water Act. Despite an impressive benefit-cost analysis that failed to generate quantifiable positive net benefits, the administration endorsed the proposal citing "significant nonquantifiable benefits" as the justification. The Great Lakes Water Quality Guidance used a similar approach.

[4]In 1983, EPA issued formal guidance on preparing RIAs that, on the basis of a review of the literature, recommended a range of $400,000 to $7 million as the preferred value of a statistical life ($1982). That range has since been updated to $1.9 to $10.1 million ($1990). A recent EPA study of the benefits and the costs of the Clean Air Act, subject to extensive peer review, has used $4.8 million as the midpoint estimate for the value of a statistical life. See U.S. EPA (1996).

REFERENCES

Arrow, Kenneth and others. 1996. *Benefit-Cost Analysis in Environmental Health, and Safety Regulation: A Statement of Principles*. Washington, D.C.: American Enterprise Institute, The Annapolis Center, and Resources for the Future.

Graham, John and Jonathan Weiner. 1995. *Risk vs Risk: Tradeoffs in Protecting Health and the Environment*. Cambridge, Massachusetts: Harvard University Press.

Hazilla, Michael and Raymond J. Kopp. 1990. Social Cost of Environmental Quality Regulations. A General Equilibrium Analysis. *Journal of Political Economy* 98(4): 853–73.

Hornstein, Donald A. 1994. Paradigms, Process, and Politics: Risk and Regulatory Design. In Adam M. Finkel and Dominic Golding (eds.) *Worst Things First?* Washington, D.C.: Resources for the Future.

ICF (ICF Resources Inc. and Smith, Barney, Harris, Upham, and Co., Inc.) 1992. *Business Opportunties of the New Clean Air Act: The Internal Impact of the CAAA 1990 on the Air Pollution Control Industry*. August. Prepared for the Office of Air and Radiation. Washington D.C.: U.S. EPA.

IPCC (Intergovernmental Panel on Climate Change). 1996. Economic and Social Dimensions of Climate Change: Working Group III Report. In *Intertemporal Equity, Discounting, and Economic Efficiency*. Cambridge: Cambridge University Press.

John Paul II. 1990. *The Ecological Crisis: A Common Responsibility*. Message of His Holiness for the Celebration of the World Day of Peace, January 1.

Kempton, Willett, James S. Booster, and Jennifer A. Hartley. 1995. *Environmental Values in American Culture*. Cambridge, Massachusetts: MIT Press.

Kopp, Raymond. 1993. Environmental Economics: Not Dead But Thriving. *Resources* 111: 7–12. Washington, D.C.: Resources for the Future.

Kopp, Raymond, Alan Krupnick, and Michael Toman. 1996. *Cost-Benefit Analysis and Regulatory Reform*. White paper prepared for the Commission on Risk Assessment and Risk Management. June 13. Washington, D.C.: Resources for the Future.

Lave, Lester. 1996. Benefit-Cost Analysis: Do the Benefits Exceed the Costs? In Robert W. Hahn (ed.) *Risks, Costs, and Lives Saved*. Washington, D.C: American Enterprise Institute.

Leonard, Harman B. and Richard J. Zeckhauser. 1986. Cost-Benefit Analysis Applied to Risks: Its Philosophy and Legitimacy. In D. MacLean (ed.) *Values at Risk*. Totowa, New Jersey: Rowman and Allanheld.

Lind, Richard C. 1990. A Primer on the Major Issues Relating to the Discount for Evaluating National Energy Options. In R.C. Lind (ed.) *Discounting for Time and Risk in Energy Policy*. Washington, D.C: Resources for the Future/Johns Hopkins University Press.

———. 1996. Analysis for Intergenerational Decisionmaking. Paper presented at conference, Discounting in Intergenerational Decisionmaking, November 14–15, cosponsored by the Energy Modeling Forum and Resources for the Future. Washington, D.C: Resources for the Future.

Miller, James C. III. 1989. *The Economist as Reformer: Reforming the FTC 1981–1985*. Washington, D.C.: American Enterprise Institute.

Oates, Wallace and Maureen Cropper. 1992. Environmental Economics: A Survey. *Journal of Economic Literature* 30: 675–731.

OMB (Office of Mangement and Budget). 1996. *Economic Analysis of Federeal Regulation Under Executive Order 12866*. January 11. Washington, D.C.: Executive Office of the President.

Palmer, Karen, Warren Oates, and Paul Portney. 1992. Tightening Environmental Standards: The Benefit-Cost or the No-Cost Paradigm. *Journal of Economic Perspectives* 9(4): 119–32.

Porter, Michael and Claas van der Linde. 1995. Toward a New Conception of the Environment-Competitiveness Relationship. *Journal of Economic Perspectives* 9(4): 97–118.

Railton, Peter. 1990. Benefit-Cost Analysis as a Source of Information about Welfare. In R.B. Hammond and R. Coppock (eds.) *Decision Making: Report of a Conference.* Washington, D.C.: National Academy Press.

Sagoff, Mark. 1982. We Have Met the Enemy and He Is Us, or Conflict and Contradiction in Environmental Law. *Environmental Law* 283(12): 286–96.

———. 1993. Environmental Economics: An Epitaph. *Resources* 111: 2–7. Washington, D.C.: Resources for the Future.

Scanlon, Thomas. M. 1991. The Moral Basis of Interpersonal Comparison. In J. Elster and J. P. Roemer (eds.) *Interpersonal Comparisons of Well-Being*. Cambridge: Cambridge University Press.

Toman, Michael A. 1994. Economics and "Sustainability:" Balancing Tradeoffs and Imperatives. *Land Economics* 70(4): 399–413.

U.S. EPA (Environmental Protection Agency). 1992. *Environmental Equity: Reducing Risk for All.* Washington, D.C.: U.S. EPA.

———. 1996. *The Benefits and Costs of the Clean Air Act, 1970 to 1990.* October (draft). Washington, D.C.: U.S. EPA.

Weitzman, Martin L. 1994. On the Environmental Discount Rate. *Journal of Environmental Economics and Management* 26: 200–209.

Wentz, Peter. 1988. *Environmental Justice.* Albany: State University of New York Press.

Zeckhauser, Richard J. 1981. Preferred Policies When There Is Concern for Probablity of Adoption. *Journal of Environmental Economics and Management.* 8: 215–37.

4

Lead in Gasoline

Albert L. Nichols

In March 1985, EPA promulgated a rule to slash the use of lead in gasoline by over an order of magnitude in less than a year. This schedule combined the most stringent elements of the two alternative schedules that had been offered only seven months earlier. EPA expressed confidence that refiners would be able to meet this tight, final schedule, especially with the aid of the flexibility afforded by the preexisting trading program and the "banking" program that it was also promulgating.

EPA's decision was supported by an extensive benefit-cost analysis. That analysis indicated that lowering lead in gasoline would reduce serious health effects related to elevated lead levels in blood; such effects were clearest for children, but new evidence suggested that adults might reap even larger benefits related to reduced blood pressure levels. The rule was also expected to reduce "misfueling"—the use of leaded gasoline in cars required to use unleaded gasoline—and its associated damage to pollution-control catalysts in new motor vehicles. Protecting these catalysts would reduce emissions of three other pollutants: hydrocarbons, nitrogen oxides, and carbon monoxide.

The RIA prepared by EPA estimated that the monetized benefits of the rule would exceed its costs by roughly a factor of three-to-one, even if one attached no weight to recent studies linking elevated lead levels to high blood pressure and consequent cardiovascular disease. If those benefits were included, the benefit-cost ratio rose to more than ten-to-one. In this case study, Albert Nichols (coauthor of the 1985 benefit-cost analysis) tells the story

---

ALBERT L. NICHOLS is Vice President, National Economic Research Associates. From 1983–85, he was Director, Economic Analysis Division, Office of Policy, Planning and Evaluation, U.S. EPA.

behind the RIA and explores the reasons that it played an unusually important role in the policymaking process.

EARLIER RULEMAKINGS

Lead in gasoline is nothing new. Refiners had started adding lead to gasoline in the 1920s as an inexpensive way to increase its octane rating. (Increasing octane reduces engine "knock" and allows higher engine compression.) By the early 1970s, when the U.S. Environmental Protection Agency (EPA) started to regulate lead in gasoline, virtually all gasoline contained lead, at an average of close to 2.4 grams per gallon (gpg) (U.S. DOE 1982, Table 1).

Initial Regulation of Lead in Gasoline

EPA first regulated lead in gasoline in the early 1970s, motivated primarily by the fact that leaded gasoline destroyed the effectiveness of catalytic converters in automobiles. These converters were needed to meet the standards for emissions of carbon monoxide (CO) and hydrocarbon (HC) emissions, and later for emissions of nitrogen oxides (NO_x). EPA also saw the standard as a way of reducing exposure to lead itself, which had long been recognized as a toxic substance.

During the 1970s and early 1980s, the amount of lead used in gasoline fell by about 80% as a result of tighter gasoline standards and the retirement of older vehicles allowed to use leaded gasoline. The rules implementing these standards set a series of steps to reduce the lead content of gasoline, with large refineries obligated to meet a limit of 0.5 gpg by 1979. Small refineries faced less stringent limits because it was more difficult for them to achieve high octane levels without lead. However, they were scheduled to reach 0.5 gpg in October 1982 (47 *Federal Register* 7812).

Lead is also regulated as a "criteria" air pollutant under the Clean Air Act, with a National Ambient Air Quality Standard (NAAQS). The lead NAAQS has had little or no effect on lead in gasoline, however, because violations of it were due primarily to stationary point sources (such as lead smelters). Moreover, the exposure routes for lead from gasoline are not addressed very well by an ambient *air* standard; lead from gasoline settles out of the air relatively quickly, with much exposure due to ingestion of lead-contaminated dust or inhalation of reentrained lead particles.

The 1982 Rulemaking

In early 1982, following a recommendation from the Vice President's Task Force on Regulatory Relief, EPA proposed to defer the 0.5 gpg lead limit for smaller refineries (47 *Federal Register* 7814) and to consider relaxing the limits on larger refineries as well (47 *Federal Register* 7812). However, the proposal encountered very strong opposition, within the agency as well as from environmental groups and public health officials. As a result, in August 1982, EPA announced that it was no longer planning to relax overall limits and that it was significantly narrowing the definition of "small" refineries that would be entitled to meet a less stringent limit (47 *Federal Register* 38078).

The final rule, issued in October 1982, set a standard for most refineries that was slightly tighter than the current one and phased out special provisions for small refineries by mid-1983 (47 *Federal Register* 49322). The final rule also changed the basis on which the regulatory limits were calculated, from all gasoline to only leaded gasoline, and allowed averaging across refineries (effectively creating an emission trading program). Small refineries challenged the rules in court, but were successful only in getting a brief extension of the time by which they had to comply with some of the limits facing them (*Platt's Oilgram News* 1983).

Averaging Method. Prior to the 1982 rule, lead-in-gasoline limits were expressed in terms of grams per gallon of gasoline produced per quarter by a refinery. Under this averaging method, refineries included both leaded and unleaded gasoline in their computations; for instance, a refinery that produced equal amounts of leaded and unleaded gasoline, with an average lead content of 1.0 gpg for its leaded gasoline, had an overall average of 0.5 gpg. Thus, a refinery could effectively increase the amount of lead used in its leaded gasoline by producing additional unleaded gasoline; for instance, if a refinery's gasoline output were only 40% leaded, its leaded gasoline could average 1.25 gpg of lead and still meet an overall target of 0.5 gpg.

This approach encouraged refineries to produce more unleaded gasoline, which was the primary goal of the early regulations. However, it also meant that even as the market share of unleaded gasoline rose, the total amount of lead used in gasoline would *not* shrink unless the regulatory limits were continually tightened. The limits varied by refinery size, with less stringent limits for smaller refineries, which generally produced a relatively high fraction of leaded gasoline, and which often lacked the more sophisticated equipment needed to create high-octane blend stocks to replace lead as an octane booster.

As of early 1982, the limit was 0.5 gpg for larger refineries, and leaded gasoline made up just over half of gasoline sales (47 *Federal Register* 49322). The final rule issued in October set a limit of 1.1 grams per leaded gallon (gplg), which was roughly equivalent to the prior limit given leaded gasoline's then-current market share. (The 1.1 gplg limit was equivalent to an overall 0.5 gpg limit if 45% of the gasoline was leaded: 0.5 = 45% of 1.1.) However, with the new averaging method, the total amount of lead used in gasoline would shrink as the use of leaded gasoline declined due to retirement of older vehicles allowed to use leaded gasoline and their replacement by catalyst-equipped cars that were required to use unleaded gasoline. Switching the basis of the averaging system also made it more feasible politically and more cost-effective to eliminate separate limits for small refineries; the new averaging method implicitly allowed any refinery that produced a high proportion of leaded gasoline (which included many small refineries) to use more lead than the old averaging method would have allowed if applied uniformly to all refineries. Conversely, refineries with leaded production less than 45% of their total gasoline output were required to use less lead under the new averaging system. Thus, although the new regime set the same limits for large and small refineries, changing the basis for the averaging helped small refineries, thus partly cushioning the end of their special treatment.

Interrefinery Averaging (Trading). The 1982 rule also allowed averaging of lead content across refineries, which both reduced costs and made it easier for smaller refineries to meet the limit. Interrefinery averaging was essentially a trading program, one in which the amount of lead a refinery could use each quarter was equal to the number of gallons of leaded gasoline produced times the 1.1 gplg limit.

These "lead rights" could be sold or traded to other refineries. The marginal cost of reducing lead in gasoline at a particular refinery depended on a wide range of factors, including its product mix and equipment. In general, however, the marginal cost of reducing lead was higher for small refineries than for larger ones. As a result, EPA expected the net direction of trading to be from larger to smaller refineries. Although small refiners were skeptical that large refiners would sell them lead rights, the net direction of trading was from large to small refiners. By late 1984, before the new phase down began, about one-half of refineries participated in the lead-rights market each quarter, and up to 20% of the lead rights changed hands (Hahn and Hester 1989), a level of activity well beyond that achieved by any other trading program before or since.

THE 1983–1985 ANALYSIS AND RULEMAKING

Once the final rule had been promulgated in 1982 and the court action had concluded, lead in gasoline lost its prominence as a public policy issue. In contrast to other areas, such as hazardous waste or carcinogenic air pollutants, EPA was *not* under significant pressure from the courts,

Congress, the media, or even environmental groups to take additional action. No suits were pending, nor were any congressional hearings scheduled.[1] Evidence of lead's harmful effects continued to accumulate, however, and officials in EPA's Office of Mobile Sources became increasingly concerned about "misfueling"—the use of leaded gasoline in vehicles that were supposed to use unleaded gasoline to preserve the effectiveness of their pollution-control catalysts. A 1982 survey (U.S. EPA 1983) indicated that over 12% of the vehicles that should have used unleaded gasoline were being misfueled, thus causing premature failure of catalysts and increasing emissions of NO_x, HC, and CO.

The Initial Analysis

In September 1983, less than a year after the previous rulemaking had concluded, EPA Deputy Administrator Alvin Alm asked Assistant Administrator for Policy, Planning and Evaluation Milton Russell to conduct a preliminary analysis of the costs and benefits of further reducing lead in gasoline, or even eliminating it altogether. Reportedly Alm's interest had been stimulated by a remark by a representative of a firm that produced alcohol-based additives for gasoline, which were potential substitutes for lead in raising octane. More importantly, lead appeared to offer an opportunity to demonstrate that the risk management principles being promoted by Administrator William Ruckelshaus and Alm were not just a sophisticated way of saying "no" to proposed regulations; they also could help identify cases in which additional regulation was justified.

Russell and Richard Morgenstern, director of the Office of Policy Analysis within the Office of Policy, Planning and Evaluation (OPPE), asked a small team led by Joel Schwartz to do the preliminary analysis. That initial analysis suggested that sharply reducing lead in gasoline would have a very favorable benefit-cost balance, and Alm told the team to proceed with a more detailed analysis. However, the study was still to be conducted quietly, without public announcement, presumably to avoid strong external pressures (both pro and con) until the analysis was complete. Unlike most studies, it was to be done primarily in house by the small team of analysts, with limited contractor support. Moreover, the study team was staffed entirely by OPPE, rather than the program office responsible for regulating lead in gasoline, the Office of Mobile Sources (OMS) within the Office of Air and Radiation. OPPE's usual roles were to conduct longer-term economic research or to review and critique analyses and regulatory proposals made by the relevant program offices, each of which was responsible for a different set of laws typically aimed at one medium. That review role often created tensions and sometimes outright hostility between OPPE and the program offices.

The team's analysis, developed during the fall of 1983, reinforced the preliminary conclusion that sharply tightening the amount of lead allowed in gasoline would yield significant benefits. In December, the team completed a preliminary draft, which was sent to a dozen outside economists and other experts for peer review. Based on positive feedback from the reviewers, Alm and the senior OPPE managers directed the team to prepare a revised version for public release and broader comment.

The March 1984 Public Review Draft

After significant additional analysis and rewriting by the study team, EPA released a revised draft benefit-cost analysis in March 1984 for public comment (U.S. EPA 1984)[2] and held a press conference. Although EPA did not propose new regulations, it indicated that it was considering doing so (U.S. EPA 1984). On the same day that the study was released, Russell addressed a meeting of the National Refiners Association, told them of the study's results, and put them on notice that stringent rules were likely to follow soon.

Benefits. The draft quantified benefits in monetary terms in three major categories:

- *Children's health effects related to lead*—Reducing lead in gasoline would reduce the number of children with levels of lead in their blood that were considered hazardous by the Centers for Disease Control (CDC) and other health agencies. The team quantified those benefits in monetary terms by estimating the avoided costs of medical treatment and remedial education.
- *Health and environmental effects related to other pollutants*—By reducing the damages to catalysts, reducing or banning lead in gasoline would lower emissions of HC, NO_x, and CO. These emission reductions were valued in two alternative ways: direct estimation and valuation of changes in health effects and other physical measures and implicit valuation based on the cost-effectiveness of mobile source rules recently issued by EPA.
- *Reduced maintenance costs*—Lead and related additives form corrosive salts in engines and exhaust systems that cause premature wear and failure of mufflers, spark plugs, and other components. Engines burning leaded gasoline also require more frequent oil changes. These benefits were expressed in dollar terms based on estimates of market prices of the relevant products and services.

Although the monetization of benefits was incomplete—for example, it did not include any willingness to pay to avoid the pain and suffering associated with lead poisoning in children—it was substantially more extensive than that of typical EPA analyses.

Costs. The draft also included cost estimates based on a linear programming model originally developed for the Department of Energy (DOE) and maintained and updated by a contractor, Sobotka and Company. This same model had been used in EPA's 1982 rulemaking, and by DOE in other proceedings. It represented the individual refining units and their interrelationships through a series of several hundred equations. Given a set of input assumptions and constraints, it computed the least-cost method of producing a given set of final products. To estimate the costs of reducing lead in gasoline, Sobotka first ran the model with the existing limits on lead use, which yielded a base cost. The contractor then reran the model with the tighter limits on lead use and computed the resulting cost. For the March 1984 draft, the cost analysis focused on the year 1988, which was chosen to be far enough in the future that refiners would have time to install any needed new equipment.

The analysis covered both a complete ban on lead in gasoline and a "low-lead" option that would limit leaded gasoline to 0.1 gplg, a reduction of more than a factor of ten from the existing limit. The primary motivation for the low-lead alternative was concern that some older engines required at least some lead in gasoline to prevent premature wear of valve seats; newer engines designed to operate on unleaded gasoline had hardened valve seats that did not pose this problem.

Net Benefits. Table 1 summarizes the estimated benefits and costs included in the March draft. For the low-lead (0.1 gplg) option, the estimated annualized refining costs for 1988 were $503 million. The corresponding benefit estimates totaled almost $1,300 million. Over half of those monetized benefits ($660 million) were associated with lower maintenance costs. Almost two-thirds of the remaining benefits were from lower emissions of HC, CO, and NO_x from vehicles with catalysts disabled by misfueling; the analysis assumed that because the 0.1 gplg limit would make leaded gasoline more costly to produce than unleaded gasoline, it would eliminate all misfueling.[3] Less than 20% of the benefits were attributed to health effects associated with lead itself, and all of those were based on reducing the number of children with lead levels above limits of concern set by the CDC.

As the analysis pointed out, however, those monetized estimates were very limited; they assumed that only children whose blood-lead levels were brought below 30 micrograms per deciliter (μg/dl) would benefit from the rule, and even for those children, the only benefits expressed in dollar terms were lower expenditures on medical care and compensatory

Table 1. Benefit-Cost Estimates in 1984 Report.

	1988 Costs/benefits (millions $1983)	
Category	*Low-lead option*	*All unleaded*
Costs		
Increased refinery costs	$503	$691
Valve damage[a]	$0	D
Total costs	$503	$691 + D
Benefits		
Reduced vehicle maintenance	$660	$755
Reduced HC, NO_X, and CO emissions from misfueling		
Monetized benefits	$404	$404
Unmonetized benefits[b]	H_1	H_3
Reduced lead-related health damages		
Avoided medical costs	$41	$43
Avoided remedial education costs	$184	$193
Nonmonetized[c]	H_2	H_3
Total benefits	$1,289+$H_1$+$H_2$	$1,395+$H_1$+$H_3$
Net benefits	$786+$H_1$+$H_2$	$704+$H_1$+$H_3$-D

Notes: The 1984 report, cited below, was released for the purpose of public comment.
[a]In the All Unleaded case, costs include potential damage to valve seats of older engines.
[b]Includes chronic health effects of ozone and CO, plus any effects from reduced sulfate particulates.
[c]Includes benefits other than avoided costs of medical care and compensatory schooling for children brought below 30 μg/dl blood lead, plus any benefits to children at lower blood-lead levels. $H_3 > H_2$ because the unleaded option would affect more children.

Source: U.S. EPA 1984, Table I-1.

education. Nonetheless, together with the other elements, the estimated benefits were more than double the costs.

The case for a complete ban was much less clear. The monetized net benefits were somewhat smaller than those with the 0.1 gplg limit. However, EPA officials believed that the unmonetized effects were also important. On the benefits side, there were several unmonetized or only partially monetized categories, the most important of which most observers believed was lead-related cognitive effects in children. On the cost side, as noted earlier, a complete ban raised concerns about damage to older engines without hardened valve seats. Although those potential damages were not quantified, they were seen as potentially large. Moreover, going for a complete ban was much more likely to stir strong political opposition, which would delay the rulemaking.

Development of the Proposed Rule

Members of the study team briefed senior officials within both EPA and the Office of Management and Budget (OMB), who would have to review and approve any new regulations that might be proposed. The key OMB staff members were enthusiastic about the analysis and were supportive during the later rulemaking. Following release of the report for public comment, EPA staff began to work in earnest to develop new rules for proposal. This effort was primarily the responsibility of OMS and the Office of General Counsel. However, the OPPE study team continued to be responsible for the analysis of costs and benefits and for preparing the preliminary regulatory impact analysis (RIA) that would accompany the proposal.

As noted earlier, the analysis leading up to the March 1984 report had focused on the costs and benefits of a rule imposed in 1988, which would give refiners three years (if the rule were promulgated by the end of 1984) to add any new equipment needed to meet a 0.1 gplg limit. However, additional analysis suggested that a significantly tighter schedule might be possible, in part because some refineries already had excess capacity to produce unleaded gasoline, having added such capacity in anticipation of sharp increases in demand for unleaded that had not fully materialized. In August 1984, EPA proposed to set a limit of 0.1 gplg effective January 1, 1986 (49 *Federal Register* 31032). However, the agency recognized that some refineries might be unable to meet such a rapid schedule and indicated that it was considering more gradual alternatives, such as a series of four steps that would begin with a limit of 0.5 gplg on July 1, 1985, and end with 0.1 gplg on January 1, 1988 (49 *Federal Register* 31032). The proposal also indicated in fairly general terms that EPA was considering a complete ban, but only in the long term (for example, 1995).

The preliminary RIA issued with the proposal was very similar to the March benefit-cost report, but included estimates for 1986–92, not just 1988. The estimated net benefits for 1988 were $803 million, as opposed to the March estimate of $786 million (49 *Federal Register* 31032). An increase in the estimated maintenance benefits was partly offset by a reduction in the estimated benefits from reducing HC, CO, and NO_x emissions from misfueled vehicles.

Evolution of the Final Rule

Reaction to the proposal was generally favorable, with the unsurprising exception of the Lead Industries Association, which strongly opposed any further tightening of the limit. Most refiners appeared reconciled to a reduction in (and eventual ban of) lead in gasoline, although many expressed concerns about the proposed schedule, particularly having to reach 0.1 gplg on January 1, 1986, less than a year after the final rule was likely to be promulgated. Numerous officials from refineries met with EPA staff to argue for a more extended schedule.

New Benefit Estimates. During the fall of 1984, the OPPE team continued to analyze the costs and benefits of the alternatives. On the benefit side, the key development was the completion of two studies that showed a link between blood-lead levels in adults and blood pressure (later published as Harlan and others [1985] and Pirkle and others [1985]). Lead had long been associated with elevated blood pressure, but until 1984, most of the studies had focused only on hypertension and relatively high lead levels (typically found only in adults who were occupationally exposed to lead).

The new studies, however, found a continuous relationship between blood lead and blood pressure using data from the Second National Health and Nutrition Examination Survey (NHANES II, the same data set used to estimate the relationship between gasoline lead and blood lead), a large representative sample of the U.S. population. The studies showed a strong statistical relationship that had no apparent threshold and that did not disappear when many possible confounding factors and alternative specifications were tested.

The primary author of the benefit-cost analysis, Joel Schwartz, was also a coauthor of one of the studies linking lead and blood pressure (Pirkle and others 1985). He used the results of the study to estimate additional benefits from the rule, both in terms of reduced incidence of hypertension and, far more important, in terms of reduced incidence of morbidity and mortality due to cardiovascular disease associated with elevated blood pressure. The blood pressure results dramatically increased the estimated net benefits of the rule; even with fairly conservative assumptions, those benefits were roughly three times larger than all of the other benefit categories combined.

However, EPA's lawyers cautioned that these new results could not be relied on as part of the legal basis for the final rule unless they were first subject to careful scientific review and an extended opportunity for public comment. As the studies were not even scheduled for publication until early 1985, such review and comment would have lengthened the rulemaking process considerably, delaying the final rule for many months if not a year or more.[4] As a result, EPA officials agreed that although the blood pressure benefits could be included in the RIA, net benefits should be calculated both with and without including the blood pressure effects, and the case for the rule had to stand even if blood pressure benefits were excluded. Nonetheless, the potentially very large benefits associated with blood pressure provided an additional sense of urgency in getting the final rule promulgated so that reductions could begin as soon as possible. It also helped push agency officials toward an even faster schedule than those proposed in August. In particular, staff proposed combining the January 1, 1986, deadline for reaching 0.1 gplg with the alternative schedule's intermediate limit of 0.5 gplg starting July 1, 1985.

Additional Cost and Feasibility Analysis. On the cost side, the primary concern was the feasibility of refineries meeting the 0.1 gplg limit without adding new equipment, which could take two years or more (including, ironically, about a year to get environmental permits for the equipment). Although it was clear that many refineries already had the extra capacity needed—in many cases because they had added capacity, before it was known how much misfueling would slow growth in demand for leaded gasoline—numerous refiners met with agency officials to discuss possible difficulties in meeting the tight schedule.

To address these concerns, the OPPE team worked closely with Sobotka to develop a series of sensitivity analyses to see if refiners could meet the tight schedule even if conditions were significantly more difficult than expected. Those analyses showed that the 0.5 gplg limit in mid-1985 would not cause problems, even under combinations of adverse conditions. Attention then focused on the 0.1 gplg deadline to ensure that it could be met with existing refinery equipment. Those sensitivity analyses suggested that the rule remained feasible (that is, product demands could be met with existing refining capacity) under most conditions. Only when unlikely combinations of multiple adverse conditions were assumed did feasibility appear to be in doubt, and then only for the peak-demand summer months. Nonetheless, because of the large stakes involved in such potential problems, senior agency officials continued to be concerned about the possibility that a rapid phase down might result in shortages of gasoline or other refined products.

"Banking" Early Reductions. These concerns were met by the development of a "banking" program, which would give refiners additional flexibility in meeting the rule. As noted earlier, since 1982 refiners had been able to meet the 1.1 gplg limit by averaging ("trading") across refineries. As part of the new phase down, however, EPA planned to eliminate inter-refinery averaging once the 0.1 gplg limit was reached; that limit had been chosen to provide enough lead to protect older engines with unhardened valve seats. Officials were concerned that averaging could lead to such valve damage because if some refiners used substantially more than 0.1 gplg, others would have to use substantially less, thus potentially placing some engines at risk.

Under the banking plan, which was initially suggested by an analyst at Sobotka and Company, refineries that reduced lead ahead of schedule would be allowed to save ("bank") the early reductions for later use or sale, when the limits would be tighter and more difficult to attain; under the existing interrefinery averaging program, averaging could take place only within a given quarter, thus precluding banking. Cost modeling suggested that if given the opportunity to bank in anticipation of the phase down, many refiners would reduce lead levels in their gasoline below 1.1 gplg virtually immediately. Moreover, with an intermediate limit of 0.5 gplg in the second half of 1985, refiners on net would continue to bank, thus building significant balances for use in 1986 and 1987, after the limit would fall to 0.1 gplg, but before most refineries could make major equipment changes. Banking would not change the total amount of lead used in gasoline, but rather would shift the timing of that use, securing larger reductions in 1985 in exchange for smaller reductions in 1986–87.

Model runs suggested that banking would save $200 million or more, roughly 20% of the estimated cost of the rule from July 1, 1985, through 1987 (U.S. EPA 1985a, Table II-16). However, officials in EPA's Air Office did not find the cost savings a compelling argument for taking an innovative approach; they were generally suspicious of trading and were happy with the initial proposal to eliminate it at the end of 1985.[5]

The link between banking and a faster phase-down schedule was essential to winning support for banking within the agency, particularly in the Air Office.[6] More important, however, for many EPA officials was the fact that banking would effectively eliminate the potential feasibility problems suggested by some of the more extreme sensitivity analyses. With banking and trading, those few refineries unable to meet the 0.1 gplg limit in 1986–87 could buy banked lead rights from refineries that had been able to reduce lead levels ahead of schedule. In addition, even those refineries for which the 0.1 gplg limit was feasible in 1986 could save money by reducing lead early (when the marginal cost of reducing lead would be relatively low) and using the banked rights later (when the marginal cost would be much higher because of the tighter limit).

The estimated cost savings assumed that banking could be implemented by January 1, 1985, so that refineries would have six months to bank from the existing 1.1 gplg limit. Unfortunately, the idea was developed too late to promulgate a banking rule by that date in light of statutory requirements for comment periods and other procedural issues. EPA's lawyers, however, concluded that the agency could in effect allow banking for all of 1985 if the final rule could be promulgated by the end of March, when the first quarter of the year would end. To the extent that refiners were willing to reduce lead use throughout the quarter, in anticipation of being allowed to bank those reductions, the banking program could effectively start on January 1, 1985. When the banking program was proposed on January 4, 1985, EPA indicated its determination to issue the final rule in time to allow banking during the first quarter (50 *Federal Register* 718).

Additional Analysis of Potential Valve-Seat Damage. As noted earlier, the primary obstacle to a complete ban on lead in gasoline was the concern that certain engines might suffer valve-seat failure if operated with unleaded gasoline. Since 1971, virtually all cars and light trucks sold in the United States have had hardened valve seats to avoid this problem. Essentially all heavy-duty truck engines also have had this protection, or even more effective hardened valve-seat inserts, despite the fact that they were allowed to use leaded gasoline.

Tests with vehicles operated on dynamometers—devices with rollers that allow researchers to simulate vehicle operation in a laboratory setting—had shown that engines without hardened valve seats could suffer rapid wear when operated at high speeds for prolonged periods of time (for instance, the equivalent of seventy miles per hour for 100 hours); the key variable appeared to be engine speed, with load and temperature playing less critical roles. These dynamometer tests, however, also indicated that a relatively small amount of lead provided sufficient protection. Several studies had found no problems at 0.5 gplg, and the one study that attempted to find the minimum amount necessary concluded that it lay between 0.04 and 0.07 gplg. As a result, EPA chose 0.1 gplg to avoid potential problems of valve-seat wear in older engines (U.S. EPA 1985a, VII-2).

In contrast to the dynamometer and track tests, studies of vehicles in actual use had generally found little or no problem with valve-seat wear with unleaded gasoline. Most of these studies involved automobiles, as did the dynamometer and track tests. The major exception was a large, three-year study that the U.S. Army conducted during the mid-1970s, in which all of the gasoline-powered engines (including those in automobiles, trucks, tactical vehicles, construction equipment, and stationary applications such as generators) on several bases were converted to unleaded gasoline. After a total of almost fifty million vehicle miles, only three engines, all in pickup trucks, suffered valve-seat failure. On the basis of that study, all of the U.S. armed services switched to unleaded gasoline in 1976, with no special problems reported in the intervening nine years up to 1985. Other large vehicle fleets, including the Post Office and more than a dozen public utilities, also converted during the late 1970s or early 1980s, with no reported problems.

The apparent inconsistency between the in-use experience and the results of the dynamometer and track tests reflected the fact that few engines normally operate at the sustained high speeds that had been found to cause problems; the laboratory and track tests showed that valve-seat wear with unleaded fuel is a highly nonlinear function of engine speed, with little or no impact at moderate speeds. These studies suggested that the 0.1 gplg standard should provide adequate protection, and that even if all lead were eliminated from gasoline, premature valve-seat wear would probably not be a significant problem. By 1988, few pre-1971 cars or light trucks would remain in use, and those that did would be unlikely to be operated at the high engine speeds that appeared to be required to induce valve-seat failure. Heavier trucks, which continued to be designed to use leaded fuel and which operated at higher average engine speeds, were potentially a more significant problem, but virtually all of them already had some form of valve-seat protection. Thus, although it was not possible to quantify the potential cost of valve-seat wear associated with a ban, it appeared unlikely to be large, at least for on-road vehicles. Tractors and other farm equipment posed potentially more serious problems because they tended to be used very intensely under heavy loads, but often for only a short time each year, so that engines were often in use for decades.

Promulgation of the Final Rule

On March 7, 1985, EPA promulgated the accelerated phase-down schedule and announced that it was considering the imposition of a complete ban on leaded gasoline, to take effect possibly as early as 1988 (50 *Federal Register* 25710). In late March, the banking rule became effective, allowing refiners to bank lead from the first quarter of the year (50 *Federal Register* 13116).[7]

The 1985 lead phase-down rule was unusual in six respects:

- The rule had been developed and promulgated much more rapidly than is typical for a major EPA regulation.
- The rule was *not* prompted by a congressional or court-imposed deadline, as is often the case with EPA rulemakings.
- The rule included an innovative "banking" program that extended the existing interrefinery averaging program to reduce compliance costs and to give refiners additional flexibility in the timing of their compliance with the rule.
- The rule was strongly supported by the benefit-cost analysis included in the RIA.
- The analysis underlying the rule had been developed by OPPE, rather than by the relevant "program office," the Office of Air and Radiation. OPPE staff normally served as reviewers of rules (and their supporting analyses) developed by program offices or conducted longer-term research that was not tied to a specific rulemaking.
- Perhaps most surprising, the benefit-cost analysis *preceded* the development of the rule and played a key role in shaping the rule, rather than being an *ex post* evaluation (or, cynics might say, rationalization) of decisions reached on other grounds.

THE BENEFIT-COST ANALYSIS IN THE FINAL RIA

Along with the final rule, EPA issued a final RIA with a revised and expanded benefit-cost analysis of the phase down (US EPA 1985a).[8] This section of the case study summarizes the final RIA's estimates of benefits and costs, along with costs and benefits of a complete ban, which was issued as a preliminary RIA at the same time (U.S. EPA 1985b).

Estimated Costs

The major cost of reducing lead in gasoline was the additional processing (primarily reforming or isomerization) or alternative additives (such as MMT or alcohols) needed to replace the octane previously provided by lead. As discussed earlier, EPA believed that the 0.1 gplg limit would still provide enough lead to protect older engines against excessive valve-seat wear.

Refinery Model. Refineries are complex operations, producing a range of products from different combinations of inputs and processing units. Because of the prevalence of joint costs, it is impossible to analyze the production of any single product in isolation; the effective cost of making gasoline, for example, depends not only on the costs of different types of crude oil but also on the production levels of other refined products (such as chemical feedstocks and residual oil) and their prices.

As with the earlier analyses, the final RIA relied on the DOE's linear programming model of the U.S. refining industry. Some observers argued that the refining model "overoptimized" because it assumed considerable flexibility in the refining industry. Several factors, however, limited the potential errors in that regard:

- The model was constrained not to use any new capital equipment before 1988, giving refiners more than two years to plan for such equipment (including obtaining environmental permits for any major additions) and to install it.
- Working with Sobotka, the contractor that maintained and ran the model, the study team specified various other restrictions designed to mimic real-world constraints.
- The model was used to estimate costs with and without the new limit, so any overoptimization was present in both the before and after estimates, thus limiting the potential error in the estimated difference.
- As noted earlier, the study team and Sobotka also performed numerous sensitivity analyses using combinations of pessimistic assumptions to determine the conditions under which refiners might find the proposed limits infeasible.

- The model had been used in earlier rulemakings and had withstood intellectual and legal challenges.

The base-case results suggested that the final rule would cost less than $100 million for the second half of 1985, when the 0.5 gplg limit would apply. For later years, when the 0.1 gplg limit would apply, the estimated costs ranged from $608 million in 1986 to $441 million in 1992.[9] (The estimated costs fell over time because of projected declines in the demand for leaded gasoline, even in the absence of this new rule.) The incremental cost of a ban was about $150 million in 1988.

Price Differentials. These cost estimates were substantially lower than one might have inferred based on then-existing retail price differentials between leaded and unleaded gasoline, which averaged over seven cents per gallon; the model estimated that it cost less than two cents per gallon more to make unleaded than leaded gasoline. This gap raised concerns among economists involved in the study that the model was underestimating the cost of replacing lead as an octane booster. The model's estimates, however, were consistent with differentials in the prices at which refiners traded gasoline among themselves, and the wholesale price differentials charged by distributors were also much smaller than the pump differentials. The model's estimate of the marginal value of lead to refiners (the "shadow price") was also consistent with the prices at which lead rights were trading. Most of the increase in the differential between unleaded and leaded gasoline took place at the retail level, where leaded gasoline was used as the "fighting grade," advertised at a low price to attract customers. Because that extra differential appeared to reflect marketing strategies, rather than real resource costs, the study team concluded that the change in manufacturing cost was the appropriate measure for estimating the social cost of the rule.

Savings from Banking. The study team also examined the likely impact on costs of the banking provisions. To project how much banking refiners would do, the study team started with economic theory, which predicted that the value of lead rights should grow with the interest rate over the period from January 1, 1985 (when banking could begin), through December 31, 1987 (when banking would end).[10] That result in turn implied that the shadow price of lead in the linear programming model for different quarters should increase with the interest rate. Trial and error was then used to find the series of prices that balanced demand for lead with the total amount allowed over the three years. This approach suggested that a great deal of banking would occur, more than the study team and the contractor thought likely to be realistic. As a result, they made some reasonable, but ad hoc adjustments, and the RIA presented estimates that refiners would bank between 7.0 and 9.1 billion grams of lead in 1985, with a net reduction in the costs of the rule (discounted at 10%) of between $173 and $226 million (U.S. EPA 1985a, Table II-16).

These estimated net savings arose from the fact that it was relatively easy for at least some refiners to reduce lead below 1.1 gplg in early 1985; the extra cost of those early reductions was more than offset by later savings from being allowed to use more than 0.1 gplg. Banking also yielded savings because it allowed trading across refineries to continue for an extra two years, but those savings were not included in the estimates.

Estimated Benefits of the Rule

The RIA estimated benefits in four major categories: blood pressure-related health effects in adult males due to lead exposure as well as the three categories covered in the preliminary RIA (children's health and cognitive effects associated with lead; damages caused by excess emissions of HC, NO_x, and CO from misfueled vehicles; and impacts on maintenance and fuel economy). Estimating the two types of lead-related health benefits required estimating changes in exposure. In all four categories, the estimates were incomplete because of gaps in the data or difficulties in monetizing some types of benefits. Nonetheless, the estimated benefits were substantial and far exceeded the costs.

Human Exposure to Lead from Gasoline. To predict the lead-related health effects of the rule, the study team began by estimating its impact on lead in individuals' blood. People are exposed to lead from gasoline through a variety of routes, including direct inhalation of lead particles when they are emitted from vehicles, ingestion of lead-contaminated dust or inhalation of such dust when it is stirred up, and ingestion of food that has been contaminated with lead. Although it was (and continues to be) difficult to estimate the separate contributions of these individual pathways, several large data sets made it possible to estimate the overall impact of lead in gasoline on concentrations of lead in human blood.

Those data sets included records of lead-screening programs from CDC, records from screening programs in individual cities, and, most importantly, NHANES II, which provided blood-lead measurements (and other important information) on a large representative sample of the U.S. population surveyed during the late 1970s. By linking these data to data on gasoline lead use, it was possible to estimate statistically how blood-lead levels responded to changes in gasoline lead.

Several studies had shown strong and consistent relationships between gasoline lead and blood lead. Figure 1 plots those two measures

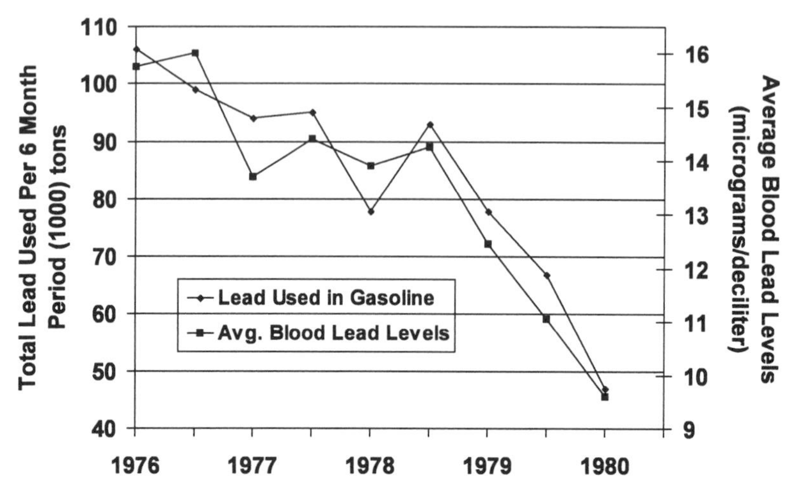

Figure 1. Correspondence between Blood-Lead Levels and Declines in Lead in Gasoline.

Source: U.S. EPA 1985a, Figure III-2.

over time using data from NHANES II. The relationship is remarkably clear; blood lead tracked both the seasonal variations in gasoline lead (rising during the summer months, when more gasoline is used) and the long-term downward trend in gasoline lead (the result of earlier EPA rules). Multiple regression analyses showed that this relationship continued to hold after controlling for other factors (such as socioeconomic status, nutritional factors, and exposure to other sources of lead). Such studies suggested that during the 1970s, gasoline was responsible on average for about half of the lead in Americans' blood. (Other sources of lead included lead paint, smelters and other stationary sources, and lead solder in cans.) An experimental study, in which the isotopes in gasoline lead had been modified so that its presence in blood could be distinguished from lead from other sources, also showed that lead from gasoline contributed significantly to blood-lead levels, although it did not quantify the relationship (U.S. EPA 1985a, III-22).

Statistical analyses indicated that gasoline lead not only raised the average level of lead in blood but also contributed substantially to the incidence of lead toxicity in children. Based on logistic regressions run on the NHANES II data set, the RIA predicted that the 0.1 gplg limit would roughly halve the number of children with blood-lead levels above those then recognized as harmful. From 1978 until 1985, the CDC had recommended that children with blood-lead levels above 30 μg/dl receive followup testing and possible treatment. In 1985, the CDC reduced that recommended level to 25 μg/dl.[11] The RIA estimated that in 1986 alone, the 0.1 gplg rule would prevent 172,000 children from exceeding 25 μg/dl blood lead.

Children's Health and Cognitive Effects. Elevated blood-lead levels have been linked to a wide range of health effects, with particular concern focusing on young children. These effects range from relatively subtle changes in biochemical measurements at low doses (such as 10 μg/dl) to severe retardation and even death at very high levels (such as 100 μg/dl). Lead can interfere with blood-forming processes, vitamin D metabolism, kidney functioning, and neurological processes. The negative impact of lead on cognitive performance (as measured by IQ tests, performance in school, and other means) was generally accepted at moderate-to-high blood-lead levels (30 to 40 μg/dl and above). One study, for example, found that children with elevated blood-lead levels were seven times more likely to be held back a grade in school or to be referred for counseling than were children in a control group matched on various measures of socioeconomic status (U.S. EPA 1985a, IV-52). Several studies also found cognitive effects at lower levels, though the existence of significant effects at low lead levels remained in dispute. Changes in electroencephalogram readings had been found at levels as low as 10 to 15 μg/dl (U.S. EPA 1985a, IV-33).

For children's health effects, the RIA estimated benefits in two categories: medical care for children exceeding the CDC cutoff and compensatory education for a subset of those children who might suffer cognitive effects from exposure to lead. Those estimates were conservative in that they did not include many benefit categories (such as avoiding lasting health and cognitive damage not reversed by medical treatment and compensatory education), nor did they attribute any benefits to reducing lead levels in children whose blood-lead levels would have been below 25 μg/dl in the absence of the rule. The study team's members recognized that many children with elevated lead levels would *not* actually receive the recommended health treatment and compensatory education. However, they reasoned that children who went untreated would suffer losses that would cost society at least as much as the recommended interventions.

To estimate reductions in medical care expenses, the RIA relied on recently published recommendations for testing and treating children with blood-lead levels above 25 μg/dl. Such treatment was estimated to cost about $900 per child over 25 μg/dl. (This average reflected lower costs for most children above 25 μg/dl, but much higher costs for the subset requiring chelation therapy.)

The estimates for compensatory education assumed three years of part-time compensatory education for 20% of the children above 25 µg/dl; that averaged about $2,600 per child above that blood-lead level. Thus, the RIA estimated a total of $3,500 in monetized benefits for each child brought below 25 µg/dl. The estimated aggregate benefits in this category ranged from about $600 million in 1986 to roughly $350 million in 1992.

Blood Pressure Effects. As discussed earlier, EPA officials concluded that although they could not rely solely on the recently developed estimates of adult blood pressure effects to justify the final phase-down rule, blood pressure–related benefits could be included in the RIA. The benefit estimates started with logistic regression equations estimated from NHANES II to predict how the rule would affect the number of hypertensives. Those estimates were applied only to males aged forty to fifty-nine, because the effect of lead on blood pressure appeared to be strongest for men and because estimates for that age range were not confounded by a strong covariance between age and blood lead. The relationship between lead and blood pressure, however, was statistically significant for other age groups as well. The RIA estimated that the rule would reduce the number of male hypertensives in the 40–59 age group by about 1.8 million in 1986. It valued reductions in hypertension based on estimates of the costs of medical care, medication, and lost wages; they yielded a value of $220 per year per case of hypertension avoided.

The RIA also estimated how reductions in blood pressure would affect the incidence of various cardiovascular diseases, based on projections of changes in blood pressure as a result of the rule and estimates of the relationships between blood pressure and heart attacks, strokes, and deaths from all causes. The latter estimates were derived from several large-scale epidemiological studies, primarily the Framingham study (McGee and Gordon 1976). Because those studies included very few nonwhites, the RIA further restricted the estimates to white males aged forty to fifty-nine.

The RIA valued reductions in heart attacks and strokes based on the cost of medical care and lost wages for nonfatal cases (the fatalities from heart attacks and strokes were included in the estimate of deaths from all causes). That procedure yielded benefits of $60,000 per heart attack and $44,000 per stroke avoided. It is important to note that those estimates did not account for any reductions in the quality of life for the victims of heart attacks and strokes (such as the partial paralysis that afflicts many stroke victims).

Valuing reductions in the risk of death is controversial, with a wide range of estimates in the literature. Based on studies of wage premiums for occupational risk, for example, EPA's RIA guidelines in 1985 suggested a range of $400,000 to $7 million per statistical life saved. The lead-in-gasoline RIA used a value from the lower end of that range, $1 million per case. Despite this relatively modest value, the benefits of reduced mortality dominated the estimates of blood pressure-related benefits.

Reductions of Other Pollutants. As noted earlier, reducing the amount of lead in gasoline was expected to decrease emissions of several pollutants in addition to lead. Most of the predicted reductions resulted from less misuse of leaded gasoline in vehicles that were required to use unleaded to protect the effectiveness of their pollution-control catalysts. EPA expected the rule to reduce misfueling significantly because it would be more expensive to produce leaded regular gasoline (at 89 octane) with 0.1 gplg than to make unleaded regular (at 87 octane). As discussed earlier, however, existing retail price differentials were much larger than the difference in manufacturing costs, apparently because of marketing strategies. Thus, while the rule was almost certain to at least narrow the price differential, it was not certain that unleaded would be priced lower than leaded at the pump.

Reducing misfueling would reduce emissions of HC, NO_x, and CO. All three of these pollutants have been associated with damages to health and welfare and contribute to ambient air pollution problems covered by NAAQS. To predict the emission reductions that would be associated with reduced misfueling, the RIA used survey data on the extent of misfueling, tests showing the effect of misfueling on emissions per mile traveled, and estimates of the numbers of miles traveled by vehicles of different ages and types.

The RIA also estimated that the rule would reduce emissions of benzene (an aromatic hydrocarbon that has been associated with leukemia) and ethylene dibromide (a suspected human carcinogen, which is added to leaded gasoline to control excess lead deposits in engines) but did not attempt to quantify those benefits. The analysis of benzene was motivated primarily by claims that the rule would *increase* benzene due to greater use of aromatics to replace octane levels previously achieved with lead. Although the potential risk from increased benzene was small relative to the gains from lead reduction,[12] the concern carried considerable weight politically because of the prominence of cancer fears among the public. Fortunately, the analysis showed that increases in benzene emissions due to increased use in gasoline would be more than offset by reductions due to reduced poisoning of catalysts as the result of misfueling.

The RIA valued reduced emissions of HC, NO_x, and CO in two ways. The first involved direct estimation of the health and welfare effects associated with the pollutants. It required several steps, including some crude assumptions about air-quality impacts, estimates of dose-response relationships, and assignment of dollar values to health and welfare endpoints. Most economists believe that damage-based methods of this type are the most appropriate approach to estimating benefits. However, these estimates were highly uncertain and did not include any values for some potentially important categories. All of the quantified benefits were from reductions in HC and NO_x emissions, and most of those were the result of projected declines in ground-level ozone, which is formed by photochemical reactions involving HC and NO_x. Although the study team made preliminary estimates of reductions in heart attacks associated with CO emissions, the Air Office and the Office of Research and Development argued successfully that the results were too uncertain to report quantitative estimates.

The second method valued the emission reductions based on the implicit cost per ton controlled of the emission control equipment destroyed by misfueling. This approach assumed that the benefits of controlling these pollutants were at least as great as the cost of the controls needed to meet congressionally mandated standards. This cost-based approach to estimating benefits suffers from some serious conceptual flaws, which have led virtually all economists to reject it in the context of efforts to estimate environmental externalities associated with electricity generation. In this case, however, this approach found ready acceptance from many observers, particularly as the costs were based on the specific equipment involved. Indeed, officials of the Air Office argued that *only* the cost-based estimates should be used, because those based on damages were too uncertain and incomplete.

The overall estimates were a simple average of the estimates of the two different approaches. The benefits in this category varied widely depending on the estimated impact of the rule on misfueling rates; the midrange estimate assumed that the 0.1 gplg rule would eliminate 80% of misfueling, with resulting benefits of $222 million in 1986.

Maintenance and Fuel Economy Benefits. As noted earlier, lead in gasoline and its associated scavengers corroded engines and exhaust systems. The RIA estimated the benefits of reducing these effects, along with expected increases in fuel economy. The estimated increases in fuel economy came from two sources: the additional refining needed to boost octane with less lead would increase the density of fuel and its energy content, and the reduced misfueling that would reduce the fouling of oxygen sensors that optimized the air-fuel mix in many newer vehicles.

The RIA estimated three categories of maintenance benefits: exhaust systems, spark plugs, and oil changes. Estimates for the first two categories were based on fleet studies of vehicles in use, which showed that exhaust systems and spark plugs lasted much longer with unleaded than with leaded gasoline. Estimates of oil-change benefits were based on studies showing that oil maintains its quality longer with unleaded than with leaded. Summing these three categories, the RIA estimated that reducing lead in gasoline from 1.1 gplg to 0.1 gplg would yield benefits of about $0.0017 per vehicle mile, or $17 per year for a vehicle driven 10,000 miles. Because of the large number of vehicles affected, the aggregate benefits were substantial, totaling about $900 million in 1986.

The fuel economy estimate, as noted earlier, had two components. To estimate the gain in fuel economy due to higher energy content, the RIA used the change in fuel density predicted by the DOE refining model and applied it to a fuel economy formula developed by the Society of Automotive Engineers. To estimate the portion due to reduced fouling of oxygen sensors, the RIA estimated the change in the number of misfueled sensor-equipped vehicles and used experimental data on how much extra fuel was consumed by vehicles with fouled sensors. Total estimated fuel economy benefits exceeded $100 million in most years.

Note that these maintenance and fuel economy benefits totaled about $1.1 billion in 1986, or almost twice the estimated cost, and that those benefits accrued to vehicle owners. Thus, in some sense, the environmental benefits appeared to be free. Such claims raise serious doubts in the minds of most economists: if the damages to vehicle owners from (mis)using leaded gasoline were greater than the cost savings, why did they continue to use leaded gasoline? Inadequate information is the most common response to such skepticism, although it is rarely supported by strong evidence. In this case, however, there was a more straightforward explanation. As noted earlier, the price gap between leaded and unleaded at the pump (which determined the cost from the vehicle owners' perspective) was much higher than the apparent "real" difference in manufacturing costs, which EPA used to measure the social costs of the rule.

Treatment of Uncertainty

The RIA contained only limited, ad hoc treatments of uncertainties in the benefit and cost estimates. Although standard errors were reported for the various regression analyses used in the study, those statistical uncertainties were not incorporated into the final benefit estimates. In some cases, the study team performed sensitivity analyses. Two key uncertainties (the inclusion of the blood pressure–related benefits and the extent to which the rule would eliminate misfueling) were carried through to the final benefit-cost comparisons, but others (such as the sensitivity analyses of costs) were not. No effort was made to assign subjective probabilities to alternative assumptions.

The limited treatment of uncertainties reflected several factors, including the following four:

- The estimated benefits exceeded the costs by such a wide margin that plausible ranges of uncertainty appeared unlikely to make a difference.
- As in most environmental cases, most of the important uncertainties were not amenable to "objective" quantification, because they concerned alternative assumptions and models rather than classic statistical uncertainties in estimators.
- For the final rule, the uncertainty about the blood pressure-related benefits dominated the other uncertainties, and that uncertainty was addressed explicitly.
- The analysis omitted some benefit categories altogether and took a fairly cautious approach to most categories, so that the study group viewed the estimates as "conservative" in the traditional scientific sense (that is, biased in a direction unfavorable to one's conclusion).

The last point deserves some additional discussion. EPA often makes "conservative" assumptions that bias estimates upward, by assuming high exposure levels, linear dose-response functions, and other conditions that lead to higher estimates of risk. These assumptions are generally most extreme for carcinogens (Nichols and Zeckhauser 1986). Thus, for example, estimates for a carcinogen might extrapolate from a study that found only benign tumors in male rats (with no increased rate of tumors in female rats or in mice of either sex) to project cancer deaths in humans of both sexes and all age groups at concentrations several orders of magnitude lower than those used in the animal studies. In contrast, consider the RIA's treatment of heart attacks, which were quantified only for white men aged 40–59. Women were excluded because both the statistical studies of humans and laboratory animal (rat) studies found that women were less likely to be affected. Other age groups of men were excluded because the strongest results were found in the 40–59 range. (However, statistically significant results were found in some other age groups.) Nonwhites were excluded because the main studies linking heart attacks to blood pressure included few nonwhites. The dollar value used for statistical lives saved ($1 million) was also low relative to most EPA analyses. As a result, although it would have been more in keeping with general principles of regulatory analysis to include a more formal treatment of uncertainties, its omission did not appear to be a serious problem in this case.

Benefit-Cost Comparisons

The RIA provided estimates of net benefits for a range of alternative phase-down rules. Estimates for a possible ban in 1988 were included in a short, supplemental RIA issued at the same time.

Net Benefits of Final Rule. Table 2 summarizes several important non-monetary measures of the effects of the rule for 1985 through 1988. The 1985 estimates are for only the second half of the year, when the 0.5 gplg rule was to apply. The estimates for the other years are for the 0.1 gplg standard. Although the change in lead emissions was relatively modest, less than 40,000 tons in any year, the estimated annual impacts are substantial, including reductions of more than 150,000 children with elevated blood-lead levels, 1.8 million hypertensives, and roughly 5,000 cardiovascular deaths.

Table 3 presents estimates of the costs and benefits quantified in monetary terms. The predicted costs and, in most categories, benefits decline over time because the use of lead in gasoline would have declined even in the absence of a new rule, as older lead-using vehicles were retired. The estimated benefits associated with reduced emissions of conventional pollutants, in contrast, rise because growth in the size and average age of the catalyst-equipped fleet would have increased the number of misfuelers without the new rule.

Table 2. Physical Measures of Estimated Benefits of Final Rule.

	1985	*1986*	*1987*	*1988*
Reductions in children above 25 μg/dl blood lead (1,000s)	64	171	156	149
Reduced emissions of conventional pollutants (1,000s tons)				
HC	0	244	242	242
NO_x	0	75	95	95
CO	0	1,692	1,691	1,698
Reduced blood-pressure effects in males aged 40–59				
Hypertension (1000s)	547	1,796	1,718	1,641
Myocardial infarctions	1,550	5,323	5,126	4,926
Strokes	324	1,109	1,068	1,026
Deaths	1,497	5,134	4,492	4,750

Source: Nichols (1985, Table 1) reports these values for the "partial misfueling" case, which was a weighted average of the "no-misfueling" results reported in the RIA (U.S. EPA 1985a) and unpublished results for the "full-misfueling" case.

Table 3. Estimated Monetized Costs and Benefits of Final Rule (in millions $1983).

	1985	*1986*	*1987*	*1988*
Monetized benefits				
Lead-related effects in children	$223	$600	$547	$502
Blood pressure-related (males, 40–59)	1,724	5,897	5,675	5,447
Conventional pollutants	0	222	222	224
Maintenance and fuel economy	137	1,101	1,029	931
Total	$2,084	$7,821	$7,474	$7,105
Costs				
Increased refining costs	$96	$608	$558	$532
Net benefits				
Including blood pressure	$1,988	$7,213	$6,916	$6,573
Excluding blood pressure	$264	$1,316	$1,241	$1,125

Source: U.S. EPA 1985a, Table VIII-7c.

Note that even without the blood pressure–related effects, the estimated net benefits exceeded $1 billion per year in 1986 and later years. If the blood pressure–related benefits were included, the estimated net benefits exceeded $5 billion per year. Although many of the individual components of these estimates were subject to uncertainty, the magnitude of the estimated benefits relative to the costs, together with the many potentially important benefits that had not been quantified in dollar terms, indicated that the rule would yield substantial net benefits.

The estimates in Tables 2 and 3 assume that the 0.5 gplg limit would have too small an impact on costs to have any effect on misfueling, but that the 0.1 gplg limit would eliminate 80% of misfueling. Those values reflected the RIA's "partial misfueling" assumption, which it characterized as a best estimate, although there was substantial uncertainty about the precise impact of the rule on misfueling. The RIA also presented estimated costs and benefits under alternative assumptions about the impact of the rule on misfueling, ranging from no impact (that is, misfueling would continue unabated even at 0.1 gplg) to its complete elimination. In addition, the RIA examined a wide range of alternative standards to see if a different phase-down schedule would be even more efficient and, in each case, computed net benefits with and without including the preliminary estimates of blood pressure–related benefits.

Figure 2 plots estimated net benefits in 1986 as a function of the stringency of the standard in that year. The higher set of curves includes the blood pressure–related benefits, while the bottom set does not. Within each set, the top line shows net benefits if the rule eliminated all misfueling, while the bottom line assumes no impact on misfueling. The middle line within each set assumes that misfueling would decline linearly from

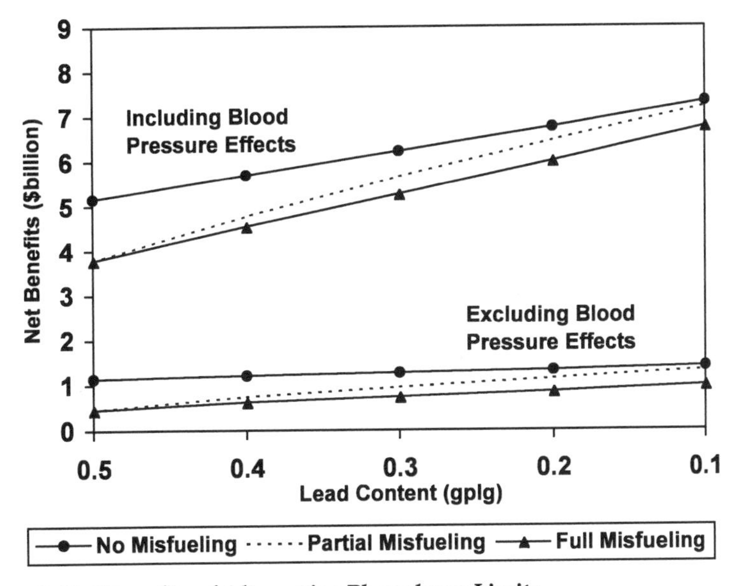

Figure 2. Net Benefits of Alternative Phasedown Limits.
Source: Computed from U.S. EPA 1985a, Tables VIII-3a,b.

100% of its current level at 0.5 gplg to no misfueling if a ban were imposed (that is, 0 gplg); it corresponds to the partial misfueling assumption discussed earlier.

As Figure 2 shows, regardless of the assumption about misfueling, and whether or not the blood pressure–related benefits were included, estimated net benefits continued to rise with the stringency of the lead limit, all the way down to 0.1 gplg. Similar results were obtained for 1985 and 1987 as well; net benefits were maximized at the tightest levels considered, 0.5 gplg in 1985 and 0.1 gplg in 1987.

Costs and Benefits of Ban. Along with the final RIA for the phase down, EPA issued a preliminary RIA for a potential ban (U.S. EPA 1985b). Despite the major lead reductions already achieved by the phase down and earlier rulemakings, the estimated incremental health and environmental effects of a ban were not trivial. Relative to the 0.1 gplg rule, the preliminary RIA estimated that a ban would reduce the number of children with blood-lead levels above 25 μg/dl by about 7,000 in 1988. Moreover, it would prevent over 100,000 estimated cases of hypertension among men aged forty to fifty-nine, and would reduce blood pressure-related fatalities among whites in that group by about 400. A ban would also eliminate whatever misfueling remained under the 0.1 gplg rule.

Under the partial misfueling assumption, the RIA estimated that a complete ban in 1988 would impose incremental refining costs of $149 million (in addition to the estimated $532 million cost of the 0.1 gplg limit in that year), while yielding incremental benefits of $193 to $635 million, depending on whether blood pressure–related benefits were included (50 *Federal Register* 9400).

Thus, in terms of monetized incremental net benefits, a ban clearly appeared to be justified. Moreover, the estimated incremental health impacts were orders of magnitude larger than those associated with most EPA regulations. However, as the preliminary RIA pointed out, the cost estimates did not include potential damage to older engines. Although the evidence available did not suggest that such damages would be major, they were the subject of much public concern.

The proposed ban was not promulgated by EPA, primarily because of continuing concerns about potential damage to older engines in farm equipment and other applications involving heavy use. Congress directed EPA and the Agriculture Department to conduct joint tests of engines used in farm equipment, and those tests suggested that unleaded gasoline could cause problems during heavy use. The two agencies recommended that owners of such engines install hardened valve-seat inserts during engine rebuilds.

Lead in Gasoline After 1985

The 1985 rule proved even more effective than anticipated in reducing lead in gasoline, despite the fact that the ban was not enacted. The combination of the phase down and declining numbers of engines on the road that were allowed to use lead rapidly reduced the availability of leaded gasoline and its sales. Many gas stations stopped carrying leaded gas, using the freed-up tanks and pumps to offer an intermediate grade of unleaded. Whereas the RIA had estimated that sales of leaded gasoline would still be over 16 billion gallons in 1990, even if all misfueling were eliminated, actual sales were less than 800 million gallons (Caldwell 1995), or one-twentieth the prediction and less than 1% of total gasoline sales. That same year, in the Clean Air Act Amendments of 1990, Congress formalized the demise of leaded gasoline by banning lead as a fuel additive effective December 31, 1995.

The banking program also proved even more popular than expected among refiners. Whereas the RIA predicted that refiners would bank 7.0 to 9.1 billion grams, banking totaled 10.6 billion grams (Hahn and Hester 1989), 16% more than the high end of the published range. More generally, although the costs of the rule were never reestimated after the rule was promulgated, it seems likely that they were even lower than anticipated, based on the fact that refiners reduced their production of leaded gasoline so much more rapidly than anticipated.

REFLECTIONS ON THE PROCESS AND THE ANALYSIS

Quantitative analysis *appears* to have played an unusually important role in the 1985 lead phase down compared to most EPA rulemakings. This section first considers whether that role was real, in the sense that the analysis played a significant role in shaping the agency's decisions, and then turns to the factors that helped determine that role and the lessons the case may offer for other rulemakings.

Did the Analysis Really Make a Difference?

Some observers may argue that the real role played by benefit-cost analysis in the 1985 lead-in-gasoline phase down was far *less* significant than it appeared to be, that EPA's actions merely represented the culmination of a long sequence of actions by the agency to reduce the amount of lead in gasoline. Indeed, far larger reductions in lead occurred *before* the analysis started, as a result of earlier rules, than as a result of the 1985 phase down. Even if the benefit-cost analysis had never been done, the argument continues, further action on leaded gasoline was inevitable to address continued concerns about lead exposure and misfueling of catalyst-equipped vehicles. All benefit-cost analysis did was put numbers on the obvious—that lead in gasoline should be reduced severely and quickly, if not banned outright.

These arguments have some truth to them, but go too far. The earlier reductions in lead were far larger and, although the benefits were never quantified, provided far larger benefits. It is also true that even without the benefit-cost analysis, the agency almost certainly would have tightened the lead limits eventually. On the other hand, it is too easy to say in retrospect that the decision was obvious even without the analysis; at the same time that the lead-in-gasoline study was proceeding, EPA's Air Office was focusing its resources on other issues, ones where there were court orders or congressional deadlines. Within the Office of Mobile Sources, NO_x standards for heavy-duty trucks were required by Congress and received more attention and effort than did lead. Within the Air Office as a whole, considerable attention and resources focused on "hazardous air pollutants," despite the fact that the risks involved were generally small, even when very "conservative" assumptions were used that overstated the likely benefits, possibly by orders of magnitude (for instance, see Haigh, Harrison, and Nichols 1984 and Nichols 1983). At the time of the lead-in-gasoline rulemaking, EPA was under court order as the result of a suit by an environmental group to promulgate rules on radionuclides emitted from uranium mill tailings, which were generally located in areas remote from populations. In meetings to discuss whether to accelerate the schedule for implementing the phase down, one member of the study team pointed out to senior officials (more in frustration than in jest) that "a week of lead is like a millennium of radionuclides," which was based on the fact that the estimated number of fatalities from lead-related heart attacks was on the order of 100 per week (5,000 per year), while the estimated risk from radionuclides was less than 1 case every 10 years (or less than 100 cases in 1,000 years). Without quantitative analysis, it would not have been possible to make a compelling case for the accelerated phase down because it would not have been possible to show how much more important lead in gasoline was relative to the vast majority of other rules competing for attention, many of which involved congressional or court-imposed deadlines, in contrast to lead.

The secrecy with which the initial benefit-cost analysis was conducted also suggests that senior agency managers were not simply looking to analysis to justify an action they had already decided on for other reasons. They wanted to retain flexibility, with their decisions depending in large part on the results of the analysis, and wanted to avoid the outside pressures that would arise once the possibility of action became widely known.

Why Did Analysis Play Such a Prominent Role?

Several factors contributed to the integral part that quantitative analysis played in the 1985 lead-in-gasoline rulemaking.

Quality and Robustness of the Analysis. Lead had long been recognized as a toxic substance, at least at sufficiently high doses, and had been studied extensively. Thus, the scientific database available for estimating the damages caused by lead was firmer than for most substances. Many of the basic issues—such as the extent to which lead in gasoline contributed to lead toxicity in children—had been addressed in earlier rulemakings. The primary author of the benefit-cost study—Joel Schwartz—had been a key participant in the earlier rulemaking and also brought an unusually broad combination of skills to bear on the problem.[13]

In addition, the analysis took a fairly cautious approach to quantifying health benefits. For example, no benefits were quantified for children with blood-lead levels below the then-current CDC level of concern (30 μg/dl in 1984, 25 μg/dl in 1985). Moreover, the study team developed benefit estimates in multiple categories and total benefits exceeded costs by a large margin. Thus, although all of the individual estimates were uncertain, the conclusion that reductions were desirable was robust; eliminating any one benefit category altogether, doubling costs, or halving all benefit estimates, for example, still left benefits larger than costs, even if one ignored the blood pressure results entirely. As a result, senior EPA officials could have reasonable confidence that the analysis would hold up under attack, as it did. During the proceedings, only the Lead Industry Association attempted to refute the conclusion that reducing lead in gasoline would have major health benefits. Some refiners disputed the cost modeling—particularly with respect to the speed at which the phase down could occur—but they were unable to provide compelling evidence to the contrary.

Analysis Preceding Decisions. Many EPA regulations are not subject to any analysis of benefits and costs. Even when such analyses are performed, they are typically conducted in parallel or subsequent to the development of regulations. Usually the primary motivation for conducting the RIA is to meet the requirements of executive orders that apply to major rules, not to inform decisionmakers.[14] Thus, the results are generally not available until most of the fundamental regulatory decisions have been made; at that point, bureaucratic momentum, often coupled with strong external pressures or deadlines, makes it very difficult for the analysis to have much influence. In contrast, the analysis of lead in gasoline was finished *before* key decisions had to be made. Moreover, as new information became available or new issues were raised, additional analyses were performed quickly, so that decisionmakers always had the relevant analysis available before they had to choose among options. The fact that the analysis was done primarily by agency staff, rather than by outside contractors, helped integrate the analysis more closely into the decisionmaking process.

Management Support. During his second term as EPA administrator, Ruckelshaus had made "risk assessment" and "risk management" major themes. He and his deputy, Alm, both emphasized the need to focus agency efforts on the most serious risks. They were also sensitive to criticisms that "risk management" was merely a sophisticated rationale for loosening environmental controls. The lead analysis was appealing on both grounds; it identified an opportunity to achieve significant health gains, and it also demonstrated that risk assessment and risk management techniques could be used to support tougher regulation, not just to justify weaker controls.

Support for an analytic approach extended to lower levels of management as well, in large part because responsibility for the analysis was lodged in OPPE, not the program office. The first-, second-, and third-level managers responsible for the analysis were all economists, all were committed to increasing the role of analysis in agency decisionmaking, and all saw lead as an important opportunity to help increase the legitimacy of benefit-cost analysis within and outside the agency.[15] Similarly, the Office of Information and Regulatory Affairs (OIRA) at OMB embraced the lead-in-gasoline rule as an example of how the RIA process could lead to improved regulation.[16] Moreover, the assistant administrator for air and radiation, Joseph Cannon, had previously been the assistant administrator for policy. In that capacity, he had been supportive of OPPE's work on lead during the 1982 rulemakings, when it helped turn back attempts to loosen the limit.

Little External Pressure. The lack of strong pressure from Congress, courts, or environmental groups gave the agency additional freedom to base decisions on costs and benefits, rather than on political considerations or legal requirements and schedules. In addition, industry response was mixed; after the analysis was released, virtually all refiners accepted the inevitability of sharp reductions in lead. On the issue of the timing of the phase down, the industry was split; although most complained that the proposed schedule was too tight and called for extending the schedule, a sizable minority actively welcomed it because they had excess octane-reducing capacity and would reap a competitive advantage from a rapid reduction. Although the lead industry was strongly opposed to the rulemaking, it had relatively little clout left by 1984.

Analysis Supporting Regulation. At the risk of sounding cynical, one of the key factors in securing support for the analysis was that it supported tight regulation. As noted earlier, that fact helped increase the enthusiasm of top agency officials, who were eager to counter charges that risk management always meant doing less. It also helped secure the support of the program office, which at the same time was arguing strongly against OPPE's use of quantitative analysis in other rulemakings in which economic analysis was unfavorable to other proposed regulations that the program office had been developing for several years to meet legal deadlines. Finally, it neutralized environmentalists, who usually oppose the use of benefit-cost analysis; here the analysis was pushing a tighter, faster rule than they had expected.

Lessons for the Future

As the foregoing discussion suggests, conditions in the case of lead in gasoline were extraordinarily favorable for analysis to influence policy. In the decade since the final rule was promulgated, no other cases have duplicated that combination of factors. The experience with lead in gasoline did not signal the beginning of some golden age for analysis. On the other hand, neither should it be viewed as a fluke that offers no lessons for those interested in making greater use of benefit-cost analysis to help improve environmental regulation.

Scientific and Technical Foundations Are Crucial. One of the reasons that the benefit-cost analysis of lead in gasoline was so compelling was that it was built on relatively strong foundations that had been laid in earlier rulemakings and research. The exposure estimates and the estimated effects of lead on blood pressure were both based on the NHANES II study, which had been conducted years before the analysis started in 1983. Without that large-scale survey study—which was *not* conducted by EPA—the exposure estimates would have been much more uncertain and much less credible.

The study also benefited greatly from the fact the refining model used to estimate costs was already developed and could be applied fairly easily. Without it, the cost estimates would have been much more uncertain and would have had much less credibility. The refinery model and the NHANES study both illustrate the value of longer-term research and development efforts to provide the building blocks for analysis.

Monetizing Benefits Is Valuable. Many people find it acceptable to quantify effects (for instance, to estimate the number of deaths or cases of illness prevented) but object to placing monetary values on those effects; that is, they accept cost-effectiveness analysis, but reject benefit-cost analysis. The lead-in-gasoline case, however, illustrates that monetizing benefits can be valuable when there are multiple benefit categories. As discussed earlier, the RIA quantified and partially monetized benefits in four broad categories, ranging from less frequent oil changes to reduced impairment of children's cognitive abilities. These disparate categories did not lend themselves to cost-effectiveness analysis because they were in very different units, with no clear effectiveness metric.[17] Adopting a benefit-cost framework made it easier to include multiple categories. It was also important, however, to keep track of physical units (number of children with elevated blood-lead levels, number of heart attacks, and so forth), so that the results could be summarized in those terms, as well as in dollars.

Management Support and Interest Is Necessary. Unless senior managers in the agency see benefit-cost and related forms of analysis as tools for helping them make better decisions, RIAs are unlikely to become an important part of the regulatory process. As noted earlier, the administrator and the deputy administrator were both keenly interested in trying to make the risk management process more deliberate and rational. Thus, they were highly supportive of the analytic effort. That interest created a sense of urgency among the members of the study team, and also helped head off potential opposition from the program office.

Analysis Can Speed Rather Than Paralyze Decisions. The lead-in-gasoline case shows that high-quality analysis need not be extremely resource intensive, nor need it delay regulatory development. As discussed earlier, the 1985 phase-down rulemaking proceeded very rapidly relative to other major rules, in significant part because the analysis made such a compelling case. Moreover, despite its unusually broad scope, the RIA did not consume unusually large amounts of staff time or contract dollars. A 1987 review of RIAs completed after 1981 estimated that the lead-in-gasoline RIA used roughly eight person-years of EPA staff time and just over $500,000 in contractor expenses, which was not significantly higher than other RIAs that attempted to measure both costs and benefits (U.S. EPA 1987).

Proactive Analysis Is Preferable. Another lesson is that analysis should be proactive, not reactive, where possible. Too often analysis is done only after an issue has achieved prominence and fundamental decisions have been made. At that point, internal bureaucratic imperatives and external political pressures leave little opportunity for analysis to have much impact. On the other hand, if analysis is part of the policy process from the start, it has a greater chance of influencing the policy in fundamental ways. The ideal is when analysis precedes the policy process and helps guide priorities. Unfortunately, EPA is often so overwhelmed by congressional and court-imposed deadlines that it has limited opportunities to set rational priorities.

Building Credibility for Benefit-Cost Analysis

A related lesson is that analysis is likely to have more influence when it identifies previously unrecognized opportunities for cost-effective health and environmental gains, rather than when it attempts to show that a proposed action fails to yield benefits commensurate with costs. Such opportunities may involve currently unregulated areas, or, as with lead in gasoline, they may entail a fresh look at an area where regulations are already on the books.

More analyses that show positive net benefits for environmental regulations would be helpful in at least three ways. First and most obvious are the direct gains that would come from implementing cost-beneficial controls. Second, in a world where environmental programs compete with one another for scarce agency resources and public attention, undertaking one rulemaking often means delaying (perhaps indefinitely) other actions. Thus, identifying desirable actions may also help indirectly by crowding out undesirable ones. Third and finally, if "negative" analyses are to have credibility, we need more examples of "positive" analyses to counter the widespread perception that benefit-cost analyses and other related techniques are inherently biased against environmental concerns. The success of the lead-in-gasoline RIA provides a powerful counterexample and illustrates the very useful role that good analysis can play in environmental policymaking.

ACKNOWLEDGMENTS

I thank Richard Morgenstern (U.S. EPA and RFF), Milton Russell (University of Tennessee), Alvin Alm (U.S. DOE), Ronnie Levin (U.S. EPA), and Alan Loeb (Argonne National Laboratory) for helpful comments on an early draft, based on their own experience in the lead-in-gasoline rulemaking while they were at EPA. I also thank Ellen Silbergeld (Johns Hopkins University) and Robert Stavins (Harvard University) who reviewed a later draft and provided important insights, both in writing and orally at a conference held at RFF. David Rowland of NERA provided his usual excellent research assistance. As always, however, responsibility for any errors rests with the author.

ENDNOTES

[1]An environmental group did file suit, but only later, when word leaked about the benefit-cost study and its conclusions. Other prominent environmentalists—such as Ellen Silbergeld of the Environmental Defense Fund—had been active for many years in efforts to reduce lead. Nonetheless, lead in gasoline was not the subject of litigation nor did it seem to be a high priority among major environmental groups.

[2]The authors of the draft report (Joel Schwartz, Jane Leggett, Bart Ostro, Hugh Pitcher, and Ronnie Levin) were all analysts within OPPE's Office of Policy Analysis.

[3]"Regular" leaded gasoline had an octane rating of 89; to reach that level with only 0.1 gplg of lead would require producing gasoline with an octane rating of more than 88 *before* lead was added. In contrast, "regular" unleaded generally had an octane rating of only 87. Thus, even if one ignored the cost of the lead itself, it was slightly more expensive to produce leaded gasoline (at 89 octane) than unleaded (at 87 octane).

[4]The link between moderate blood-lead levels and cardiovascular disease remains controversial more than a decade later.

[5]The division in the Air Office that was responsible for enforcing gasoline regulations had some problems enforcing the lead rules, with protracted litigation and some refineries (especially small ones) going bankrupt before fines could be collected. At least some of the division's staff and management attributed these problems to trading, although the problems encountered were really due to other factors, primarily limited enforcement resources. Some refineries, for example, submitted reports that clearly showed they violated the rules; they counted on slow enforcement to delay action to a time when they were going to be out of business anyway. (Many of these refineries were artifacts of the oil entitlements program, which had allocated extra amounts of price-controlled domestic oil to small refineries. With the removal of price controls by the Reagan administration, many such refineries were no longer economically viable.) Such strategies for violating the lead rules would have been just as feasible if there had not been any trading allowed.

[6]Even with this link, some officials in the Air Office were still reluctant to allow banking and continued trading; they did not see the intermediate limit of 0.5 gplg in 1985 as very valuable because it was unlikely to have much impact on misfueling (as opposed to direct health effects of lead itself), which continued to be their primary goal in the rulemaking.

[7]The banking rule was not published until April 2, 1985, but its effective date was March 25, 1985, so that it took effect before the end of the first quarter of the year. Thus, in effect, refiners were able to bank reductions below 1.1 gplg from the beginning of the year, as EPA had assured them they would be able to do.

[8]The authors of the final RIA were Joel Schwartz, Hugh Pitcher, Ronnie Levin, Bart Ostro, and Albert Nichols (the author of this case study). Jane Leggett, who had been part of the original team, had left Washington for an assignment with the Organization for Economic Cooperation and Development in Paris. Nichols, who had become director of the Economic Analysis Division, had become deeply involved in the analysis after the release of the draft report.

[9]As discussed in more detail in the section on benefit-cost comparisons, the cost and benefit estimates depended on how much the rule would reduce misfueling. Unless otherwise indicated, the estimates in the chapter reflect the "partial misfueling" assumption in the RIA, which assumed that the 0.1 gplg limit would eliminate 80% of misfueling.

[10]This is a basic result from the theory of depletable resources and applies to all trading programs involving banking. If the price rose at a rate less than the interest rate, refiners would find it profitable to use lead rights early, investing the savings elsewhere. If the price rose faster than the interest rate, refiners would find it profitable to invest in early reductions in lead, thus delaying the use or sale of the lead rights until their value was higher.

[11]In October 1991, the level of concern was lowered to 10 μg/dl (CDC 1991).

[12]*All* benzene emissions from gasoline resulted in less than fifty cancer cases per *year*, based on the usual conservative (high) estimation methods used for car-

cinogens (U.S. EPA 1985a, VI-47), less than half of the estimated number of cardiovascular deaths associated with lead per *week*.

[13]Schwartz was a physicist by training but had become well versed in energy economics and epidemiology. Thus, he was able to take the lead both in the analysis of health effects and in working with the outside contractor on the linear programming model of the refining industry. He recently left the government to become a professor in the Environmental Sciences Department of the Harvard School of Public Health.

[14]At the time of the 1985 rulemaking, the applicable executive order was 12291. The current order is 12866.

[15]Albert Nichols, an economist on leave from Harvard's John F. Kennedy School of Government, was director of the Economic Analysis Division and was closely involved with the lead-in-gasoline analysis. His superior was Richard Morgenstern, then director of the Office of Policy Analysis, who had also been an academic economist prior to joining the government. Milton Russell, the assistant administrator for policy, planning and evaluation, had previously directed the energy program at Resources for the Future.

[16]OMB was supportive of the rule throughout the process. The relationship with OMB was helped by the fact that Nichols and Christopher DeMuth, the director of OMB's OIRA, had been colleagues at Harvard's Kennedy School until DeMuth joined OMB in 1981. DeMuth's support of the 1984–1985 rulemaking stood in sharp contrast to his earlier role as the director of the Vice President's Task Force on Regulatory Relief, which had urged EPA to *loosen* the lead-in-gasoline rules in 1982.

[17]In theory, one could develop relative weights for a composite effectiveness measure. However, monetary values seem more transparent than trying, for example, to estimate the trade-off between fouled spark plugs, lower NO_x emissions, and reduced cognitive effects.

REFERENCES

Caldwell, J. 1995. Telephone conversation with D. Rowland regarding average lead content. September. Ann Arbor, Michigan: U.S. EPA Office of Mobile Sources.

CDC. 1991. *Preventing Childhood Lead Poisoning*. Atlanta: Centers for Disease Control.

Hahn, R.W. and G.D. Hester. 1989. Marketable Permits: Lessons for Theory and Practice. *Ecology Law Quarterly* 16(2): 261–406.

Haigh, J., D. Harrison, and A. Nichols. 1984. Benefit-Cost Analysis of Environmental Regulation: Case Studies of Hazardous Air Pollutants. *Harvard Environmental Law Review* 8.

Harlan, W.R., J.R. Landis, R.L. Schmouder, and others. 1985. Relationship of Blood Lead and Blood Pressure in the Adolescent and Adult U.S. Population. *Journal of the American Medical Association*. January 25.

McGee, D. and T. Gordon. 1976. The Results of the Framingham Study Applied to Four Other U.S.-based Epidemiologic Studies of Coronary Heart Disease. Section 31, DHEW Pub. No. NIH76-1083. Washington, D.C.: Government Printing Office.

Nichols, A.L. 1983. The Regulation of Airborne Benzene. In T. Schelling (ed.), *Incentives for Environmental Protection*. Cambridge, Massachusetts: MIT Press.

———. 1985. The Costs and Benefits of the New U.S. Phasedown of Lead in Gasoline. In M. Baier (ed.) *Energy and Economy Global Interdependencies*. Volume 9 of *Implications of Environmental Issues*. Bonn, Germany: IAEE Conference of International Association of Energy Economists, June.

Nichols, A.L. and R. Zeckhauser. 1986. The Dangers of Caution: Conservatism in Assessment and the Mismanagement of Risk. In V.K. Smith (ed.) *Advances in Microeconomics*. Greenwich, Connecticut: JAI Press.

Pirkle, J.L., J. Schwartz, J.R. Landis, W.R. Harlan. 1985. The Relationship Between Blood Lead Levels and Blood Pressure and Its Cardiovascular Risk Implications. *American Journal of Epidemiology* 121(2): 246–58.

U.S. DOE (Department of Energy). 1982. *Trends in Motor Gasolines: 1942–1981*. DOE/BETC/RI-28/4. June. Bartlesville, Oklahoma: Bartlesville Energy Technology Center.

U.S. EPA (Environmental Protection Agency). 1983. *Motor Vehicle Tampering Survey, 1982*. April. Washington, D.C.: National Enforcement Investigations Center.

———. 1984. *Costs and Benefits of Reducing Lead in Gasoline: Draft Final Report*. EPA-230-03-84-005. March. Washington, D.C.: Office of Policy Analysis.

———. 1985a. *Costs and Benefits of Reducing Lead in Gasoline: Final Regulatory Impact Analysis*. EPA-230-05-85-006. February. Washington, D.C.: Office of Policy Analysis.

———. 1985b. *Supplementary Preliminary Regulatory Impact Analysis of a Ban on Lead in Gasoline*. February. Washington, D.C.: Office of Policy, Planning and Evaluation.

———. 1987. *EPA's Use of Benefit-Cost Analysis: 1981–1986*. August. Washington, D.C.: Office of Policy, Planning and Evaluation.

5

Water Pollution and the Organic Chemicals Industry

Peter Caulkins and Stuart Sessions

In November 1987, EPA promulgated final effluent guideline regulations for the organic chemicals, plastics, and synthetic fibers (OCPSF) industry. Among the most costly and complex effluent guidelines issued under the Clean Water Act, the OCPSF guidelines were supported by a full regulatory impact analysis. This RIA had three main components: a direct-cost analysis (called the "cost analysis" in this chapter), an employment and industry impact analysis (here called the "economic impact analysis"), and one of the first attempts at a water-related benefits analysis (here termed the "benefits analysis"). In this case study, Peter Caulkins and Stuart Sessions summarize and critique all three components of the RIA and proceed to show how the cost and economic impact analyses were most influential in the final rulemaking. Although the benefits analysis did not affect this particular regulatory decision directly, it did identify the issue of cross-media transfer, which was incorporated into subsequent Clean Air Act rulemakings. After reflecting on various reasons for its limited impact, the authors suggest that EPA did not view benefits issues as central to making decisions about a technology-based standard because the

---

PETER CAULKINS is an environmental economist with the U.S. EPA. His work on the effluent guidelines for the OCPSF industry concentrated on the benefits of the regulations. STUART SESSIONS is an environmental consultant. For four years during development of the OCPSF effluent guidelines he was director of the Regulatory Policy Division in EPA's Office of Policy, Planning and Evaluation and was involved in various aspects of the rulemaking. The views expressed in this paper are those of the authors and do not represent official EPA positions.

law was specific about requiring "best available" technologies. Furthermore, the water quality benefits analysis might have been seen as challenging the merits of nationally uniform technology-based standards and thereby contesting the regulatory decisionmaking paradigm already established by the agency. Finally, an effluent guideline rulemaking process that gave serious consideration to water quality benefits would have required the development of large databases on current and potential uses of water bodies as well as the socioeconomic characteristics of potential user groups.

INTRODUCTION

To understand the OCPSF effluent guidelines, one first should know some details of their statutory and historical basis as well as of the industry itself. This background is provided below.

Statutory Basis for the OCPSF Effluent Guidelines

The Clean Water Act (Federal Water Pollution Control Act Amendments of 1972) established a comprehensive program to "restore and maintain the chemical, physical and biological integrity of the nation's waters" (Section 101[a]). States were to establish water quality standards for individual water bodies. Both point and nonpoint sources were to be controlled in order to meet the water quality standards, with differing approaches prescribed for each. Point sources were to be governed by technology-based effluent standards; that is, effluent guidelines for industrial point source dischargers and secondary treatment and other requirements for municipal dischargers. These technology-based standards were to be established category by category: a set of standards for all organic chemical plants, another set for all foundries, another set for all publicly owned treatment works (POTWs), and so forth. For this reason, these standards are now referred to colloquially as "categorical standards." The appropriate categorical standards were to be incorporated into a discharge permit for each point source, converting the broad categorical standards into enforceable specific limits for each individual discharger. The categorical standards provided a technology-based floor for the permit limits developed for each discharger. If compliance with the categorical standards by all point sources affecting a water body would be insufficient to achieve the ambient water quality standard for that water body, the permits for the relevant dischargers were to include more stringent water quality based control requirements in addition to the technology-based requirements.

The Clean Water Act directed EPA to establish several different sets of technology-based requirements for industrial dischargers:

- One set of requirements for existing dischargers and another for new dischargers.
- One set of requirements for facilities that discharge directly to surface waters—"direct dischargers"—and another set for facilities that discharge to sewers that connect to POTWs before discharging to surface waters. (The standards applicable to these "indirect dischargers" were to take account of the treatment provided by the subsequent POTW, addressing only those pollutants that pass through or interfere with the operations of the POTW.)
- Two sets of requirements for control of conventional pollutants (biological oxygen demand [BOD], suspended solids, oil and grease, fecal coliforms, and pH) and another set for control of toxic and nonconventional pollutants. ("Toxic" pollutants were listed as such in the act, while "nonconventional" pollutants are any others in addition to conventional and toxic pollutants.)

These sets of requirements are collectively referred to as effluent guidelines. The act required EPA to develop guidelines for twenty-seven listed categories of industrial sources, at a minimum. EPA generally chose to develop all the different sets of requirements for a single category of industrial dischargers simultaneously. Thus the effluent guidelines for the organic chemicals industry included all the standards applicable to plants in the industry—for existing and new plants; direct and indirect dischargers; and conventional, toxic, and other pollutants. The specific requirements in the act for effluent guidelines are as follows.

Section 301(b)(1)(A) set a deadline of July 1, 1977 for existing industrial direct dischargers to achieve effluent limitations requiring the application of the *best practicable technology* (BPT) for controlling discharges. BPT effluent limitations are generally based on the average of the best existing performance by plants of various sizes, ages, and unit processes within the industrial category for the control of conventional pollutants. In establishing BPT effluent limitation guidelines, EPA considers the total cost of the requirement in relation to the benefits from the resulting effluent reductions. The age of equipment and facilities involved, process changes required, engineering aspects of the control technologies, and nonwater quality environmental impacts are also taken into consideration.

Section 301(b)(2)(A) of the act set a deadline of July 1, 1983 for industrial direct dischargers to achieve effluent limitations requiring the application of the *best available technology* (BAT) economically achievable. BAT effluent limitations are generally based on the best existing performance in the industrial category for the control of toxic and nonconventional pollutants. While the reasonableness of costs and their economic impacts may be taken into consideration in the determination of these limits, the primary determinant of BAT is the effluent reduction capability of the control technology.

The 1977 Clean Water Act Amendments added Section 301(b)(2)(E), establishing *best conventional technology* (BCT) requirements for the discharge of conventional pollutants from existing industrial point sources. BCT was intended not to be an additional limitation, but to replace BAT as the incremental level of control beyond BPT for conventional pollutants. In establishing requirements for incremental control beyond BPT, the act directed EPA to consider a different mix of factors in addressing conventional pollutants than those used in addressing toxic pollutants. BAT for toxics was to be based largely on technological capabilities, whereas BCT for conventional pollutants was to be subject to a test of whether the effluent reduction was worth the costs.

Section 306 required EPA to develop new source performance standards (NSPS) applicable to newly constructed direct dischargers. NSPS are based on the performance of the best available demonstrated technologies for all pollutants.

Sections 307(b) and (c) required that *pretreatment standards* for *existing sources* and *new sources* (PSES and PSNS) be set for indirect dischargers to prevent the discharge of toxic pollutants that pass through, interfere with, or are otherwise incompatible with the operation of POTWs. (*Passthrough* is generally determined to exist for a pollutant if the average removal percentage realized by well-operated POTWs achieving secondary treatment is less than the percentage removed by the BAT for the industry for that pollutant.)

Table 1 summarizes the various requirements that were to be established for plants in each industrial category and that collectively comprise the effluent guidelines for an industry.

History of the OCPSF Effluent Guidelines

EPA began developing effluent guidelines in the early 1970s. EPA initially interpreted the effluent guidelines as being technology-based limitations

Table 1. Requirements Comprising an Effluent Guideline.

		Requirements for	
Type of discharger	*Pollutants controlled*	*Existing sources*	*New sources*
Direct	Conventional pollutants	BPT, BCT	NSPS
	Toxic and nonconventional pollutants	BAT	—
Indirect	Pollutants that pass through or interfere with POTWs	PSES	PSNS

requiring a balancing of treatment effectiveness with potential economic dislocations within the affected industry. This led to a first generation of guidelines addressing mostly conventional pollutants that were promulgated as quickly as possible in an attempt to meet statutory deadlines. The guidelines were based on limited and reasonably available information. Numerous court challenges to the guidelines ensued, filed by both industry and environmental groups. Many of the initial guidelines were remanded to the agency for reconsideration.

For the organic chemicals sector, EPA promulgated the original effluent limitation guidelines in two phases. Phase I, covering forty product processes, was promulgated in 1974, and Phase II, covering twenty-seven additional product processes, was promulgated in 1976. Effluent limitation guidelines for the plastics and synthetic fibers sector were also promulgated in two phases—Phase I, covering thirty-one product processes, in 1974 and Phase II, covering eight additional product processes, in 1975.

These regulations were challenged. In the 1976 *Union Carbide v. Train* case, the Court remanded the Phase I organic chemical regulations, and the agency withdrew the Phase II regulations. The Court also remanded the Phase I plastics and synthetic fibers regulation (*FMC v. Train*), and the agency withdrew the Phase II regulations in 1976. Only the regulations covering butadiene production in the organic chemicals sector and the pH regulations for the plastics and synthetic fiber sectors were not part of the lawsuits and were left intact.

EPA was unable to meet many of the deadlines specified by the 1972 Federal Water Pollution Control Act. As a result, EPA was sued in 1976 by several environmental groups (*Natural Resources Defense Council, Inc. v. Train*). The settlement agreement required EPA to develop a program and meet a court-ordered schedule for controlling sixty-five "priority" toxic pollutants through the promulgation of BAT and PSES effluent limitation guidelines for twenty-one major industries, including the OCPSF industry. Much of the settlement agreement was incorporated into the 1977 Clean Water Act. EPA then began a second generation of effluent guidelines to address both toxic and conventional pollutants, under tighter statutory standards, and subject to a court-ordered timetable.

In this new round of effluent guidelines activity, EPA eventually proposed OCPSF effluent guideline regulations in March 1983, and promulgated final regulations in November 1987. The analyses EPA conducted to support these OCPSF guidelines, and the influence of these analyses on EPA's key regulatory decisions, are the focus of this study.

Overview of the OCPSF Industry

As EPA began a new round of effluent guidelines, the agency recognized the OCPSF industry, producing a large volume of effluents with significant levels of toxic pollutants, as one of the largest sources of water pollution among the manufacturing industries regulated. In 1982, the OCPSF industry produced 185 billion pounds of product valued at $59 billion. Approximately 183,000 workers were employed. In 1984, the industry accounted for 5% of U.S. exports, and 1% of U.S. imports.

At the time the regulations were being developed, more than 25,000 different organic chemicals, plastics, and synthetic fibers were being manufactured, but less than half of the products were produced in amounts exceeding 1,000 pounds per year. Seven hundred and fifty plants were considered to be primary producers of OCPSF products, and 190 other plants were considered secondary producers. Secondary OCPSF facilities were often part of other industries producing other chemicals or products, such as petroleum refining, inorganic chemicals, pharmaceuticals, pesticides, adhesives and sealants, paint and ink, and plastics molding and forming.

Some plants produced chemicals in large volumes, while others produced only small volumes of specialty chemicals. Large-volume production tended to use continuous processes that were more efficient than batch processes in minimizing pollution and water use. Different products were made by varying the raw materials, the chemical reaction conditions, and the chemical engineering unit processes. A single large plant could produce many different products simultaneously using a variety of continuous and batch operations. The products manufactured at a plant and the effluents that resulted could vary on a weekly or even on a daily basis.

The approximately 940 plants in the OCPSF industry in the early 1980s were distributed among regulatory categories as noted in Table 2. *Direct dischargers* were typically large plants, often located on large water bodies in order to meet needs for shipping access and ample water supplies. *Indirect dischargers* were typically smaller but in total also constituted a large source of toxic pollutants. The nondischarging plants (*zero dischargers*) relied on deep well injection, incineration, contract hauling, or evaporation and percolation ponds to dispose of their wastewaters.

Table 2. Categories of Dischargers among OCPSF Plants (1987).

Category	*Number of OCPSF plants*	*Average daily wastewater flow per plant (mgd)*
Direct dischargers	300	1.31
Indirect dischargers	395	0.25
Zero dischargers	245	0
Total	940	0.52

Source: U.S. EPA 1987c, pp. 42525–26.

The nature of the OCPSF industry made EPA's task in developing effluent guidelines for the industry a very difficult one. The industry was very large and diverse. Across the industry a wide array of products was produced using different raw materials, chemical reaction conditions and engineering processes, resulting in differing effluents that were treatable to varying degrees. Effluents were also variable over time for a single plant as the plant's product mix shifted. Developing efficient, cost-effective, and equitable regulations for such variable waste streams was exceedingly difficult.

Other difficulties for EPA stemmed from the legal context for these guidelines. The 1976 decision in the case of *Union Carbide v. Train*, remanding the earlier OCPSF effluent guidelines, required EPA to develop extensive new technical information about the industry. As well, the time available to the agency to perform this work was sharply constrained by the court-ordered deadlines imposed by the subsequent settlement agreement in *NRDC v. Train*.

EPA'S OCPSF DATA COLLECTION AND ANALYSIS

At approximately the same time as the court remanded the original regulations in 1976, EPA began a large new effort to collect and analyze data on the industry. This work continued for a decade, ultimately resulting in many thousands of pages of data and studies supporting the final rule in 1987.

The analytical effort was organized in a way that had become largely standardized across all of the several dozen effluent guidelines developed by EPA during this period. EPA interpreted its legal mandate as developing effluent limitations that would be achievable by plants in the industry if they were to install the control technologies judged to be "best practicable," "best available," "best conventional," and so forth. The standards issued by EPA were to be in the form of numerical effluent limitations that a regulated source could choose to meet in any way the source wished. The regulations did not require that specified control technologies be installed—they were written as performance standards rather than as design standards. However, the effluent limitations specified in the regulations were those that EPA believed could be achieved by the technologies selected by EPA. Thus, the key decisions for EPA in every effluent guideline were to determine the technologies on which the regulation was to be based (known colloquially as the "technology basis" for the regulation) and the performance level that the chosen technologies could be expected to provide when applied within the industry being regulated. Most of EPA's analytical work to support development of effluent guidelines therefore focused on potential control technologies and their effectiveness.

Activities in Developing the Regulation

For the OCPSF industry, as for other effluent guidelines, the basic technology-oriented data were collected as follows:

- surveying establishments in the industry to obtain basic background on products manufactured, processes employed, toxic chemicals used, control technologies employed, economic status, and additional available data sources;
- collecting sampling information from the industry on the levels of pollutants in raw (untreated) wastewater and the reductions that are achieved in wastewaters that have been treated using different technologies; and
- further evaluating the performance of promising technologies, including obtaining more targeted data from the industry and sometimes performing bench tests or pilot plant tests or obtaining performance data from a related industry.

This information was used in a regulatory development process that included the following seven activities.

1. *Define the scope of the regulated industry.* EPA determined which sorts of facilities to consider as being in the industry and subject to the OCPSF regulations and which should be left to other regulations.
2. *Determine which pollutants to regulate.* These would include, for direct dischargers, any conventional, toxic, or other pollutants discharged from plants in the industry in amounts that could be harmful. For indirect dischargers, the regulated pollutants would include any that pass through or interfere with the operations of the POTWs receiving the industry's waste water.
3. *Evaluate the available control technologies for these pollutants and their performance in the industry.* As noted, this step involved very extensive data collection from the industry as well as sampling by the agency.
4. *Develop subcategories within the industry.* To the extent that portions of the OCPSF industry differed significantly from each other in terms of products made, effluent characteristics, effluent treatability, economic capability, and so forth, EPA aimed to develop effluent limitations tailored for each subcategory.
5. *Develop regulatory options.* For each of the standards (BPT, PSES, and so forth) and for each OCPSF subcategory, several alternative technology bases were developed. A candidate technology basis typically included not just a single control technology, but an entire treatment train, including combinations of specific in-plant measures (water use

reduction, stream separation, lime precipitation for metal-bearing streams, steam stripping for streams with volatile organic pollutants) and end-of-pipe measures (such as biological treatment, and polishing filters). EPA's major regulatory choice then involved deciding which among these alternative treatment trains to select as the basis for establishing effluent limits.

6. *Evaluate the impacts of different regulatory options.* This step included three different analyses: cost analysis, economic impact analysis, and analysis of effluent reduction benefits. Each is described below.
   - *Cost analysis.* Engineering costing procedures were developed to project the capital and operating/maintenance costs of applying each candidate technology at a regulated facility. The cost-estimating approach used by EPA varied from guideline to guideline. Industrywide compliance costs were sometimes estimated by costing the technology at all facilities, then reducing the estimate by a "credit" amount at each facility that represented the value of the treatment technology already in place at that facility. Treatment in place was determined from the baseline industry surveys. Sometimes compliance costs were estimated by determining for each facility the additional treatment steps needed for compliance and then costing them. Sometimes compliance costs were estimated for each of several "model plants" rather than for actual facilities in the industry. An industrywide cost estimate was then developed by assigning each actual facility to a most closely representative model plant and aggregating costs across the models. (For the OCPSF guideline, several of these different costing approaches were employed at various times during regulatory development.)
   - *Economic impact analysis.* Economic analyses were performed to project the impacts on the industry and the economy from the estimated compliance costs. The impacts investigated included closure of plants and production lines, local community impacts, and changes in product prices, firm profitability, employment, international trade, and industry concentration. The economic impact analysis began with information from the industry survey on each plant's baseline financial operating condition prior to incurring compliance costs for this regulation. Various financial and economic models were then used to estimate the impacts of the compliance costs. A novel aspect of the economic impact analysis for the OCPSF industry was incorporation into the analysis of the expected costs of several environmental requirements under statutes other than the Clean Water Act (CWA). For both the OCPSF industry and other effluent guidelines, EPA also

performed a regulatory flexibility analysis (RFA) that was closely related to the economic impact analysis. The RFA was required to evaluate whether the regulation created a disproportionate effect on small businesses.

- *Analysis of effluent reduction benefits.* The CWA required EPA to establish effluent guidelines by considering, among other things, compliance costs in relation to effluent reduction benefits. Therefore, for each effluent guideline, EPA estimated the nationwide amount of discharge of each pollutant that would be abated through compliance with the regulation. EPA also performed a dilution analysis to assess the significance of the loading reduction that was expected from each direct discharger. For each discharger, the precompliance effluent load for each pollutant was compared with the flow level in the receiving water body to determine whether the discharge was likely to result in violations of ambient water quality criteria for any pollutants. Similar calculations were performed for the postcompliance effluent load for each discharger. A comparison of the precompliance and postcompliance results then showed how many violations of water quality criteria across the country could be eliminated by the regulation.

  For nearly all the effluent guidelines issued by EPA, these two simple analyses (amount of effluent reductions and dilution analysis regarding elimination of violations of water quality criteria) were the only ones performed that evaluated the benefits of the regulation. However, for effluent guidelines with a projected compliance cost exceeding $100 million annually, EPA also performed a more detailed benefits analysis as required by Executive Order 12291. As of 1987, the OCPSF guideline and the guideline for the iron and steel industry (promulgated in 1982) were the only two for which full RIAs had been performed.

7. *Consider the impacts of the different regulatory options and adopt limitations for each of the industry subcategories.* By the time the OCPSF guideline was promulgated in 1987, EPA had implemented for all effluent guidelines a regular procedure for coordination among offices in regulatory development and a streamlined process for making final agency decisions and review by the Office of Management and Budget (OMB). Past decisions on numerous guidelines had also resulted in an informal and unwritten, yet important, rough sense across the agency of how stringent effluent guidelines should be. Past guideline decisions had generated precedents on issues such as: how widely demonstrated a technology must be within an industry for it to be adoptable as a basis for regulation; what degree of economic disloca-

tion might be acceptable in a regulated industry; and what minimum level of cost-effectiveness in reducing pollution loadings should be necessary for a technology to be considered as a basis for regulation. This last issue concerning cost-effectiveness was evaluated quantitatively for every guideline. For each guideline, EPA estimated the incremental cost per "pound-equivalent" for removal of the pollutants controlled by the regulation. A pound-equivalent was calculated by multiplying the number of pounds of a pollutant by a toxic weighting factor for that pollutant. The weighting factors gave relatively more weight to more highly toxic pollutants. In this manner, the relative cost-effectiveness of guidelines for different industries abating different amounts of a wide variety of different pollutants could be compared. As a general rule of thumb, candidate effluent guideline technologies that abated toxic pollutants at a cost exceeding $100 per pound-equivalent were viewed skeptically.

For each effluent guideline, all the information and analyses supporting the final guideline decisions were assembled into a set of three packages:

- A *development document* (U.S. EPA 1987a), establishing the technical foundation for the regulations by describing the industry; dividing it into appropriate subcategories; and covering the nature of the discharges from the industry, the control technologies available to abate the discharges, the actual and expected performance of the control technologies, and the costs of installing and operating the control technologies.
- An *economic impact analysis* (U.S. EPA 1987b), projecting the economic impacts of the compliance costs estimated in the development document.
- An *environmental analysis*, projecting the likely effect of different effluent guideline options on water quality. Typically, the number of expected violations of water quality criteria due to discharges from the industry were estimated before and after the regulation. For the OCPSF industry, a much broader analysis of regulatory benefits was conducted pursuant to Executive Order 12291, but the various studies comprising the benefits analysis were not compiled into a single document.

Technology-Related Data and Analysis

EPA initially required OCPSF facilities to provide detailed information related to plant locations and age, products manufactured, processes used, production rates, mode of discharge, treatment technologies employed, wastewater flows, and effluent characteristics in a 1976 BPT questionnaire and a 1977 BAT questionnaire mailed to every company in the industry. The data from these questionnaires were computerized and sent to the plants for their review and revision at the end of 1979.

Over the period from 1977 to 1981, EPA also conducted three major sampling studies to characterize raw and treated wastewaters in the industry. In total, samples were collected and analyzed from 213 plants, covering 176 product/process wastewater streams. Some of the sampling at these plants was short term (typically three days), while some was long term (often four to six weeks). Much of the data collected by EPA and that submitted by industry were generated pursuant to a separate EPA regulation prescribing sampling and analysis methods for the organic pollutants generated by the industry.

EPA analyzed these data and issued proposed effluent guidelines (BPT, BAT, BCT, PSES, NSPS, PSNS) for the industry in March 1983. The industry was divided into four subcategories for the BPT and BCT regulations and two subcategories for the other regulations, based on the general types of products manufactured at each plant. Proposed effluent limitations were established by averaging the effluent quality achieved by existing plants that were selected as being well controlled. In selecting from its database the plants that had suitable and well-operated treatment in place, EPA used a series of "editing rules." For example, in establishing proposed BPT limits for BOD, EPA averaged the performance of the seventy-one plants that achieved either 95% BOD removal or a 40-mg/l long-term average. Different editing criteria were used for BAT. EPA estimated the cost of the proposed regulation based on estimates of compliance costs for model plants, referred to as "generalized plant configurations" (GPCs). The GPCs represented typical combinations of 176 products/processes used in the industry.

EPA received significant comments on the proposed regulation, including some major criticisms: the agency' s database was inadequate to cover such a large and diverse industry; the cost of compliance was underestimated due to limitations in the coverage provided by the GPC models; the BPT subcategorization scheme was arbitrary, as minor shifts over time in the products a plant produced could move the plant into a different industry subcategory with entirely different effluent limitations; and the editing rules for selecting plants on which effluent limitations would be based were inappropriate.

To respond to the first two criticisms, the agency issued in 1983 a new survey questionnaire to all manufacturers of OCPSF products and a supplemental questionnaire to eighty-four facilities known to have installed selected wastewater treatment unit operations for which EPA sought additional information. The two surveys generated new and updated technical information on processes, production levels, raw wastewater characteristics, and treatment performance. The surveys also provided additional cost, economic, and financial data, including detailed cost information regarding capital and operating costs for specific treatment technologies. EPA also conducted another detailed sampling program at twelve plants to improve and expand the database on toxic pollutants and extend monitoring to cover the performance of in-plant controls such as steam stripping, chemical precipitation, and activated carbon adsorption.

Using the data from the new surveys, EPA revised the approach to estimating regulatory compliance costs. Models were fit to the data to project the capital and operating costs of specific treatment technologies. The agency then abandoned the GPC approach and instead estimated compliance costs for each plant in the industry, as follows:

1. EPA compared the individual plant's current effluent concentrations with those in the proposed regulation.
2. If the plant's current discharge concentrations exceeded the target levels, EPA then selected the additional treatment units or operational upgrades that would be needed at the plant.
3. EPA estimated the costs for the additional units or upgrades by applying the engineering cost models scaled to the particular plant's size and circumstances.

EPA also revised its procedure for subcategorizing the industry with regard to BPT. In the proposal, the industry had been broken into four subcategories each with different sets of BPT effluent limits. Commenters complained that minor changes in plant operations could cause a plant to shift among BPT subcategories, resulting in major changes in regulatory requirements. The agency responded by developing seven BPT subcategories, each with its own effluent limits, reflecting alternative sorts of products produced in the industry. However, rather than assign a plant exclusively to one or another of the seven subcategories based on the predominant production at the facility, EPA decided to calculate the limits applicable to a plant by combining the limits for the seven subcategories in proportion to the fraction of the plant's production in each of the subcategories. If, for example, half of a plant's production was in one subcategory and half in another subcategory, the plant's effluent limitation would be the sum of half of the limit of the first subcategory and half of the limit of the second.

During the four years following proposal of the guidelines in 1983, EPA collected information on and analyzed many other technology-oriented issues, for example:

- Developing additional candidate technology bases for BAT, PSES, NSPS, and PSNS

- Revising the editing rules for determining the effluent quality that the chosen technologies could obtain
- Evaluating controls that could prevent treatment technologies from stripping volatile pollutants out of the wastewater and into the ambient air
- Pursuing options for monitoring toxic pollutants
- Analyzing pollutant passthrough and interference at POTWs. This was necessary in establishing limits for indirect dischargers
- Transferring data from other industries on treatability of some waste streams and control technology performance

On three occasions during 1985 and 1986, EPA provided its new data and analyses for public review and comment in Federal Register notices of data availability. In total, the proposed rule and three notices of data availability generated over 2,000 technical comments from the public.

Overall, the agency was very thorough in attempting to understand and analyze technology-related issues for the OCPSF effluent guidelines. Extensive data were collected pertaining to the key issues, and the data were analyzed sensibly and objectively. The regulated industry and interested public were kept informed of the agency's activities through frequent meetings and workshops, wide distribution of key documents for comment, and formal notice and comment periods. EPA appeared quite responsive to reasonable concerns raised in public comments.

Economic Impact Analysis

Analyses were performed to estimate the economic impacts of moving from then-current levels of treatment in place in the industry to BPT, BAT, or PSES. These economic analyses measured impacts on existing plants, small plants, new plants, employment losses and associated community effects, and international trade effects. Economic impact analyses for many other effluent guidelines also assessed impacts on product prices, but this was not done for the OCPSF industry—presumably because of the analytical difficulty of dealing with the large number of different OCPSF products.

Three measures were used to assess the impacts on existing plants: closure, profitability, and cost to sales. The *closure analysis* compared the net present value of cash flow for each plant after regulatory compliance to the plant's salvage value. A closure was projected if the salvage value exceeded the net present value of cash flow. The *profitability analysis* indicated the extent to which compliance costs affected plant profitability. A significant impact was counted if the plant's profitability was reduced to the lowest decile for all plants in that particular three-digit SIC (standard industrial classsification) code. The *cost-to-sales analysis* compared compliance costs to plant sales. A significant impact was counted for a plant when the projected compliance cost exceeded 5% of sales.

Potential community impacts were assumed to result primarily from local losses in employment and earnings if a plant were to close. Since 44% of the national population was employed in 1980 (99.3 million employed out of a total population of 226.5 million) and a decline in employment of 1% nationally was considered significant, EPA counted a significant community impact as occurring in cases where the employment lost from plant closures exceeded 0.44% of the surrounding community population.

An upper bound for the impacts on international trade was estimated by assuming that all the production lost as a result of plant closures would have been exported. The resulting impact on the U.S. balance of payments was calculated.

Regulatory Issues and Some Answers from Analysis

Some major issues were faced by EPA in issuing the final regulation, These issues, and how the various analyses performed by EPA contributed to resolving them, are discussed below. The specifics of the issues are not of concern here; rather, the utility and sufficiency of the analyses in informing decisions about the issues are considered.

Technology Bases for the Regulations. Choosing the specific treatment trains that are to be the basis for the various standards is the critical decision in the effluent guidelines development process. Once the technology basis is established, the limitations themselves follow almost mechanically as editing rules are applied to the collected database on effluent quality across plants in the industry.

BPT Limits. For BPT for the OCPSF industry, EPA considered the following major technology options.

- *Option I*—Biological treatment. This typically involved either activated sludge or lagoons, followed by clarification and preceded by appropriate process controls and in-plant treatment to ensure optimal performance of the biological system.
- *Option II*—This included the Option I technology plus the addition of a polishing pond after biological treatment to achieve additional removal of total suspended solids (TSS).
- *Option III*—This included Option I plus multimedia filtration as an alternative basis (in lieu of Option II ponds) for additional TSS removals.

EPA selected Option I for BPT primarily because the treatment performance data collected from the industry did not demonstrate that polishing ponds or multimedia filtration significantly enhanced the treatment of OCPSF discharges beyond the levels achieved by Option I. Polishing ponds or filters were also not used extensively in the industry. Technical arguments were made as to why the successful performance of ponds or filters in other industries should not be assumed transferable to the OCPSF industry. EPA also noted that polishing ponds and filters had rarely been selected by EPA as a BPT technology for other industries in other effluent guidelines.

BPT Option I was projected to require 214 OCPSF plants to incur compliance costs totaling $76.6 million ($1986) on an annualized basis. No plant closures or product line closures were projected. Eight plants were projected to experience significant impacts as measured by changes in profitability or the ratio of compliance cost to sales. No job losses were expected to occur as a result of BPT.

It is clear that EPA's decision on the BPT option was based on technological concerns. The database that EPA had developed on treatment technologies used in the industry and their performance provided the information on which the decision was made. The findings of other analyses considering costs, economic impacts, cost-effectiveness, benefits (see the next section of this chapter), and legal issues were much less important to the decision. These other analyses essentially raised no issues to cause EPA to reconsider the decision EPA intended to make on sound technological grounds.

BAT Limits. For the BAT limitations, EPA divided the OCPSF industry into two subcategories. The end-of-pipe biological treatment subcategory included plants that already had or would install biological treatment to comply with BPT limits. The non-end-of-pipe biological treatment subcategory included plants that either generated such low levels of BOD that they did not need to use biological treatment or that would choose to use physical/chemical treatment alone to comply with the BPT limits. For the two subcategories, EPA considered the following major BAT technology options.

- *Option I* (BAT the same as BPT)—For the first subcategory (end-of-pipe biological treatment), the limits for toxic pollutants would be set equal to the performance expected to be achieved by the BPT end-of-pipe biological treatment. For the second subcategory (non-end-of-pipe biological treatment), the treatment basis would be in-plant controls (physical/chemical treatment and in-plant biological treatment) on appropriate waste streams to achieve the same toxic pollutant limits as would be achieved by BPT end-of-pipe treatment.

- *Option II*—This included the Option I technology plus additional in-plant controls to remove toxic pollutants from particular waste streams prior to discharge to any end-of-pipe treatment systems. The in-plant controls included steam stripping to remove volatile organic compounds, activated carbon to remove various base/neutral priority pollutants, chemical precipitation for metals, alkaline chlorination for cyanide, and in-plant biological treatment for removal of other priority pollutants.
- *Option III*—This included Option II and the addition of activated carbon adsorption at the end-of-pipe to follow biological or physical/chemical treatment.

EPA selected Option II for BAT. Option I was rejected on technological grounds without undergoing full analysis for costs, economic impacts, benefits, and so forth. The Option I technologies were capable of removing some toxic pollutants to some extent, but were far from constituting the "best available" technologies. In particular, biological treatment was thought to be insufficiently effective in removing volatile organic compounds (instead primarily transferring them to the air) and metals.

Option II would remove a very significant 1.1 million pounds per year of toxic pollutants incrementally beyond BPT. Its incremental cost over BPT was estimated at $224.2 million, affecting 289 of the estimated 300 direct dischargers in the industry. Five plants and six product lines were projected to close, resulting in 1,197 job losses.

EPA rejected Option III for BAT. While the addition of end-of-pipe carbon adsorption would achieve further reduction over Option II in discharge of less biodegradable organic compounds, the incremental costs and economic impacts associated with Option III were significant. The annualized costs of Option III were estimated at $802.6 million. Twenty-six plants and sixteen product lines were projected to close resulting in 6,475 job losses. Forty-four additional plants were projected to experience significant profitability impacts.

EPA's BAT decision thus appears to have been based both on technological grounds and on concerns about costs, cost-effectiveness, and economic impacts. Option I was rejected on a legal and technological basis: it did not constitute the best available treatment. Option III, although technically and legally acceptable, failed economic tests. It is clear that each of EPA's major OCPSF analytical efforts—characterization of available technologies and their performance, estimating costs, and projecting economic impacts—was important in guiding EPA's BAT decision.

It is also notable that neither EPA's environmental analysis nor the benefits analysis appeared to have contributed to the BAT decision, even though they would have supported the choice of Option II over Option

III. The environmental analysis showed that Option III would result in almost no increase in the number of stream segments meeting water quality criteria. The benefit-cost comparison could have been interpreted as showing that the incremental benefits of Option III were far short of the incremental costs. Perhaps curiously, these findings were not cited in the final rule preamble to support the choice of Option II, nor do we remember them playing any role in the agency debate at the time.

PSES Limits. In establishing PSES limits for the OCPSF industry, EPA did not fully and formally consider explicit technology options as the agency did for BPT and BAT. The most important PSES issues involved determining which pollutants pass through or interfere with POTWs and were therefore to be regulated. EPA believed that indirect dischargers generated wastewaters with the same pollutant characteristics as did direct discharging plants, and that the same technology options that were discussed previously for BAT were appropriate to consider for PSES. EPA decided to base PSES on the BAT Option II technologies with the exception that end-of-pipe biological treatment would not be considered necessary. End-of-pipe biological treatment for an indirect discharger was judged to be redundant to the biological treatment also provided by the subsequent POTW.

EPA estimated that PSES would impose an annualized cost of $204.3 million on 365 indirect dischargers in the OCPSF industry. Large adverse economic impacts were projected, with about 14% of these plants closing and another 17% incurring significant profitability reductions or cost-to-sales impacts. EPA considered and rejected several ways of mitigating these impacts by relaxing the PSES limits for some or all indirect dischargers.

Again, in deciding on PSES as well as BPT and BAT, EPA relied primarily on its technology-related databases and analyses, but also developed and thoroughly considered information about costs, cost-effectiveness, and economic impacts. Table 3 summarizes the industry costs and impacts of the most important regulatory options.

Adjustment of the Standards to Mitigate Economic Impacts. A second important issue considered by EPA in developing the final regulations was whether the effluent limitations resulting from the chosen technology bases should be modified to reduce the projected adverse economic impacts.

The selected BAT option (Option II) was projected to have significant impacts on small direct discharger plants that had annual production of less than five million pounds. Of the nineteen plants in this group, 80% were projected to incur significant adverse economic impacts (closures or reductions in profitability to the lowest decile). EPA determined that for

Table 3. OCPSF Costs and Economic Impacts.

Category of impact	*BPT*	*BAT II*	*BAT III*	*PSES*
Number of plants incurring costs	214	289	289	365
Annualized cost of compliance (million $1982)	$76.6	$224.2*	$802.6*	$204.3
Plant closures	0	5	26	25
Product line closures	0	6	16	27
Significant profit or sales impacts (nonclosure only)	8	11	44	62
Employment reduction	0	1,197	6,475	2,190
(Percentage of total employment in category)		(0.7%)	(3.5%)	(1.2%)

Notes: *BAT costs are incremental to BPT.

BPT (best practicable technology); BAT (best available technology), Options II and III; PSES (pretreatment standards for existing sources)

these nineteen plants BAT at the Option II level was not economically achievable; therefore, for this segment of the industry, EPA set BAT equal to BPT. This prevented the loss of 162 jobs. The incremental amount of toxic removals forgone in making this adjustment represented less than 0.07% of the reduction in toxic discharges being obtained from all direct dischargers.

The PSES requirements were also expected to cause substantial economic dislocation, as 31% of all indirect dischargers were projected to close or suffer significant impacts. EPA considered various less stringent technology options to determine whether it would be possible to afford relief to some indirect dischargers while still obtaining most of the very large reduction in pollutant discharges to POTWs that these regulations would achieve. For example, EPA considered the options of regulating only metals, or metals and cyanide, or of reducing monitoring frequency. However, no options short of total exemption reduced projected closures or other impacts substantially, and EPA was not willing to forgo regulating this sector because of the large amount of toxic discharge at stake.

EPA also considered graduating the stringency of the PSES requirements by size of plant as the agency had done for BAT. However, the fraction of small plants significantly affected by PSES was not much larger than the fraction of large plants significantly affected by PSES (in contrast to BAT, which was expected to cause much worse impacts among small dischargers than among large dischargers). EPA concluded that the impacts on small indirect dischargers, while significant, were not disproportionate to the impacts on other indirect dischargers. EPA decided not to relax the PSES requirements for small plants.

EPA's deliberations regarding possible adjustments to the standards for categories of adversely affected plants again demonstrate that the agency carefully considered the results of its cost and economic impact analyses.

Benefits Analyses

For most effluent guidelines, very little attention was devoted to estimating benefits. The statutory requirement for EPA to consider the "effluent reduction benefits" of its guidelines was interpreted literally by EPA—the agency simply projected the reductions in pollutants discharged that would result from compliance with the regulations.

The Environmental Analysis and Its Shortcomings. Only a modest effort was made to evaluate the importance of the effluent reductions achieved by a guideline. We have previously mentioned the "environmental analysis" that was performed for most guidelines. In this analysis, EPA would project for each direct discharger whether its discharge was sufficient to cause violations of water quality criteria in the receiving water. This modeling was done assuming effluent quality, first at the preregulation level and then at the postregulation level. By comparing the before and after cases, EPA projected the extent to which compliance with the guideline was likely to reduce violations of water quality criteria. The results of the analysis were typically viewed as disappointing—the effluent guidelines were rarely projected as reducing potential water quality problems significantly.

For the OCPSF guideline, for example, 170 direct dischargers affecting 134 receiving stream segments were assessed. At preregulation discharge levels they were projected to cause water quality criteria violations in 32% of the stream segments. After compliance with the regulation, the affected stream segments were projected to decline by only three percentage points, to 29%.

A similar analysis of indirect dischargers showed slightly more impact from the regulations. Before regulation, impacts were projected for five of the forty-one stream segments analyzed, and after compliance only one segment would remain affected.

Several shortcomings to this sort of analysis were recognized by the agency. For one, the analysis relied on a simple dilution calculation rather than on more sophisticated water quality modeling. The receiving water flow (in volume per unit time) was assumed to be pristine. The pollutants in the discharger's effluent flow (also expressed in volume per unit time) were then considered to be diluted into the receiving water stream—mathematically in proportion to the ratio of the receiving water flow to the effluent flow. If, for example, the effluent flow was 1 million gallons per day (mgd) and the receiving water flow was 2 mgd, the effluent was diluted with twice as much clean water and the calculated resulting instream concentration of pollutants was one-third of the concentration in the effluent. Two limitations resulted from this simple approach: no analysis could be done for dischargers into nonflowing water bodies (such as lakes and estuaries), and analysis could be done only for "conservative" pollutants that did not degrade (such as BOD) or volatilize. Both of these limitations could have been overcome if EPA's analysis had incorporated a capability for water quality modeling that was more sophisticated than simple dilution.

Another shortcoming of the analysis was that it did not reflect the use values that the receiving water body provided. Some river segments are far more important than others (ecologically, for recreational purposes, for water supply) and violations of water quality criteria in these segments should be viewed with more concern than violations in other segments. EPA's environmental analysis did not reflect these differential values, though.

In addition, the method was applicable only for pollutants for which water quality criteria existed—typically toxic pollutants—since the metric for assessing impacts involved water quality criteria. The major pollutants controlled by BPT and BCT are BOD and TSS, and water quality criteria did not exist for either of them. The result was that the environmental analysis was applicable only to the parts of the effluent guidelines addressing toxic pollutants (BAT and PSES).

As a result, the environmental analyses were viewed as having limited significance. We are not aware of any instance in which one played an important role in influencing an effluent guideline decision. Surprisingly, in retrospect, for nearly all the effluent guidelines there was no attempt other than the environmental analysis to evaluate the benefits to be gained from the regulation. There was no risk analysis, no evaluation of whether the dischargers affected highly valued water bodies, no assessment of impacts on drinking water intakes, and so forth.

Industry-Specific Benefits Analysis. For the OCPSF guideline, however, the situation was different. Executive Order 12291 required federal agencies to perform an RIA, which included a benefit-cost comparison, for all major regulations that imposed more than $100 million in annualized costs on the economy. The OCPSF effluent guideline was the second effluent guideline estimated by EPA to cost more than the $100 million threshold and the second to include a full benefits analysis. The first had been the effluent guideline for the iron and steel industry, promulgated in 1982. The benefits analysis for the iron and steel guideline included a national aggregate assessment of the benefits of BAT relative to current treatment, and several detailed case studies of particular river segments comparing the BAT compliance cost for steel plants along these segments with the benefits from the resulting improvements in water quality. Both the national aggregate assessment and most of the case studies found that the benefits of the guideline exceeded the costs. In the view of the official in EPA's Office of Water who was primarily responsible for developing the guideline, the iron and steel benefits analysis had no effect on the regulatory decisions EPA made. However, the fact that, on balance, the studies found the benefits of the regulation to exceed its costs was helpful in obtaining OMB approval for a controversial regulation.

The benefits assessment conducted for the OCPSF industry pursued three principal lines of investigation: three in-depth site-specific case studies of water quality benefits from the regulation; a national water quality benefit assessment; and a national assessment of health and ozone-related benefits that would result from controlling intermedia transfer of volatile organic compounds (VOCs) from wastewater treatment processes into the air.

While the OCPSF benefits assessment attempted to identify all relevant beneficial outcomes, only a subset could be quantified and monetized due to deficiencies in the data and/or limited methods for valuing nonmarket commodities. Monetized water quality benefits were estimated for recreational boating and fishing, commercial fishing and intrinsic values (nonuser benefits including option and existence values). Monetized air quality benefits were estimated for cancer reduction and smog reduction. Significant impacts that were not monetized included:

- Reduction in health risks from air emissions of noncarcinogenic pollutants. (It was estimated that air emissions from OCPSF direct discharger effluents exposed 45,000 people to levels of these pollutants in excess of thresholds for chronic health effects, but the monetary value of reducing these exposures could not be estimated.)
- Reductions in health risks associated with air emissions of nonpriority pollutants from both direct and indirect dischargers.
- Reductions in health risks from consumption of fish or drinking water contaminated by OCPSF effluents. (These health risks were partially quantified but were not monetized.)
- Reductions in health risks from dermal exposure of swimmers to OCPSF effluents.
- Ecological benefits. The likely improvements in the condition of aquatic ecosystems could provide both important services (such as energy capture, nutrient cycling, habitat provision) and enhance existence values. Such benefits, at best, were only partially addressed in the monetized estimates of recreational and commercial fishing and intrinsic benefits.

Consequently, the economic benefits to society of the OCPSF effluent guidelines likely exceeded the monetized estimates provided in the RIA.

Three Site-Specific Benefits Case Studies. Three in-depth case studies were conducted on river segments where there were large concentrations of OCPSF direct dischargers: a 9-mile segment of the lower Houston Ship Channel in Texas, a 74-mile segment of the Kanawha River in West Virginia, and a 130-mile segment of the Delaware River from Trenton, New Jersey to Liston Point, Delaware. The case studies estimated water quality benefits for direct uses including both in-stream uses (recreational fishing, boating, commercial and subsistence fishing) and withdrawal uses (drinking water), as well as intrinsic benefits from potential use (option value) and no use (existence value).

The recreational water quality benefits portions of the case studies were performed in stages. First, the stream segments were delineated, and the OCPSF plants directly discharging to those segments identified. Then, the total benefits of improving water quality from current to "clean" were estimated for each of the delineated stream segments. Next, the OCPSF share of these total benefits attributable to moving OCPSF dischargers from current levels to zero discharge were calculated based on the OCPSF dischargers' share of total pollutant loadings to each affected river segment. Finally, to estimate the water quality benefits of moving OCPSF dischargers from in-place treatment to BAT, the percentage removal achieved by BAT was used to derive the appropriate level of benefits. As a result, EPA arrived at benefits estimates that were effluent-weighted percentages of the total benefits estimated for water quality improvements on each water body.

For the Kanawha and Delaware River case studies, recreational fishing and boating benefits were each estimated by two different methods. One was a participation approach that used off-the-shelf models. These generated ranges of benefit estimates based on projected increases in the number of recreation days resulting from improvements in water quality. The other method used was a more direct approach derived from the use of historical data on recreation levels observed on each river. The results from each approach were interpreted to generate a best judgment range of recreational benefits.

Intrinsic benefits were projected on the basis of a proportional relationship to recreational benefits. The conceptual and empirical rationale was put forth in the iron and steel effluent guidelines economic benefits assessment.

Health risks from exposure to OCPSF discharges into receiving water bodies were believed at that time to result from three different exposure routes: ingestion of contaminated fish, ingestion of contaminated drinking water, and dermal exposure of swimmers/water skiers. A screening analysis that provided upper bound ingestion-related exposures from fish consumption was performed by assuming no treatment in-place at

OCPSF facilities. The resulting discharges were converted to instream concentrations and translated through fish uptake and bioconcentration to ingestion levels corresponding to the amount of fish consumed. The analysis could not account for uptake of toxic materials in sediments. The resulting risk estimates were so low that further refinement of the estimates was not warranted.

Only the Delaware River case study had drinking water intakes in the delineated river segments studied. An existing regional risk assessment was used there. Finally, no assessment of risks to swimmers and water skiers was conducted due to a lack of critical data.

Findings of the Case Studies. Each of the case studies bore specific findings.

The principal findings of the Houston Ship Channel case study were:

- Recreational fishing, boating, commercial shellfishing, and intrinsic water quality benefits of moving OCPSF direct dischargers from in-place treatment to BAT were estimated at less than $1 million annually. This low benefit estimate was due largely to the predominant use of the channel for commercial shipping, which deterred other water uses and was not water quality dependent.
- The excess lifetime cancer risk among those consuming contaminated fish from the Houston Ship Channel was estimated to be less than 1.8×10^{-7}, though the model was not able to consider fish uptake of toxics from sediments, which may have resulted in an underestimate of risk for consumption of bottom-feeding fish.
- No drinking water intakes were present in the study area, so there were no drinking water benefits.
- The annualized costs to the OCPSF dischargers of moving from in-place treatment to BAT were estimated at $8.8 million.

For the Houston Ship Channel case study, the monetized water-quality benefits were insignificant and considerably less than the costs.

The principal findings of the Kanawha River case study were:

- Recreational fishing, boating, and intrinsic water-quality benefits were estimated to range from $0.1 to $2.7 million annually.
- An assessment of cancer risks resulting from the consumption of contaminated fish suggested cancer risks less than 1.7×10^{-7} but subject to the same limitations as described above for the Houston Ship Channel assessment.
- The annualized costs to the OCPSF plants of moving from in-place treatment to BAT were $1.5 million.

For this case study, the monetized water-quality benefits were commensurate with the costs.

The principal findings of the Delaware River case study were:

- Recreational fishing, commercial fishing, boating and intrinsic water quality benefits were estimated to range from $2.1 to $9.1 million annually.
- An assessment of cancer risks from the consumption of contaminated fish suggested moderate risks, but an incongruity between modeled fish tissue concentrations and the results from actual samples cast doubt on the results.
- A 1983 EPA ambient water quality monitoring study near the intake of Philadelphia's Baxter Drinking Water Plant indicated the presence of ten priority pollutants in the intake water, and one—bis (2-chloroethyl) ether, a suspected carcinogen—was present in the finished water from the plant as well. The source of this pollutant was an OCPSF indirect discharger to one of Philadelphia's POTWs. A pretreatment system was installed at this plant, and pollutant levels were reduced. Assessment of potential contamination from OCPSF plants at other drinking water intakes indicated low risks, which did not warrant a formal assessment.
- The annualized costs of moving from in-place treatment to BAT were estimated at $18.7 million.

The monetized water quality benefits of achieving BAT for the plants along the Delaware River were significant but less than half as much as the costs.

Comparisons and Limitations of the Case Studies. The three case studies provided somewhat different results. In the Kanawha River case study, the annual monetized water quality benefits were commensurate with the costs; in the Delaware River case study, the annual, monetized water quality benefits were significant but clearly less than the costs; and in the Houston Ship Channel, the annual monetized water quality benefits were insignificant and an order of magnitude less than the costs. The different benefit-cost results in the three case studies prompted the obvious question of how to interpret site-specific findings in the context of a national regulation. The critical issue involved the representativeness of the case studies and the degree to which their results could be extrapolated to the larger affected universe. Extrapolation would depend on the character of the industry and the receiving water bodies with respect to parameters that significantly influence the magnitude of potential benefits. These factors might include:

- characteristics of the OCPSF dischargers and their effluent,
- characteristics of other local dischargers and their effluent,
- current and potential future uses of the water body,

- availability of substitute sites in the vicinity potentially providing for these uses, and
- size, proximity, and socioeconomic characteristics of the local user population.

Other factors may also be important, but any credible attempt to extrapolate from case studies to the nation as a whole would require consideration of those mentioned above.

A further question that arose in assessing the national implications of these case studies involved the importance of the nonlocal benefits that might arise in each case. Some of the pollutants discharged by OCPSF plants are long lived, and they could eventually be transported significant distances and cause damages well beyond the geographic areas included within the case studies.

The agency first explored representativeness in late 1983, following completion of the Houston and Kanawha case studies. These efforts focused mostly on characterizing the industry and the use designations of receiving water bodies. Unfortunately, insufficient information was developed to provide a basis for projecting the extent to which the case studies were typical or atypical in comparison with other locations affected by OCPSF dischargers. Consequently, the results of the two case studies could not be extrapolated to other OCPSF dischargers and the remainder of the country.

National Water Quality Benefits Estimate. Just as it had done in the iron and steel effluent guidelines, EPA pursued a second approach to estimating water quality benefits—a national, aggregate share approach. This second approach began with preexisting estimates of the national water quality benefits if all sources of water pollution were completely controlled, and then allocated benefits to control of OCPSF sources on the basis of the OCPSF industry's share of total national pollution loadings.

This analysis used Mitchell and Carson's (1984) national estimates of willingness to pay for recreational and intrinsic benefits augmented by Freeman's (1982) estimates of national water quality benefits for commercial fishing and withdrawal uses.

The national aggregate share approach required first that Mitchell and Carson's and Freeman's benefit estimates be adjusted downward to reflect the improvement in current water quality conditions since the time when the benefit estimates were originally developed. Second, a share of these benefits was allocated to the OCPSF industry on the basis of the industry's percentage contribution to total pollutant loadings. Finally, the benefits that corresponded to the percentage reduction in OCPSF loadings that would occur as a result of the regulation were determined.

The national aggregate water quality benefits of moving OCPSF direct dischargers from in-place treatment to BAT were estimated to range from $178.1 million to $330.2 million annually.

The key assumption in this estimation procedure was that the OCPSF industry's share of national benefits was identical to its share of national pollutant loadings. No evidence was assembled either to support or to question this assumption. One might speculate that abatement of OCPSF discharges would yield a share of aggregate benefits more than in proportion to the industry's share of loadings, since the OCPSF industry's pollutant loadings are probably disproportionately toxic and dangerous relative to other industries. On the other hand, one could note that OCPSF dischargers are commonly located on large, heavily trafficked, and industrialized waterways that might not be used extensively for recreation, fishing, or water supply even if pollution levels were significantly reduced. Given how critical this assumption was to the magnitude of the benefits estimates that were derived, and given the lack of corroborating evidence to support this assumption, the aggregate shares approach generated national water quality benefits estimates of uncertain validity.

National Benefits from Reduction of VOC Air Emissions. Information indicated that biological treatment systems failed to treat substantial amounts of volatile and semivolatile pollutants and instead transferred them to the air. In light of this, a national assessment of the magnitude of these emissions and their associated human health and environmental impacts was conducted.

The level of emissions of priority pollutants was derived on a chemical-by-chemical, product process line-by-product process line and plant-by-plant basis. EPA estimated volatile releases from the biological treatment system ponds for eighty-eight priority pollutants. Only priority pollutants were included because they were the only compounds for which the amended CWA required expedited regulations for and the only compounds for which detailed data had been collected.

Nonpriority pollutants, which also volatilized from OCPSF plants and posed potential health risks, were estimated on the basis of a proportional relationship to the priority pollutants as derived from an analysis of a separate database, the Industrial Studies Database, Office of Solid Waste and Emergency Response (Raucher 1987). Over 700 OCPSF plants with wastewater streams containing potentially volatile compounds were considered. Atmospheric concentrations of volatile pollutants and resulting human inhalation exposure were estimated for populations within a fifty-kilometer radius of the plants using standard air quality models, local climatic conditions and population data. Adverse environmental (smog formation) and human health (cancer and noncancer) impacts were estimated. The principal findings of this study were:

- Five hundred and ninety-nine direct and indirect OCPSF dischargers emitted priority pollutant VOCs from wastewaters. The emissions totaled 7,901 metric tons per year at current control levels.
- Of these air emissions, 6,864 metric tons contributed to tropospheric ozone formation (smog). About 77% of these releases were estimated to occur in areas where the national ambient air quality standard for ozone was not attained.
- Exposure to these air emissions was estimated to result in seventy-nine additional cancer cases over a seventy-year period of exposure (1.1 excess cancers per year nationally). Sixty-six of these cancers were estimated to result from air emissions from direct dischargers, and thirteen cancers were estimated to result from air emissions from indirect dischargers and POTWs.
- Noncancer health impacts might also be possible, as 43,000 people residing in six different geographic areas in the United States were estimated to be exposed to levels of benzene or chlorobenzene exceeding health-based thresholds.
- On average, if the recommended BAT technology were installed, an 84% reduction in these emissions would result. (The recommended technology was steam stripping of VOC-bearing aqueous waste streams in the plant before the streams were combined with other streams and subjected to treatments that would cause the pollutants to volatilize. Steam stripping would capture the VOCs, and they could be either recycled or incinerated. Note that although EPA recommended use of this technology and calculated the effluent limits assuming that it would be used, the regulation did not actually require its use.)
- Adequate characterization of nonpriority pollutants in OCPSF wastewaters was not available and nonpriority pollutants could not be included in air emission estimates. This was due to the NRDC settlement agreement and court-ordered deadlines that focused agency attention on a specific list of "priority" toxic pollutants. Considerable uncertainty existed on the magnitude of nonpriority pollutant air emissions, but extrapolations suggested that 10,200 to more than 64,000 metric tons per year of nonpriority pollutants might be emitted from OCPSF wastewaters.

Several of these VOC-related benefits were monetized. The reduction in cancer cases was valued using a range from $1 million to $7 million per cancer case avoided, and ozone (smog) reduction benefits were valued using a range from $720 to $1,250 as the value per metric ton for abating VOC emissions.

Summary of Benefits and Costs

The RIA included a benefit-cost comparison only for direct dischargers and compared the costs and benefits for moving from in-place treatment to BAT. Table 4 summarizes the comparison. The monetized water quality and air quality benefits were estimated at $188 million to $393.1 million annually. The annualized costs of achieving BAT were estimated at $300.8 million, which fell well within the benefits range. Of the national monetized benefits, 84%–95% were attributed to improvements in water quality, while 5%–16% of the benefits were associated with air quality improvements.

Criticisms of the Benefit-Cost Assessment

Four main points of criticism of the benefit-cost assessment can be made: the lack of extrapolation of site-specific case studies, the incompleteness of the case studies, the assumption of linearity that was used to apportion

Table 4. Summary of OCPSF Benefits and Costs (millions of $1987).

Total benefits	*$188.0–$393.1*
Monetized annual benefits	
National aggregate water quality benefits	$178.1–$330.2
Incremental cancers avoided from priority pollutant air emissions	$0. 8–$5.9
Ozone (smog) reduction due to reduced air emissions of priority pollutants (5,045–14,832 millions of tons (MT)/yr)	
—valued at $1,250/MT removed	$6.3–$18.5
—(valued at $720/MT removed)	($3.6–$10.7)
Ozone (smog) reduction due to reduced air emissions of nonpriority pollutants (7,626–30,771 MT/yr)	
—valued at $1,250/MT removed	$9.5–$38.5
—valued at $720/MT removed	$5.5–$22.2
Nonquantified benefits	
Reductions in human health risks involving air emissions of nonpriority pollutants, dermal exposures of swimmers, and subsistence fishing	
Ecological benefits	
Addressed only in terms of the likelihood of affected stream segments recovering from failure to meet water quality criteria	
Quantified but nonmonetized benefits	
Noncancer health risks from air emissions of noncarcinogenic priority pollutants (43,000 people exposed at levels exceeding health-based thresholds)	
Total costs	*$300.8*

Notes: These figures are for direct dischargers using BAT relative to current treatment in place. Costs are annualized costs required to move from current treatment in place to BAT.

Source: U.S. EPA 1987d.

some of the national water quality benefits, and the limited scope of the benefits assessment.

Lack of Extrapolation from Case Studies. A major concern about the three site-specific case studies involves the absence of a method for extrapolating the case study results to all OCPSF direct dischargers. No database had been developed that could provide some perspective on whether the case study water bodies were typical or atypical of the remainder of the country in terms of the magnitude of benefits that could result from water quality improvements. EPA had not developed a socioeconomic database that could help to suggest the amount of beneficial use that each water body currently supported or might support in the future if it were cleaned up.

One might imagine, for example, a database that would indicate for each water body such information as the socioeconomic characteristics and size of the user population, presence of drinking water intakes and the number of people using it as a drinking water source, the estimated total value of commercial fish and shellfish catch, the level of recreational use (perhaps measured through proxies, such as the number of boat ramps and miles of public beaches), the presence of unusual features that could contribute to a high ecological value for the water body, the prevalence of nearby alternative sites for water-based recreation, and so forth.

EPA appears to have initiated the OCPSF benefits case studies without having given much thought to how to extrapolate their results to the national level. One impetus for the case studies was the success of the site-specific case studies in supporting the iron and steel effluent guideline, where the case studies had resulted in a positive benefit-to-cost comparison and no national extrapolation seemed necessary. For the OCPSF guideline, though, EPA was left in a quandary when the case studies gave disparate results and generally did not support the guideline. EPA did not know what to do with the case study results and, ultimately, essentially ignored them.

In hindsight, it is clear that a well-designed approach for selecting benefits case studies for the OCPSF guidelines and extrapolating their results to the nation would have been highly desirable. The case studies could then have contributed much more than they did in forming the OCPSF regulatory decisions. And, together with the national aggregate shares benefits analysis, the case studies might have encouraged more discussion within EPA of ways to improve the ratio of regulatory benefits to costs. When two different approaches, each with its own shortcomings, provide benefit estimates that overlap, there is more confidence in the general magnitude of the benefits estimated. Given the sensitivity of the national water quality benefits estimates to the assumed linearity between loadings and benefits, a second approach (that is, an approach that generated national water quality benefits from the extrapolation of case study results) would have significantly enhanced the credibility of the estimates.

Incompleteness of Case Studies. An advantage of conducting in-depth, comprehensive case studies is to identify all relevant outcomes and then pursue quantification and monetization of these outcomes to an extent that is often not possible when the scope of the analysis is national. In fact, the intermedia transfer of VOCs from OCPSF wastewater treatment processes into the air was evaluated analytically first in the Delaware River case study. Once the significance of this intermedia transfer was determined, a national assessment of air emissions from OCPSF wastewaters and their human health and environmental impacts was undertaken. Given the court-ordered deadlines, there was insufficient time after completing the national air emissions assessment to revisit each case study and incorporate the air quality benefits. Consequently, the benefit-cost comparisons depicted by these case studies were incomplete and less favorable than they would have been had the air emission reduction benefits been incorporated.

This omission had no direct effect on the final regulation since the benefit estimates used in the RIA and cited in the rulemaking package were taken from the national water quality and air emission studies and not from the case studies. But the credibility of the national assessments would have been enhanced by more complete and more benefit-cost favorable case studies. Furthermore, since a case study is able to translate general, nationally occurring beneficial outcomes into concrete, site-specific enhancements that the public can both understand and identify with (that is, improved recreational fishing, boating, and swimming opportunities in a specific nearby river; less smog in the city; fewer cancers in the community), the incompleteness of the case studies represented a missed opportunity to portray more effectively for the public the specific ways in which it would benefit from the regulation.

Assumption of Linearity. The linearity assumption that was used to apportion some of the national water quality benefits estimated in the Mitchell and Carson (1984) and Freeman (1982) studies to the OCPSF effluent guidelines had been used before (in the iron and steel industry effluent guidelines) and in subsequent applications as well. Empirical testing of this assumption was certainly warranted and overdue.

Limited Scope of Assessments. Finally, the scope of the benefits assessment was too limited. Indirect dischargers were not included in the monetized economic benefits assessments, nor were the benefits and costs of controlling indirect dischargers included in the benefit-cost comparison provided in the RIA. The cost to indirect dischargers of moving from in-place treatment to PSES was $204.3 million annually, or 40% of the total annual costs of the regulation. The economic benefits from further controls on indirect dischargers might be expected to be rather different from the benefits from control of direct dischargers. In effect, the benefits to be gained from a major share of the regulatory costs were unknown.

An economic benefit assessment for indirect dischargers would have been difficult to perform since a significant portion of the benefits would accrue to POTWs in the form of treatment costs avoided. A slug of OCPSF toxic pollutants discharged into a sewer leading to a POTW, at sufficiently high concentrations, could inhibit biodegradation in the POTW operation, causing potentially severe disruption of treatment resulting in the passthrough not only of OCPSF discharges but also of other wastes from other sources as well. Furthermore, high concentrations of OCPSF priority pollutants in the POTWs' sludge would limit the POTWs' options for disposing of the sludge. Some of the lowest cost options, including the beneficial use of sludge on agricultural lands, might be precluded. Despite the analytical and data difficulties associated with performing a benefits assessment to address such issues, one should have been attempted given that the pretreatment costs for indirect dischargers represented 40% of the costs of the OCPSF regulation.

THE IMPACT OF THE BENEFITS ANALYSIS ON REGULATORY DECISIONS

The benefits analysis clearly had much less effect on the OCPSF effluent guideline decisions than did the other analyses. The rule was shaped primarily by the technical data and analyses developed by the agency. The cost analysis, economic impact analysis, and response to public comments also left clear imprints on the final rule. In contrast, we cannot point to any significant aspect of the final rule that would have been different if the water quality benefits analysis had not been performed. The water quality benefits analysis was not a factor as the agency made its final regulatory decisions.

Limited Impact of the Benefits Analysis

The three case studies suggested that the benefits of the rule might not exceed its costs, at least for two river segments with major concentrations of OCPSF dischargers. The case study findings had no effect in inducing the agency to consider a less stringent rule that might bring costs into a better balance with benefits or a more targeted rule that would focus more intensively on the locations where regulatory benefits might be greatest. The impact of the case studies was largely defused by the inability to extrapolate the results to the entire nation.

The environmental analysis showed generally that BAT controls on OCPSF dischargers were likely to have little impact in reducing the frequency with which water bodies affected by OCPSF dischargers violated water quality criteria. Parallel environmental analyses developed for other effluent guidelines typically had reached similar conclusions. An expectation had developed that the environmental analysis for any effluent guideline would show disappointing results (that is, results not supporting the need for the regulation), and these analyses were generally ignored. The typical finding that an effluent guideline for a single industry would have a limited effect on receiving water quality was interpreted by EPA not as suggesting that the effluent guideline might not be worthwhile but instead as suggesting that many effluent guidelines covering all industries were necessary in order to accumulate a meaningful impact on water quality.

The national aggregate shares benefits analysis appeared to support EPA's choice of Option II for BAT. The analysis projected the benefits of BAT to exceed the costs of Option I, to be roughly comparable to the costs of Option II, and to fall well short of the costs of Option III. Despite this support for one regulatory option over others, this benefits analysis was hardly mentioned during EPA's internal debate over option selection, nor was it referred to in the preamble to EPA's final rule as supporting the choice of Option II.

Identification of VOC Air Emissions as an Issue

The benefits assessment did, however, contribute importantly to the agency's identifying as an issue the cross-media transfer of VOCs to the air from OCPSF wastewater treatment processes. The air emissions benefits assessment was the first analysis to investigate and quantify the significant adverse impacts these emissions had on human health and the environment.

The impact this assessment had on the regulatory decision process (but not the outcome) was significant. When the OCPSF guidelines were proposed in March 1983, before the benefits assessment was conducted, the potential for air emissions was mentioned and dismissed in one paragraph in the *Federal Register* notice. After the national OCPSF air emission assessment was completed, the agency included steam stripping in the selected BAT option of recommended technologies to control the volatiles. Also, the agency changed the way it calculated cost-effectiveness (the cost per pound-equivalent removed) to include the tons of VOCs controlled by the steam strippers. This resulted in a very significant lowering of the cost-effectiveness numbers to $34 per pound-equivalent removed for BAT and $5 per pound-equivalent removed for PSES. This reduction of the cost-effectiveness numbers to well below the informal cutoff of $100 helped to ensure OMB approval of a controversial regulation. Furthermore, the agency considered requiring the installation of in-plant steam stripping to remove and recover VOCs from wastewater before it reached the settling ponds and aeration lagoons where the VOCs would volatilize into the atmosphere. EPA considered requiring control of VOC air emissions through any of three approaches:

- *BAT VOC removal*—Technologies, such as steam stripping with recovery, that must be employed as BAT to remove VOCs should be specified. This approach would be contrary to both the legislative history of the Clean Water Act and EPA's traditional preference for performance standards, in which effluent limitations are established with which facilities are free to comply by whatever means they think best.
- *Non-BAT VOC removal*—Technologies involving significant levels of air stripping that are *not* BAT should be specified. This approach would raise similar concerns about dictating control technologies as the previous approach.
- *In-plant limitations on VOCs in wastewater*—Such in-plant limitations that apply at a point prior to any unit or process that is capable of transferring VOCs to air should be established. Compliance monitoring would be required at this point. This would presumably force dischargers to install steam stripping upstream of this point. This approach of setting in-plant limitations to control air emissions, however, was not explicitly authorized by the CWA. The CWA's requirement that EPA consider potential adverse nonwater-quality environmental impacts in promulgating effluent limitations was intended to encourage EPA to temper effluent limitations to prevent such effects. It was not intended to authorize EPA to use CWA authority to regulate these nonwater-quality effects.

After considerable internal debate, EPA decided that each of these approaches to controlling VOC air emissions was unprecedented and not appropriate under CWA. EPA concluded that the Clean Air Act (CAA) provided more appropriate authority for directly regulating the air emissions, and no requirements to control air emissions or employ steam strippers were included in the OCPSF guidelines. Nevertheless, the end-of-pipe effluent limitations for VOCs promulgated by the agency were calculated based on the assumed use of steam stripping with product recovery or destruction rather than techniques that would allow air emissions, and the costs EPA estimated for compliance with BAT assumed that this more expensive technology would be used. Furthermore, the agency recommended that plants adopt this technology to limit air emissions in order to avoid costly retrofits that might subsequently be required under CAA or other appropriate statutes, possibly the Resources Conservation and Recovery Act (RCRA).

In 1994, EPA finalized a national standard for hazardous air pollutants that addresses a subset of the emissions from organic chemical producers. EPA also published in 1994 a guidance document for states, which under CAA have the authority to write rules to control another subset of emissions from the organic chemicals producers. In an inventory of treatment in place at organic chemicals plants that was conducted to support these regulatory efforts, the Office of Air found that few steam strippers had been installed—indicating that the OCPSF industry had not voluntarily followed the agency's recommendations on this issue contained in the 1987 effluent guidelines preamble.

CONCLUDING OBSERVATIONS

The effluent guidelines for the OCPSF industry were developed over a ten-year period of intensive work by EPA. Large amounts of information on the industry and on treatment technologies and their effectiveness were developed, analyzed, published several times for public review and comment, revised, and reanalyzed. EPA evaluated applicable technologies and organized the major regulatory options it considered around alternative combinations of control technologies. Compliance costs for the different regulatory options were estimated for each plant in the industry and then totaled, as were the reductions in pollutant loadings expected with compliance. An extensive economic impact analysis was conducted to evaluate the potential effects of these costs on plant closures, profitability, employment, and so forth. All this information provided the basis for the OCPSF effluent guidelines. The regulatory issues were structured in a manner consistent with how this information was organized (that is, choosing among control technologies), and the agency's regulatory decisions were consistent with what this information showed. In sum, EPA very thoroughly considered cost, effectiveness, and economic impacts in developing the OCPSF effluent guidelines.

Limited EPA Interest in the Benefits

In contrast, EPA paid much less attention to the question of what was to be gained from the regulation. EPA did not, for this or any other effluent guideline, prepare a baseline analysis of what the environmental problem was that might warrant regulatory action. EPA did eventually perform an extensive benefits analysis as required by Executive Order 12291, but the analysis had significant limitations:

- The results of the site-specific benefits case studies could not be extrapolated to the national level.
- The national aggregate shares benefits analysis was subject to great uncertainty because of the untested critical assumption that benefits were directly proportional to loadings reductions.
- The environmental analysis employed a highly simplified methodology that left the results far short of representing the real effects of the regulation on receiving water quality.
- The benefits analysis addressed only direct dischargers, leaving aside 40% of the rule's cost and the rather different issues associated with indirect dischargers.

Each of these limitations was foreseeable before EPA began the OCPSF benefits work and each might have been overcome. The problems with the site-specific case studies and the national aggregate shares analysis had both been identified previously in the work for the iron and steel effluent guideline. The shortcomings in the methodology for the environmental analysis had been recognized in many previous effluent guidelines. Approaches for analyzing the benefits of controlling indirect dischargers existed and had been applied to other water issues (such as RIA work for municipal sewage sludge regulations). In short, EPA did not work as intensively on the benefits analysis as it did on the other data and analyses supporting the OCPSF regulations.

In the final analysis, the water quality–related benefits information that EPA generated had no demonstrable impact on the OCPSF regulatory decisionmaking. The air emissions benefits assessment did affect EPA's decision process and what technologies were selected as the basis for BAT (such as steam stripping to control volatiles) as well as how the cost-effectiveness numbers were calculated and their magnitude. (They were significantly lower as a result of controlling for the VOCs). However, EPA's preference that OCPSF facilities install this technology was not written into the regulatory requirements, and plants appear to have adopted steam stripping only rarely.

One possible interpretation of this limited impact from the benefits analysis might be simply that the benefits analysis was so uncertain as to be unconvincing. The contradictory results of the water quality case studies, the inability to extrapolate from the case studies to the nation, and the substantial uncertainty of the aggregate shares approach made it difficult to argue strongly for influencing the regulation one way or another.

Another perspective might be that EPA did not view benefits issues as central to making decisions about a technology-based standard when the law was quite specific about requiring control technologies that were the "best available." The dozens of previous effluent guidelines had all been established with technology concerns in the forefront and benefits concerns virtually nonexistent. The most recent changes to the CWA and litigation by environmental groups had generated tremendous pressure on EPA to issue effluent guidelines promptly and to base them on straightforward, uniformly applied technologies. The water quality–based approach to determining tailored effluent control requirements for dischargers had been discredited as slow and ineffectual.

In retrospect though, we suggest that much of the water quality benefits analysis that was done for the OCPSF rule—the case studies and the environmental analysis—could have been seen as challenging the rule promulgated by EPA rather than supporting it. In particular, the analysis could have been read as suggesting that stringent, nationally uniform technology-based standards for industrial point-source dischargers represented an economically inefficient way of addressing the nation's widely varying water quality problems, particularly for advanced control requirements directed specifically at toxic pollutants (BAT). Other information was available in the mid-1980s that pointed in the same direction:

- EPA's experience in developing many effluent guidelines showed that there was great variability across dischargers within an industry in their cost of controlling a unit of pollution. Other data suggested that there was likely even more variability across dischargers in the benefits of controlling a unit of pollution. The variability on the benefits side stemmed from variations in the degree of dilution of the effluent afforded by the receiving water, the background level of water quality in the receiving water, the desired uses for the receiving water, and the number of people wishing to make use of the receiving water. The result was that the most economically efficient level of water pollution control—the level of control at which the sum of control costs plus damages is minimized—would vary widely across dischargers.
- Based on the biennial compilations of state assessments of surface water quality, EPA found nonpoint sources to be responsible for far more of the nation's impaired waters than were point sources (U.S. EPA 1988). Studies by the U.S. Geological Survey had found that a decade of concentrated effort on control of point sources had resulted in little demonstrable improvement in water quality across the country (Smith and others 1987). A large national modeling effort by researchers at Resources for the Future found that only about 7% of all the stream reaches in the country were sufficiently influenced by

point sources so as to show detectable improvements in water quality if point-source controls were improved (Gianessi and Peskin 1981).
- The compilation of state assessments also showed that far more of the water bodies not supporting designated uses were impaired by conventional pollutants (BOD, suspended solids, pathogens) or nutrients than by toxic pollutants.

The OCPSF case studies and these other sources suggested that a more efficient approach to water quality improvement might have been to conduct a comprehensive, national baseline assessment of water quality, identifying the sources and types of pollutants adversely affecting specific water bodies and their controllability, and the magnitude of benefits that could result from improved water quality. With such an assessment, better-tailored control programs, whether in the form of uniform national standards or site-specific standards, could have been developed. We envision for most water bodies that control efforts would have focused on nonpoint sources, conventional pollutants, and nutrients. Effluent guidelines to control toxic pollutants from industrial discharges would probably not have been among the highest priorities within such an overall strategy for improving water quality.

The mid-1980s probably represented the high point in EPA's concern for uniform standards, point sources, and toxic pollutants. The Clean Water Act of 1977 and the NRDC settlement agreement demanded this emphasis. EPA needed nearly a decade to accomplish most of what was legally required. Since the mid-1980s, though, EPA's surface water program has clearly shifted in other directions, consistent with the sorts of findings from the OCPSF benefits case studies. Amendments to the Clean Water Act in 1987 significantly increased the attention given to nonpoint sources and strengthened requirements for state development and achievement of water quality standards for individual water bodies. EPA responded by delegating greater authority to states to deal with site-specific problems on a case-by-case basis and by emphasizing the watershed approach, in which management attention focuses on developing tailored controls for particular important water bodies. Recently, EPA has moved further away from uniform requirements by supporting trading approaches to reduce the cost of meeting water quality goals by reallocating control responsibilities efficiently among individual dischargers and between point and nonpoint sources.

Relative Importance of Economic Impacts versus Benefits Issues

Our second general observation about EPA's OCPSF effluent guidelines is also apparent in a historical context. We have argued that the technology-related data and analysis (available technologies, performance, and cost) provided the foundation for virtually all of EPA's decisions. In a few instances, the economic impact analysis raised concerns about the affordability of the basic technology requirements that EPA had chosen, and the requirements for some OCPSF sectors were modified to reduce the economic dislocations likely to result from the rule. The water quality benefits analysis, though, had no apparent impact on the regulation. This pattern—the rule was fundamentally determined by technology issues, with affordability concerns resulting in marginal adjustments and benefits not really considered—was repeated in other effluent guidelines. And, we contend, this pattern prevailed broadly throughout the early and middle 1980s as EPA established other technology-based standards: standards for RCRA treatment, storage, and disposal facilities; NSPS, RACT, and BACT under the Clean Air Act; the secondary treatment requirements under the Clean Water Act; standards for underground injection wells under the Safe Drinking Water Act; and so forth.

In more recent technology-based regulations, though, it seems to us that economic affordability has declined as a factor affecting EPA's decisions, while concern about benefits has increased. Risk and benefit analyses appear to have played important roles in recent EPA decisions about RCRA facility standards and maximum available control technology (MACT) for toxic air pollutants, and some of the resulting standards are significantly risk based as well as technology based. It has also been infrequent that the chosen technology basis for these regulations has been adjusted and made less stringent in order to avoid projected large adverse economic impacts. We pose for the reader the question of why the concern for economic affordability of technology-based standards seems to have declined somewhat over time and concern for benefits has increased.

EPA Analysis to Support Similar Rules. A final observation arises about EPA's analysis supporting the OCPSF effluent guidelines if you view this regulation as a later one in the long series of effluent guidelines that EPA has issued over the past twenty years. Whenever EPA must issue many parallel rules under a single statutory mandate for a series of different categories of regulated entities, there are great opportunities for efficiencies in analysis across the multiple rules. Major expenditures to improve agency decisionmaking may be justified when these costs can be spread across a series of rules rather than be incurred for a single rule. In issuing a series of similar rules, EPA can take useful steps both prospectively and retrospectively:

- EPA can afford prospectively to develop large databases or sophisticated analytical methods that will contribute to decisionmaking across many rules.

- After issuing some of the rules, EPA can retrospectively evaluate experience with them and use the results to help with future rules in the series.

The OCPSF rule illustrates several respects in which EPA took advantage of opportunities of this sort. EPA developed many methods for investigating the availability and performance of treatment technologies that were used across multiple effluent guidelines. For example, EPA developed a common format for survey questionnaires to elicit basic industry information and promulgated standard methods for pollutant sampling and analysis. EPA developed a sophisticated, consistent approach to performing economic impact analyses. Following the NRDC consent decree, EPA recognized the necessity of developing many effluent guidelines quickly, and adopted a streamlined procedure for participation by offices other than the Office of Water in guideline development and review. Finally, EPA developed a procedure for evaluating the relative cost-effectiveness of the different technology options considered for different effluent guidelines. A major difficulty in comparing cost-effectiveness across effluent guidelines was that the different guidelines abated different mixes of pollutants. One guideline might address mostly toxic metals, while another addressed toxic organic compounds. EPA developed a simple but reasonable procedure for weighting each pollutant abated by its relative toxicity. The agency then calculated the incremental and total "pound-equivalents" abated by each regulatory option. Over several years a rule of thumb developed that helped to ensure some consistency across effluent guidelines in the stringency of the technology options that were chosen as the basis for BAT and PSES: candidate effluent guideline technologies that abated toxic pollutants at an incremental cost exceeding $100 per pound-equivalent were viewed skeptically. For the OCPSF guideline, BAT and PSES were estimated to cost $34 and $5 per pound-equivalent, respectively, well below the informal $100 limit.

Some Missed Opportunities: Databases and Analysis. EPA also missed many opportunities to develop data or methods that would have helped importantly across many effluent guidelines. To continue with the last example, EPA never took the seemingly obvious step of asking whether $100 per pound-equivalent was a reasonable cutoff level to adopt. What are the typical benefits of abating a pound-equivalent of toxic discharge from industrial point sources in the United States? Clearly, as the OCPSF benefits case studies indicated, there would be wide variation in the benefits per unit of pollution abated across different settings. But some general conclusions could have been drawn if EPA had investigated this question, and the answer could have significantly enhanced decisionmaking across all of the effluent guidelines. (We recollect that during the early 1980s EPA's Air Office established and used such benefits-based cost-effectiveness cutoffs to organize EPA's decisions across multiple technology-based rules for air emissions. Benefits studies roughly quantified the dollar amounts that abating emissions of a ton of different pollutants (particulate matter, sulfur dioxide, VOCs) might be worth. These values were then adopted as loose limits on the stringency of, for example, air new source performance standards across many industrial categories.

As it was, the $100 per pound-equivalent figure was technology based and had no relation to the benefits of abating toxic discharges. The $100 figure had emerged as a rough pattern across the earlier effluent guidelines developed under the NRDC consent decree—the technology options selected across these guidelines generally showed incremental cost-effectiveness figures of less than $100, while technology options that had been rejected for these guidelines often showed cost-effectiveness figures of more than $100. Furthermore, the adopted technologies that cost less than $100 per pound-equivalent typically seemed routine and reasonable, while the rejected technologies costing more than $100 often seemed exotic or aggressive (such as polishing filters, activated carbon adsorption). The $100 figure was then applied in decisionmaking on subsequent guidelines to provide some consistency with past decisions.

There were at least two other instances where EPA missed important opportunities to perform analytical work that would have been valuable across many effluent guidelines.

First, for many years, EPA did little to organize a multi-industry database on treatability of different sorts of wastewaters. One of the key steps in developing any effluent guideline occurred in documenting the performance of different candidate, well-operated treatment technologies on the waste streams typical of the industrial category to be regulated. EPA initially interpreted this task as requiring treatment effectiveness data specifically from the industry in question. The result was that EPA incurred a high cost for intensive sampling and data acquisition from each industry. Even so, for an industry such as OCPSF with hundreds of different sorts of production processes and associated wastewater streams, EPA was rarely able to generate a sufficient volume of treatability data specific to the industry that covered all the processes. EPA eventually then began to transfer from other industries to the industry being regulated data on the treatability of wastewater streams that had similar characteristics. For the OCPSF guideline, the proposed rule was specifically criticized as being based on insufficient data on the treatability of the many waste streams produced by the industry. EPA responded by conducting more sampling of OCPSF plants, but also by transferring data on treatability of metal-bearing wastewater streams from several metals industries that had been studied earlier in more detail. This was a sensible approach, but it occurred rather late among the series of effluent guidelines that were issued. This approach could also have been used for many earlier guidelines and could have been planned for and organized more appropriately from the start.

Second, EPA never created a database on current and potential uses of water bodies that would have facilitated benefits analysis for all the effluent guidelines. For most effluent guidelines, EPA's consideration of benefits stopped with simply an accounting of the amount of pollutants that would no longer be discharged and the environmental analysis that projected the potential impact of the reduction in discharge on violations of water quality criteria in receiving waters. EPA did not go farther to analyze the site-specific significance of the loadings reduction that would occur from each discharger. Such analysis would have been possible if EPA had generated a database on receiving waters—for each water body, what is the current, baseline water quality; what uses are currently made of the water; what uses might be made of the water if its quality were to improve; what demands exist for these uses; and so forth. These data could have been added to EPA's existing database on water bodies that consisted mostly of location and flow information. With the augmented database, EPA could locate for the appropriate receiving water each discharger affected by an effluent guideline, and EPA could then evaluate more fully the use and nonuse benefits that might result from reduced loadings from the discharger. Had such a database been developed for the OCPSF guideline, EPA would have been able to judge the respects in which the settings for the three benefits case studies were typical or atypical of those for the remainder of the plants in the industry.

Retrospective Evaluations. To our knowledge, EPA has also made no extensive or systematic effort to evaluate the results of the promulgated effluent guidelines. Analyses to support future effluent guidelines could profit from evaluation of where analyses to support past effluent guidelines have proven accurate or inaccurate. A retrospective evaluation of effluent guideline compliance experience might have been possible and useful by as early as the mid-1980s. By then, some industrial dischargers covered by early effluent guidelines had made their compliance choices and had begun significant investments. By the late 1980s, the experience that could have been evaluated was much richer. However, to our knowledge, EPA began only two very limited retrospective analyses of effluent guidelines, and completed neither of them.

EPA's Office of Policy, Planning and Evaluation performed a pilot retrospective study of the effluent guidelines in 1990–91. The study agenda was ambitious. Industry had had sufficient time to comply with many effluent guideline requirements, and EPA set out to compare the nature of the compliance activities that had occurred with the projections that had been made in the analyses supporting the original regulations. The comparison of what actually happened with what had been expected to happen was to cover all the important issues: choice of compliance technologies, effectiveness of control technologies, cost of compliance, economic impacts, water quality benefits, cross-media impacts, and risk reductions. Detailed case studies were to be conducted for several plants in each of three industrial categories.

Because of funding limitations, only three case studies of plants in one industry were completed. The findings were quite interesting, although one hesitates to generalize because of the very limited sample size. Perhaps the most controversial finding was that EPA had greatly overestimated the extent to which industry would achieve compliance by making in-process changes (such as altering production techniques, substituting inputs, or recycling or recovering wastes for reuse). Instead, the three plants complied nearly entirely by installing distinct end-of-pipe treatment facilities. In some cases, the plant managers had attempted process changes but abandoned them because of adverse effects on product quality or yield. They ultimately chose end-of-pipe approaches often because they did not want to tinker with an apparently delicate production process.

If this OPPE study had been larger and more representative, such findings could have suggested important changes in how EPA developed and supported effluent guidelines.

A second retrospective study was begun by EPA's Office of Water, focusing specifically on the accuracy of EPA's economic impact analysis. The porcelain enameling industry was chosen for a case study, as projected adverse economic impacts had played an important role in shaping the final rule for this industry. The study was to compare actual impacts (plant closures, price changes, trade impacts, and so forth) from this guideline with projected impacts and ascertain the reasons for any differences. The study was not completed, and no results are available.

In sum, we view EPA's record as mixed in the extent to which it took steps that would facilitate making good decisions across the whole two-decade-long series of effluent guidelines. Some opportunities to create broadly useful databases and analytical tools were seized, some were missed. With specific regard to retrospective evaluation of the actual experience as industry complied with promulgated effluent guidelines, EPA has done very little.

We believe that concern about these issues is still timely. EPA continues to issue and revise effluent guidelines. And, some of the databases, tools, or approaches that we argue could improve EPA's ability to issue good effluent guidelines may also be applicable to other series of technology-based regulations that EPA must issue. For example, EPA is now only several years into a likely decade-long process of issuing MACT standards for emission of toxic air pollutants from many industries. As yet, EPA has not developed for the MACT rules any counterpart to the "pound-equivalent" scheme for estimating the cost-effectiveness of different effluent guideline options. We are not aware that any method has been developed yet to weight and then aggregate the many different toxic air pollutants at issue in the MACT rules. There appears to be no generally accepted way of judging the relative cost-effectiveness of different MACT rules nor quantitative approach that can provide some consistency across MACT decisions by distinguishing cost-effective from cost-ineffective technologies.

REFERENCES

Freeman, A. Myrick III. 1982. *Air and Water Pollution Control: A Benefit-Cost Assessment.* New York: John Wiley.

Gianessi, Leonard and Henry Peskin. 1981. Analysis of National Water Pollution Control Policies. A National Network Model. *Water Resources Research* 17: 796–821.

Mitchell, Robert C. and Richard T. Carson. 1984. *Willingness to Pay for National Freshwater Quality Improvements*. Prepared for U.S. EPA. Washington, D.C.: Resources for the Future.

Raucher, Robert. 1987. *A Summary of the Benefit-Cost Analyses of the Final Effluent Limitations Guidelines for the Organic Chemicals, Plastics, and Synthetic Fibers Industry.* Energy and Resources Consultants, Inc.

Smith, Richard A., Richard B. Alexander, and M. Gordon Wolman. 1987. *Analysis and Interpretation of Water Quality Trends in Major U.S. Rivers, 1974–1981*. Water Supply Paper 2307. Washington, D.C.: U.S. Geological Survey.

U.S. EPA (Environmental Protection Agency). 1987a. *Development Document for Effluent Limitations Guidelines and Standards for the OCPSF Point Source Category.* (Volumes 1 and 2). EPA 440/1-87-009a and b. NTIS PB88-171335. Washington, D.C.: U.S. EPA, Office of Water.

———1987b. *Economic Impact Analysis of Effluent Limitations Guidelines and Standards for the OCPSF Industry*. EPA 440/2-87-009. NTIS PB89-11430. Washington, D.C.: U.S. EPA, Office of Water.

———1987c. *OCPSF Effluent Limitations Guidelines, Pretreatment Standards, and New Source Performance Standards. Federal Register* November 5. Vol. 52, no. 214.

———1987d. *Regulatory Impact Analysis of the Effluent Guidelines Regulation for the OCPSF Industry*. Washington, D.C.: U.S. EPA, Office of Water.

———1988. *National Water Quality Inventory: 1986 Report to Congress.* Washington, D.C.: U.S. EPA, Office of Water.

6

Stratospheric-Ozone Depletion

James K. Hammitt

On September 16, 1987, representatives from nations around the world adopted the Montreal Protocol on Substances that Deplete the Ozone Layer, an international agreement seeking to limit the production of ozone-depleting chemicals. In this case study, James K. Hammitt, a veteran researcher in the field, reviews the scientific, analytic, and regulatory issues surrounding U.S. adoption of this agreement. Hammitt concludes that while the global and potentially catastrophic consequences of CFC emissions may have been the principal factor in the U.S. decision to sign the Montreal Protocol, economic analysis clearly played an important role in both shaping the specifics of the agreement and providing assurance that the policy's benefits did in fact outweigh the costs. Moreover, in a rare case where one can compare ex ante *and* ex post *information, Hammitt shows that the initial analysis in support of the proposed rule considerably overstated marginal control costs. Subsequent analysis, in support of the final rule, proved to be more accurate and may even have underestimated costs somewhat.*

INTRODUCTION

CFCs and Ozone Depletion

Chlorofluorocarbons (CFCs) are a family of synthetic chemicals developed in the 1930s for use in home refrigerators. Composed of chlorine, fluorine, and carbon, CFCs provide an unusual combination of desirable

JAMES K. HAMMITT is Associate Professor of Economics and Decision Sciences and Director of the Environmental Science and Risk Management program at the Harvard School of Public Health. Several CFC studies he co-authored at RAND for EPA provided input to both negotiation of the Montreal Protocol and the CFC RIA.

chemical and physical properties, including low toxicity. As a result, CFCs were adopted for use in a wide range of applications. By the 1970s, the largest quantities were used in aerosol spray cans; home and commercial refrigeration; automobile and room air-conditioning; and rigid and flexible plastic foams for insulation, cushioning, and packaging and as solvents for cleaning electronic parts and dry-cleaning clothing. A related family of compounds, the Halons (which contain bromine), are used in fire-extinguishing systems and are especially useful for valuable equipment in enclosed spaces, such as computer centers and telephone exchanges, museums, and aircraft.

The number of firms using CFCs as inputs to their production processes is large, befitting the wide range of applications. In contrast, production has been restricted to a relatively small number of firms. In 1986—the year before international regulation was adopted—there were five U.S. producers. E. I. Du Pont de Nemours & Co. (Du Pont), the first company to produce CFCs, had maintained its market leadership and accounted for about half of U.S. production. An additional fifteen firms accounted for all CFC production outside the Soviet Union and its satellite states: four in Japan, two each in the United Kingdom and West Germany, and one each in France, Italy, the Netherlands, Canada, Australia, Greece, and India (CMA 1986).

CFCs and Halons are nearly inert: they do not react readily with other compounds but persist in the environment on release. James Lovelock, an independent British scientist, noted in 1969 that this feature made CFCs potentially valuable as a chemical tracer of atmospheric motions, since virtually all of the CFCs produced to date remained in the lower atmosphere (Lovelock and others 1973). Not until the early 1970s did scientists determine the eventual fate of CFCs released to the environment: on drifting up to the stratosphere (the region of the atmosphere from about eleven to fifty kilometers above the earth's surface), CFCs are dissociated by the intense solar ultraviolet (UV) radiation of that region. Through a chain of chemical reactions, the chlorine so released transforms ozone, a relatively unstable molecule of three oxygen atoms, into stable molecular oxygen, which contains only two atoms. Although ozone is in turn produced by photodissociation of molecular oxygen, the chlorine destroys ozone faster than it is produced and consequently the concentration of ozone in the stratosphere declines.

Depletion of stratospheric ozone could have a wide range of consequences. Perhaps most important is the effect on UV radiation. Ozone is the primary atmospheric constituent that absorbs UV radiation, and most atmospheric ozone is concentrated in the stratosphere. Reduction in stratospheric ozone concentrations is likely to increase the quantity of solar UV radiation reaching the earth's surface. Increases in UV radiation at the earth's surface are expected to lead to increases in human skin cancers;

increases in cataracts; possible reductions in immune function in humans and other animals; damage to terrestrial crops and other plants and to marine life, particularly plankton and fish larvae; and increased weathering of plastics. In addition, by reducing heat input to the stratosphere associated with UV absorption, ozone depletion may alter the heat balance of the atmosphere and so affect atmospheric circulation and global climate.

CFCs also contribute to the global "greenhouse effect," the enhancement of which is expected to lead to substantial global changes in climate over the next century. The direct contribution of CFCs to radiative forcing—the "heat trapping" effect that results because CFCs allow incoming (visible) radiation to penetrate but prevent outgoing (infrared) radiation from escaping—is many times stronger, per molecule, than the effect of carbon dioxide. CFCs, however, are much less important than carbon dioxide to global climate change for two reasons. First, the quantities of CFCs released to the atmosphere are several orders of magnitude smaller than the quantities of carbon dioxide released through fossil-fuel combustion. Second, because ozone contributes to radiative forcing, CFCs approximately offset their direct contribution to climate change by depleting stratospheric ozone.

Within a few years of the discovery that CFCs could deplete ozone, the United States and a few other countries had imposed restrictions on their use, primarily through bans on CFC use in most aerosol dispensers. During the late 1970s and early 1980s, global CFC production and use of the two primary CFCs (CFC-11 and CFC-12) declined, due to these restrictions and to weak economic growth.[1] Concern about the environmental consequences of CFC use moderated, as substantial control measures had been undertaken, scientific understanding of the magnitude of their effect on ozone suggested that ozone depletion might not be as great as previously believed, and the antiregulatory Reagan administration came to power in the United States. By the mid-1980s, however, pressures developed for additional control measures, as production of CFC-11 and -12 began to increase after the aerosol ban and early 1980s recession. In addition, production of CFC-113, a solvent, was increasing rapidly, in part because of its use in the rapidly expanding electronics industry (see Figure 1). Recognizing that CFCs presented a global commons problem, diplomats negotiated toward international control measures, eventually leading to adoption in 1987 of the Montreal Protocol, an agreement that limited production and consumption of the major CFCs and Halons.

Historical Background

Initial interest in the effects of human actions on stratospheric ozone began before it was recognized that CFCs might have significant environmental effects. In the late 1960s, scientists identified the possibility that

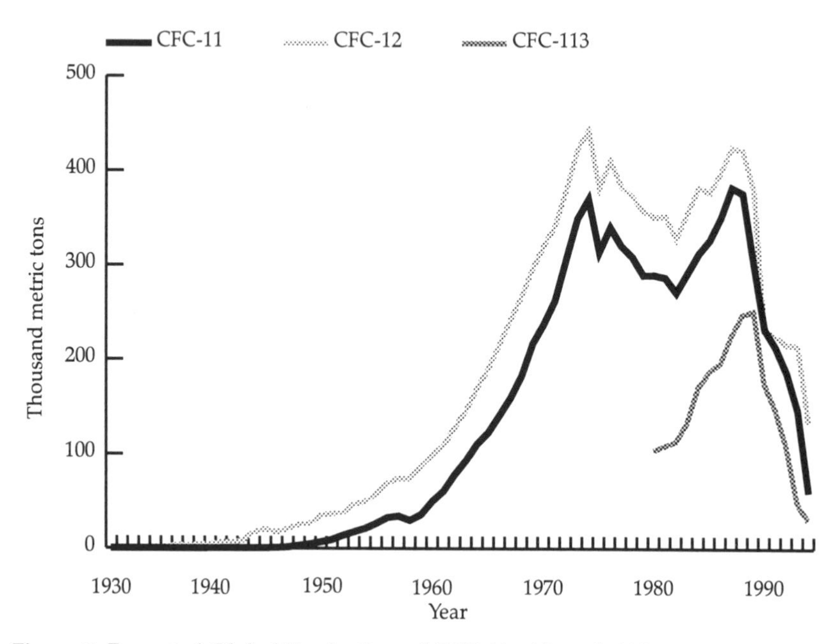

Figure 1. Reported Global Production of CFC-11, -12, and -113.

Note: These data exclude communist countries.

Source: AFEAS 1995.

introduction of water vapor and nitrogen oxides from a proposed fleet of commercial supersonic transports (SSTs) and chlorine from solid-fuel launches of the NASA space shuttle could lead to the destruction of stratospheric ozone. Stolarski and Cicerone (1974) concluded that hydrogen chloride emitted by the space-shuttle boosters would break down in the stratosphere and release chlorine atoms, which would in turn react with ozone in a self-perpetuating cycle similar to a cycle Crutzen (1970) had earlier proposed for nitric oxide released from SST exhaust. These results generated relatively little concern because scientists were focusing on the small quantity of chlorine that would be released in the stratosphere by spacecraft and volcanoes, and other means of transporting chlorine to the stratosphere were as yet unrecognized.

In late 1973, Molina and Rowland identified a pathway for the destruction of stratospheric ozone by CFCs (Dotto and Schiff 1978) and pointed to the importance of the large quantities accumulating in the stratosphere. They concluded that because CFCs are inert and insoluble, they would be unaffected by the natural sinks (for example, oceans, rainwater, and particles) that trap other chemicals and facilitate their breakdown in the lower atmosphere. Instead, CFCs would diffuse in the atmosphere and be transported to the stratosphere with the global circulation. The only significant mechanism for removal of CFCs from the atmosphere would be dissociation by UV radiation in the stratosphere. Photodissociation would release chlorine that would in turn convert ozone into oxygen. By the 1970s, millions of tons of CFCs had already been released and were on their way to the stratosphere. But more troubling was the fact that production of CFCs appeared to be increasing rapidly, doubling every five to seven years (see Figure 1). Molina and Rowland (1974) predicted that, even if current emission levels continued, the earth's ozone layer would be depleted by 7% to 13% before reaching a steady state at the end of the next century.

This work received widespread public attention, which led consumers to boycott aerosol products and aerosol manufacturers to substitute alternative propellants or product dispensers (such as pump sprays). While market reactions and economic recession led to drops in CFC demand, the scientific understanding of the threat posed by CFCs became more complicated, because interactions between chlorine and nitrogen released from fertilizers could interfere in a way that dampened the effects that might be caused by either substance alone (Cogan 1988).

The U.S. National Academy of Sciences (NAS) released its first of what became a series of four reports on CFCs and ozone depletion in September 1976. The report gave a best estimate of ozone depletion of 7% (the low end of the original Molina and Rowland estimate) but indicated that the uncertainties in atmospheric chemistry were so large that actual depletion could range between 2% and 20%. Assuming that most deaths from melanoma, a frequently lethal form of skin cancer, are due to exposure to solar UV radiation, the report estimated that for a 7% depletion there might be up to a 15% increase in melanoma deaths. Although the 1976 report discussed the evidence suggesting that less lethal nonmelanoma (basal- and squamous-cell) skin cancers are also related to exposure to ultraviolet radiation, the committee did not provide quantitative estimates for the incidence of these cancers. The report also discussed agricultural effects (crop damage), effects on natural terrestrial and aquatic ecosystems (harm to phytoplankton), and possible effects of climatic change on the biosphere, and emphasized that any nonhuman biological effects could lead to significant indirect impacts on humans.

Throughout the 1976 report, a great deal of discussion focused on the uncertainties complicating assessment and particularly on uncertainty about the magnitude of depletion that would occur. The report concluded that:

> The selective regulation of [CFC] uses and releases is almost certain to be necessary at some time and to some degree. Neither the

> needed timing nor the needed degree can be reasonably specified today because of remaining uncertainties. However, measurement programs now under way promise to reduce these uncertainties quite considerably in the near future. The prospect for narrowing uncertainty and the finding that the rate of ozone reduction is relatively small engender in the committee the conclusion that a one- or two-year delay in actual implementation of a ban or regulation would not be unreasonable. (NAS 1976: iii)

Early U.S. and International Efforts. Following these conclusions, congressional interest in regulating CFCs waned. In the 1977 Clean Air Act Amendments, Congress delegated the responsibility for evaluating the issue to the Environmental Protection Agency (EPA). Section 157(b) required the EPA administrator to issue "regulations for the control of any substance, practice, process or activity which in his judgment may be reasonably anticipated to affect the stratosphere, if such effect ... may reasonably be anticipated to endanger public health or welfare."

In the spring of 1977, the government announced a two-phase plan to limit the use of CFCs. Phase One banned all "nonessential" uses of CFCs in aerosol products after December 15, 1978, while Phase Two aimed at reducing nonaerosol uses of CFCs. While Phase One essentially completed the phase out that market processes had initiated in the United States, only three other nations (Canada, Sweden, and Norway) banned CFC use in aerosols (Parson 1993: 36). Thus, much of the rest of the world continued to consume substantial quantities of CFCs as aerosol propellants. Phase Two of the government's plan was not seriously implemented, partly because of the apparent lack of more international concern, partly because the public and most policymakers perceived that they had taken care of the problem by eliminating aerosols, and partly because the search for CFC substitutes in other applications had not yielded attractive replacements.

Between about 1978 and 1985, scientific research continued to yield varying estimates of ozone depletion. Figure 2 illustrates the changing estimates of equilibrium ozone depletion for a standard CFC emission scenario (generated by one of the primary atmospheric models used to simulate ozone depletion) as it was continually revised to incorporate new understanding. In a report issued in 1979, the NAS increased its estimate of eventual ozone depletion to 16%. This increase reflected new evidence that suggested that nitrogen oxides would convert to an inactive form more rapidly than expected and would consequently impede the effects of chlorine less than Molina and Rowland had suggested in 1974. Continuing the theme of the 1976 NAS report, the 1979 NAS report also pre-

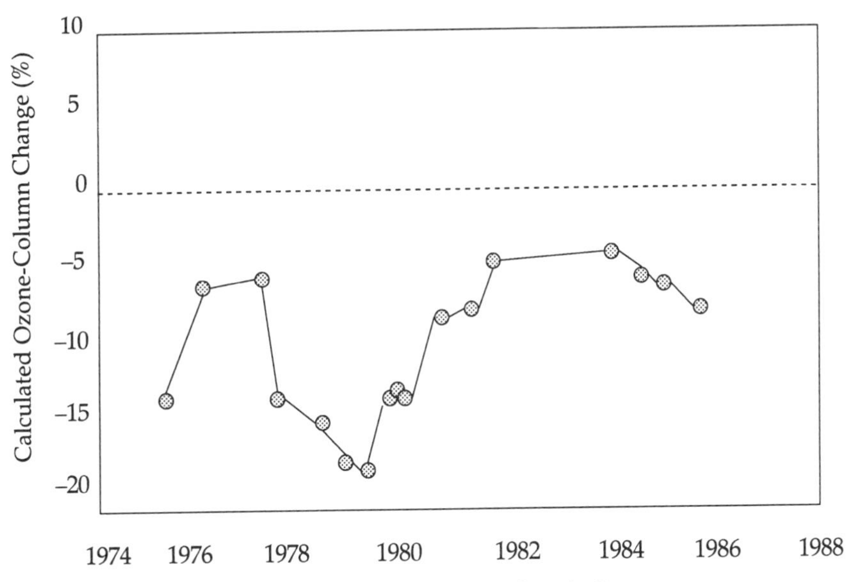

Figure 2 . Calculated Ozone: Column Change to Steady State.

Note: Equilibrium column ozone calculated by the then-current version of the Lawrence Livermore National Laboratory Model for a scenario with CFC-11 and -12 emissions constant at 1974 rates.

Source: WMO 1985: 773.

dicted sharp increases in human skin-cancer rates and serious damages to plant and marine life that could occur if the CFC buildup continued.

In the early 1980s, international concern motivated many nations to discuss freezing their CFC production capacities and reducing consumption by existing uses. Some nations, notably France and the United Kingdom, continued to assert that the theories about ozone depletion were still too speculative to warrant restrictions on CFCs. Nevertheless, in March 1980 the European Community Council of Ministers voted to freeze production capacity and reduce consumption in aerosols by 30% from 1976 levels (Litfin 1994: 70; Maxwell and Weiner 1993: 23). In October 1980, EPA published an advanced notice of proposed rulemaking to control domestic CFC emissions (U.S. EPA 1980), but this proposal was not pursued because of the antiregulatory stance of the newly elected Reagan administration combined with strong industry reaction, smaller model-predicted ozone depletion, and the shifting of public attention to hazardous waste (for example, Love Canal).

Refining the Models, Moderating the Message. In the early 1980s, Wuebbles (1981, 1983) developed the Ozone Depletion Potential (ODP) for comparing the abilities of chemicals to deplete stratospheric ozone. ODP measures the cumulative ozone depletion induced by emission of a kilogram of a specified compound relative to that induced by emission of a kilogram of CFC-11. The ODP provided a means for evaluating possible regulations of different compounds. It also suggested that Halons could become significant contributors to ozone depletion, despite their relatively low production levels, because they are about ten times as efficient ozone destroyers as CFCs (see Table 1).

In March 1982, the NAS released its third report on ozone depletion. It estimated global ozone depletion would eventually reach 5% to 9%, approximately a factor of two below its 1979 estimate. This drop gave the impression that the problem was less serious than had been previously thought. In addition, the 1982 NAS report indicated that the etiology of skin cancer was more complex than previously believed, and consequently the committee was unwilling to make a quantitative estimate of the incidence of melanoma that might result from reduced stratospheric ozone concentrations. The 1982 report also added concerns about possible damage to human vision and immune systems.

In 1984, the NAS released its fourth report, further dropping its estimates of global ozone depletion to between 2% and 4%. These lower estimates were based on the recognition that increasing atmospheric concentrations of methane, carbon dioxide, and nitrous oxide would temper ozone depletion, in part by accelerating global warming that, by trapping more heat in the lower atmosphere, cools the stratosphere and slows ozone-destruction reactions (Wuebbles and others 1983). The NAS report

Table 1. Atmospheric Lifetime, 1986 Emissions, and Ozone-Depletion Potential for CFCs and Related Compounds.

Compound	*Chemical formula*	*Atmospheric lifetime (yrs)*	*Model-derived ozone-depletion potential*	*1986 emissions (million kg)*
CFC-11	$CFCl_3$	50	1	342
CFC-12	CF_2Cl_2	102	0.82	444
CFC-113	$C_2F_3Cl_3$	85	0.90	163
Methyl chloroform	CH_3CCl_3	5.4	0.12	545
Halon 1211	CF_2ClBr	20	5.1	10.8
Halon 1301	CF_3Br	65	12	10.8
HCFC-22	CF_2HCl	13.3	0.04	146
HCFC-123	$C_2F_3HCl_2$	1.4	0.014	0
HCFC-141b	$C_2FH_3Cl_2$	9.4	0.10	0
HFC-134a	CH_2FCF_3	14	$< 1.5 \times 10^{-5}$	0
Carbon tetrachloride	CCl_4	42	1.20	1,029

Sources: Lifetime and ODP from WMO (1994). Emissions from Hammitt and others (1986) except HCFC-22 from AFEAS (1995).

focused attention on the potential contribution of CFCs to global warming and suggested that efforts to reduce ozone depletion would have important ramifications for global warming and vice versa.

Following release of the 1984 NAS report, the Natural Resources Defense Council (NRDC) sued EPA to compel the agency to issue final regulations for the proposed 1980 rulemaking. In March 1985, while negotiations continued between EPA and NRDC, delegates from twenty-one nations including the United States met in Vienna to adopt the Vienna Convention for the Protection of the Ozone Layer. The Vienna Convention established a program for cooperative international research and monitoring and scheduled a second meeting for spring 1987 to discuss international measures to limit CFC use.

Beyond Models: Discovering the Ozone Hole. Two months later, in May 1985, a British Antarctic Survey Team reported the first observed damage to the ozone layer. Measuring ozone above Antarctica, the team had first observed a springtime decrease in ozone above its research station in 1981 (Somerville 1996). However, these findings were not consistent with satellite measurements, which did not seem to show a decrease. After confirming the observations over several springs, Farman and others published their report in 1985. Shortly afterward, NASA scientists confirmed the existence of the Antarctic "ozone hole" using satellite data.[2]

News of the ozone hole and the graphic satellite images NASA produced revived public concern about ozone depletion and its possibly catastrophic consequences. It also demonstrated a fundamental lack of understanding of upper atmospheric chemistry and physics. None of the models that had been developed had predicted so much ozone depletion so quickly, and none were able to explain the development of the hole over Antarctica, which appeared only in the austral spring. Several competing hypotheses were advanced, implicating either chemical or dynamical processes.

A consent decree settling the NRDC lawsuit was issued in October 1985. It required EPA to decide by 1987 whether to regulate nonaerosol uses of CFCs. EPA announced in January 1986 that it was preparing a Stratospheric Ozone Protection Plan.

In late 1986, several teams of scientists assembled in Antarctica to study the ozone hole and try to explain its development. The teams were unsuccessful in deciphering the origins of the hole, but they were able to rule out dynamical hypotheses, which required transport in the lower stratosphere that was opposite that observed (personal communication, D. Wuebbles).

A major shift in negotiating position occurred during 1986. An association of U.S. producers announced its support for a reasonable global limit on CFC growth and, subsequently, Du Pont announced its support for global limits on production (Parson 1993: 41). This change in position was apparently due to the increasing scientific evidence of ozone depletion (notably the Antarctic ozone hole) and the unwillingness of producers to tarnish their public images for products representing a small share of their sales and profits. In addition, because of the NRDC lawsuit, it seemed increasingly likely that new regulations would be imposed in the United States. Producers began to see the choice as one between domestic and international regulations, rather than one between new regulations and none. Since the U.S. industry had already suffered from the aerosol ban, it was not eager to accept further unilateral regulation (Cogan 1988; Rowlands 1995). Moreover, producers may have seen greater commercial opportunity in developing new substitutes than in continuing to sell CFCs, which required relatively low-technology production processes and were becoming low-margin commodity chemicals.

Although the source of the ozone hole remained uncertain, delegates convened in Geneva in April of 1987 as scheduled under the Vienna Convention. During the Geneva meetings, the delegates reached an agreement to freeze CFC production at 1986 levels beginning in 1990, to reduce CFC production 20% by 1992, and to discuss options for controlling Halons and further reductions of CFCs. A meeting was scheduled for September 1987 in Montreal at which the delegates planned to sign a protocol agreement. Shortly after the groundwork for the Montreal Protocol had been laid, U.S. Interior Secretary Donald Hodel and presidential science adviser William Graham suggested that CFC production cuts might not be necessary as people could adopt "personal protection" like hats, sunglasses, and suntan lotion to reduce UV exposure. Environmentalists dubbed this the "Rayban Plan," noted that it would do little for crops and wildlife ("It's very hard to get fish to wear sunscreen"—David Doniger of NRDC, quoted in Cogan 1988: 52), and estimated the costs of such a plan to be $10 billion a year, at least ten times higher than the projected costs of a freeze in production.

In June 1987, President Reagan formally endorsed the State Department's position supporting government controls on CFCs. The Council of Economic Advisers was conducting an ongoing benefit-cost analysis, with extensive sensitivity analysis about the planning horizon and discount rate, among other factors, which supported CFC regulation. Although not published, results from the analysis were presented in administration meetings and were apparently influential in the decision (Benedick 1991: 63; personal communication, S. DeCanio).

The Montreal Protocol. The Montreal Protocol on Substances that Deplete the Ozone Layer was signed in September 1987. Under its terms, delegates were to meet periodically to review the adequacy of the protocol. In December, EPA published its regulatory impact assessment (described below) that estimated that millions of cases of skin cancer among Americans would result over the next century if CFC and Halon emissions were not controlled. The agency concluded that regulations to implement the Montreal Protocol were highly cost-effective. It issued draft regulations in December 1987 and final regulations in August 1988 (U.S. EPA 1987a).

Following the events in Montreal, scientists returned to Antarctica. The 1987 ozone hole was both larger and deeper than had been previously observed. Solomon and others (1986) postulated that surface reactions on the polar stratospheric clouds present during the Southern Hemisphere's winter months might explain its development. Polar stratospheric clouds develop when the air mass over the Antarctic rotates in the "polar vortex," limiting mixing with outside air, and extremely low temperature in the vortex causes water vapor and other chemicals to freeze into ice crystals. Laboratory measurements subsequently demonstrated that the reaction of two important sinks for chlorine, chlorine nitrate and hydrogen chloride, was greatly increased in the presence of ice crystals like those present in the polar stratospheric clouds (Molina and others 1987). The products of this reaction are two chlorine atoms, which react with ozone, and nitric acid, which prohibits the inhibiting reaction between nitrogen dioxide and chlorine atoms. In short, the polar stratospheric clouds create some of the best possible conditions for ozone depletion, and lower levels of ozone (which mean cooler temperatures) create some of the best possible conditions for polar stratospheric clouds.

In March 1988, NASA released the Executive Summary of the Report of the Ozone Trends Panel, an international group of more than 100 scientists (NASA 1988). This report provided the first observational evidence of large-scale ozone loss, which was at a rate exceeding that predicted by contemporary atmospheric models. The report concluded that between 1969 and 1986, average ozone levels dropped between 1.7% and 3% in the temperate Northern Hemisphere, with wintertime losses averaging between 2.3% and 6.2%. Although these observations could not establish causality, the losses were consistent with the theory of CFC-caused depletion, albeit somewhat larger than predicted by current models. The report also implicated CFCs as a primary cause of the Antarctic ozone hole and recognized the hole's hemispheric effects, documenting that ozone-depleted air moves toward the equator when the polar vortex breaks up in the Southern Hemispheric spring, causing ozone declines of 5% or more in southern latitudes poleward of sixty degrees. Following release of the report, Du Pont announced that it would discontinue production of CFCs and that it expected to be able to provide substitute compounds in commercial quantities within about five years, if adequate incentives were provided. Du Pont also recommended that the Montreal Protocol be amended to require a virtual phase out of CFC production by early in the next century (Cogan 1988: 74–75).

The Montreal Protocol took effect in July 1989. The protocol required that signatory nations freeze CFC consumption at 1986 levels until 1993, cut them 20% beginning in 1997, and cut them another 30% by mid-1998 (developing countries are allowed a ten-year grace period before meeting similar conditions). The protocol also froze annual consumption of Halons at the 1986 level beginning in February 1992 (the protocol is summarized in Benedick 1991 and reprinted in U.S. EPA 1987a). EPA is charged with implementing the requirements of the Montreal Protocol in the United States, and it issued regulations in August 1988 to initiate the process. There have been three subsequent revisions to the protocol: the 1990 London, 1992 Copenhagen, and 1995 Vienna amendments, which have accelerated the phase out of some compounds and added restrictions on additional ozone-depleting substances.[3]

THE 1987 RIA: AN OVERVIEW

While the 1987 ozone-hole research was under way, EPA was completing its regulatory impact analysis (RIA). Although the RIA was not published until after the Montreal Protocol was signed, much of the input to the analysis was presented during the two years leading up to the protocol. Indeed, EPA played a significant role in promoting negotiations, including joint sponsorship with the United Nations Environment Programme (UNEP) of a series of scientific and economic workshops.

The RIA quantified the costs and benefits to the United States of national and international regulation of ozone-depleting substances. It concluded that the roughly 50% cut in CFC emissions and stabilization of Halon emissions proposed by the Montreal Protocol was extremely cost-effective, with U.S. benefits of $6.5 trillion and costs of only $27 billion. Because of the long time horizon appropriate to this analysis (due to the century-long atmospheric residence times of the primary CFCs), these estimates are necessarily subject to large uncertainties. The RIA addressed some of the important uncertainties, finding positive net benefits in all cases. Consideration of other important uncertainties could have substantially widened the range of plausible net benefits, but the probability that net benefits would have been negative appears to be small.

The RIA begins with a set of baseline scenarios representing future CFC and Halon emissions through 2075, assuming no additional controls beyond the aerosol bans implemented by the United States and a few other countries in the 1970s. Based on several studies of current trends and the broad economic determinants of CFC and Halon use, "low," "medium," and "high" emission scenarios were developed. EPA invested substantial resources in overcoming the post–aerosol ban presumption that CFC use would not grow in the future, commissioning studies of future emissions by the environmental consulting firm ICF (Gibbs 1986), Nordhaus and Yohe (1986), and RAND (Hammitt and others 1986, 1987) (EPA 1987b; personal communication, J. Hoffman).

In the RIA, emissions of CFC-11, -12, and -113 and Halon 1301 were projected to grow between about 2% and 6% per year for the period 1985–2000, and between about 1.3% and 4% per year for the period 2000–2050 (see Table 2). Halon 1211 was projected to grow at a somewhat greater rate. Emissions of all compounds were conservatively and arbitrarily assumed to stop growing in 2050 and to remain constant thereafter.

Policies Considered

The 1987 RIA considered a variety of options with respect to which compounds would be covered, the stringency of controls, and the regulatory mechanism to be employed. Under all of the control options, the fully

Table 2. CFC and Halon Emission Scenarios and Calculated Ozone Depletion.

	Low growth	*Medium growth*	*High growth*
Projected emissions growth rate (%/yr)			
1985–2000			
CFC-11, 12	2.1	3.6	5.1
CFC-113	2.1	4.0	6.1
Halon 1301	–0.5	1.1	2.0
Halon 1211	5.5	8.8	12.0
2000–2050			
CFC-11, 12	1.3	2.5	3.8
CFC-113	1.3	2.5	3.8
Halon 1301	1.6	3.2	4.7
Halon 1211	1.5	2.9	4.4
2050–2075			
All gases	0.0	0.0	0.0
Projected ozone depletion (%)			
1990	0.27	0.27	0.28
2010	1.25	1.71	2.33
2030	2.55	4.86	9.13
2050	4.32	12.3	>50
2075	7.09	39.9	>50

Source: U.S. EPA 1987a.

halogenated CFCs (primarily CFC-11, -12, and -113) would be controlled; under some scenarios, other fully halogenated compounds (primarily Halons 1211 and 1301) would also be controlled. Controls on compounds not fully halogenated (in particular, HCFC-22 and methyl chloroform) were not proposed. These chemicals are roughly an order of magnitude less efficient ozone depleters than the fully halogenated CFCs (Table 1) and were expected to be used as transitional substitutes until nonozone-depleting alternatives could be brought into production. In addition, carbon tetrachloride was not proposed for regulation. Even though it is fully halogenated and approximately as destructive to ozone as the CFCs, carbon tetrachloride use in the United States and other developed countries was very small (except as an input to CFC production) because of its toxicity to humans.

All of the control strategies begin with a freeze on ODP-weighted use of fully halogenated CFCs at the 1986 level, to become effective in 1989. A series of more stringent control options supplemented this initial freeze with increasingly stringent reductions from the 1986 level of 20%, 50%, and 80%, to become effective in 1993, 1998, and 2003, respectively. Adherence to the Montreal Protocol is represented by the initial freeze followed by 20% and 50% reductions in ODP-weighted CFC use, plus a freeze on ODP-weighted Halon use at the 1986 level, beginning in 1992. Two additional variations on this scenario were considered: in one, the United States would go beyond the Montreal Protocol by unilaterally adopting the 80% reduction in CFC use effective in 2003; in a second, the United States alone would follow the Montreal Protocol; other countries would impose no controls on CFC use.

CFCs and Halons both deplete stratospheric ozone, but neither the Montreal Protocol nor the control options considered in the RIA allowed trading between these two classes. Although ODPs were available for both types of compounds, the ODPs for Halons depend on atmospheric chlorine concentrations and were also less accurately estimated because of greater uncertainty about bromine chemistry.

Compliance and Regulatory Considerations. The benefits of CFC control depend on global emission levels, and consequently the RIA required assumptions about international accession to the Montreal Protocol and compliance with its terms. In all cases, U.S. compliance was assumed to be complete, in part because the small number of CFC producers and importers was thought to make enforcement relatively easy. In fact, some relatively small-scale black-market imports and mislabeling of virgin production as recycled CFCs appeared in the 1990s under the more stringent provisions of the London amendments to the protocol (Cook 1996: 12; UNEP 1994: 7–14). Based on anticipated ratification of the Montreal Protocol, countries representing 94% of developed-country CFC use, and 65% of developing-country CFC use, were assumed to comply with the protocol. (In sensitivity analysis, these fractions were varied between 75% and 100% for the developed countries and between 40% and 100% for the developing countries.) Moreover, CFC use in countries not acceding to the Montreal Protocol was assumed to grow more slowly than it would have absent the protocol because of a "rechanneling" of technological development. This rechanneling reflects the expectation that research and development of new applications would concentrate on non-CFC alternatives, and fewer new applications of CFCs would be developed. As a result, CFC use in the countries not acceding to the protocol was assumed to grow under the control scenarios to only half the level it would under the baseline scenario. "Rechanneling" is the opposite of the "leakage" often anticipated under environmental regulation, in which producers migrate from countries with more stringent regulations to countries with less stringent regulations, and reduced demand for regulated materials may lower world prices and thereby increase use elsewhere (Felder and Rutherford 1993). The incorporation of restrictions on trade with countries not signing the Montreal Protocol was intended to prevent importation to the signatory countries, and so reduce leakage.

Four alternative regulatory mechanisms were considered: a regulatory fee on CFC production or use (a pollution tax), permits for CFC use that would be sold by the government at auction, marketable production/import quotas assigned to CFC producers, and command and control restrictions on process engineering and product bans. The mechanisms were evaluated with respect to the certainty with which the desired level of control could be obtained, economic efficiency, equity considerations, administrative costs and enforceability, legal certainty, and impacts on small business. Command and control approaches were rejected, as they would be extremely cumbersome to implement, given the large number of firms using CFCs in their production processes, and the large number of product and process applications. Moreover, command and control approaches could not be guaranteed to achieve the control levels specified under the Montreal Protocol. Similarly, the regulatory fee was discarded because of uncertainty about the level of fee that would be required to yield the emission reduction required by the protocol and about EPA's legal authority to levy such a fee. Production and import quotas assigned to existing producers were viewed as preferable to use permits that would be auctioned by the government because of administrative simplicity, given the small number of producers and importers (estimated as no more than fifteen or twenty) and legal uncertainty about whether EPA could auction permits. Taxes on CFC sales and inventories were subsequently introduced in 1989. These taxes have multiple uses, including capturing part of the rents created by the limited number of production rights. Barthold (1994) and Hoerner (1995) conclude that these taxes reduce CFC consumption below the allowable quantity, thereby acting as additional control measures.

Policy Limitations and Difficulties. A significant limitation to the policies considered in the RIA is the failure to consider adaptive or sequential decision strategies, in which CFC controls are modified as new information about growth in CFC emissions, control costs, ozone depletion, and damages becomes available (Hammitt 1986, 1987). The RIA frames the decision problem as a one-time choice between alternative regulatory levels (including no additional regulation) and forecasts the consequences of following each choice for the next century. The failure to consider sequential decision strategies is somewhat peculiar, as the Montreal Protocol was structured with the recognition that a one-time choice of CFC restrictions for the next century was inappropriate. The protocol calls for periodic meetings of the parties to assess the state of the science and economics and to revise its conditions as appropriate; as noted earlier, three such revisions have occurred to date. Incorporating the possibility of altering policies in response to new information would be expected to sharply reduce the difference in expected net benefits between alternative near-term control levels, since it would be assumed that, if the selected degree of control were subsequently shown to be highly nonoptimal, it would be adjusted. Acknowledging the fact that current policymakers cannot fix CFC control policy for succeeding generations would obviate the need for the unrealistic assumption of the RIA that, if additional CFC controls were not adopted in the mid-1980s, emissions would grow in accordance with "business as usual" assumptions for the next century.

Incorporating sequential policy choices in the RIA would have required EPA to address a number of complicated and uncertain interactions in economic and social processes. Although it is relatively straightforward to determine optimal control policies using a dynamic-programming framework, a number of factors may be difficult to represent. For example, opportunities to revise control levels may present themselves only intermittently, often when some exogenous event highlights the issue. The discovery of the Antarctic ozone hole was one such event that focused attention on stratospheric-ozone concerns; if international regulations had not been adopted then, other events could have captured governmental attention and made it difficult to obtain an international agreement. In addition, research and development of improved control technologies depend in part on regulatory pressure; CFC producers had devoted substantial resources to developing substitute compounds in the late 1970s when additional regulations appeared likely to follow the aerosol ban, but curtailed these programs in the early 1980s as the likelihood of additional regulation waned. The phenomenon of technological "lock-in," where a new technology becomes widely adopted, can also have significant long-term implications for the cost of substituting an alternative technology.[4]

Estimated Costs of Control

Social and private control costs, for the United States, were estimated using a detailed engineering economic analysis.[5] The analysis accounted for several types of possible control actions, including product substitution (such as switching from CFC-blown foam to cardboard egg cartons), use of production processes that are less CFC-intensive (such as CFC recapture and reuse in foam blowing), and use of alternative CFC-free production processes (substitution of methylene chloride in foam blowing).

Nearly 900 technical control measures were identified for a total of 74 applications of CFCs. Of these, about 350 options were excluded from further consideration because of potential difficulties with toxicity or flammability, technical feasibility, cost, effectiveness in reducing CFC emissions, enforceability, or inadequacy of available data for assessment. For the remaining 550 technical controls, the potential reduction in CFC use was estimated by considering the quantity of CFCs used in the application, the share of the application to which the alternative technology could be applied, and the extent to which it would reduce CFC use in that share of the application. Because many of the control options were not yet available, the date of introduction of the technology and the time required for it to penetrate the applicable market were also assessed.

The annualized marginal costs of the options were based on estimates of capital and operating costs, other costs or cost savings (for example, changes in energy use), salvage value of existing capital equipment, and transitional costs (for example, research and development, product reformulation, retraining). Offsetting risks posed by the alternatives were not formally incorporated, except when such risks led to exclusion of an option or when additional costs associated with controlling the risk (for example, fire protection) could be quantified. An important assumption is that the marginal costs of controlling CFC-11 and -12 in foam-blowing and refrigeration applications was capped by the availability of a "backstop" technology: the newly developed chemicals HCFC-123 and HFC-134a were assumed to substitute for CFC-11 and -12 in these applications at a price increase of $5.48 per kilogram (a substantial increase from the unregulated price of about $1.50 per kilogram). (This estimate seems remarkably prescient: upon completion of the U.S. phase out of CFC-11 and -12 in spring 1996, the price of these compounds was about $11 per pound or $5 per kilogram; personal communication, E. Parson.)

Underlying Assumptions of the Analysis. The economic analysis made several simplifying assumptions. First, CFC production was assumed to be perfectly elastic, with marginal cost equal to price. Under this assumption, producer surplus was zero under both baseline and control scenarios, and so welfare changes could be evaluated in terms of changes in consumer surplus alone. The assumption of constant marginal cost was justified by analysis of CFC production technology; the possibility that CFC producers could be earning rents, because of the concentration of production in a few firms, was acknowledged but not incorporated in the analysis. Any producer surplus was probably small, however, since CFCs had become commodity chemicals with substantial international trade.

Demand for products incorporating or manufactured with CFCs was assumed to be perfectly inelastic, so that CFC regulation would not affect quantities of these products. This assumption allowed for calculation of changes in consumer surplus as the product of the price increase and the (unchanging) quantity produced. It was justified by the recognition that CFCs accounted for less than 5% of costs for most products in which they were used, so that product prices would be insensitive to CFC price increases. In applications where CFCs accounted for a larger share of costs, demand for product services was considered inelastic (such as eggs being packaged using either CFC-blown foam or cardboard).

Costs were estimated under alternative assumptions about the rate at which substitute technologies would become available and would be adopted. The primary cost estimate represents a "least-cost" simulation in which new technologies are adopted as soon as they become available and cost-effective. This estimate assumes CFCs would be regulated using some type of market mechanism (or a command and control mechanism that mimics a market mechanism). To account for real-world "stickiness" due to imperfect information, imperfect capital markets, and other sources of inertia, the RIA also considered "stretch-out" scenarios incorporating more pessimistic assumptions about the rate of adoption of new technologies.

Present-value compliance costs for the period 1989 (the date of initial control) through 2075 are reported in Table 3. The uncertainty ranges shown in Table 3 reflect alternative assumptions about adoption of new technology. Control costs are based on the medium growth scenario for future CFC emissions. Sensitivity of the control costs to alternative assumptions about general economic growth and specific growth in demand for CFC-using goods was not reported, although these uncer-

Table 3. Estimated Present Value Costs of Control Options, 1989–2075 (millions of $1985, discounted at 2%).

		Stretchout	
Control scenario	*Least cost*	*Moderate*	*Major*
No controls	0	0	0
CFC freeze	6,778	7,050	17,240
CFC 20% cut	12,070	16,590	27,460
CFC 50% cut	24,440	26,360	35,550
CFC 80% cut	31,350	41,820	55,370
CFC 50% cut, Halon freeze	27,040	29,220	38,140
CFC 50% cut, Halon freeze, U.S. 80% cut	33,950	44,410	57,960
U.S. only CFC 50% cut, Halon freeze	27,040	29,220	38,140

Notes: CFC control options are cumulative, that is, 80% cut includes freeze in 1989, 20% cut in 1993, 50% cut in 1998, and 80% cut in 2003. Stretchout cases reflect delays in adoption of cost-effective control measures. RIA also includes an additional stretchout case intermediate to the moderate and major cases.

Source: U.S. EPA 1987b, Exhibit 9-6.

tainties are primary determinants of the alternative baseline emission paths (Table 2).

Estimated Benefits of Control

Because CFCs' atmospheric lifetimes are on the order of a century (Table 1), the benefits of controlling CFC emissions persist for a similar time period. As a result, the choice of population whose benefits are to be included in the analysis is more important than in many other regulatory contexts. The RIA used as its target population the U.S. population living at the time of the assessment plus all additional U.S. residents born before 2075. Health benefits incurred by members of this population are apparently counted only if they occur before 2075; thus, there are presumably no averted skin cancers for people born after about 2060 or so, as these would be unlikely to manifest before 2075.

Benefits are estimated by simulating time paths of ozone depletion, UV flux at the earth's surface, and effects on humans, wildlife, and materials. Time paths of total global ozone depletion corresponding to the three baseline (no control) emission scenarios shown in Table 2 were calculated using a simplified representation (Connell 1986) of contemporary atmospheric models that represent a globally averaged "one-dimensional" atmosphere—that is, the models represent an idealized column of the atmosphere, with solar insolation and other input conditions representing some average of the latitudinally and temporally varying conditions in the real atmosphere. Although two-dimensional models (representing the effects of latitude and altitude, and incorporating seasonal variation) were available at the time, EPA judged them to be insufficiently reliable for its analysis. Even holding the atmospheric model constant, the range of future emissions was so large that projected ozone depletion ranged from about 4% to more than 50% by 2050 (Table 2). As illustrated in Figure 3, ozone depletion for the medium emission scenario was projected to increase modestly until about 2030 or so, when nonlinearities in the atmospheric response would lead to a "cliff." In contrast, depletion was projected to be minimal under the 50% reduction in CFC emissions and freeze on growth in Halon emissions called for by the Montreal Protocol. Only the medium emission scenario seems to have been used in the benefit assessment.

The effects of ozone depletion on a variety of endpoints, including human health, materials damage, plants, and aquatic organisms, were considered. (Because projected ozone depletion was near zero under the primary control scenario, corresponding to the Montreal Protocol restrictions, nearly all the damages due to ozone depletion would be offset by controls.) The estimated benefits are summarized in Table 4. Despite the range of endpoints considered, the estimated monetary benefits are dominated by the effect on deaths from skin cancer. To some extent, this reflects the partial and incomplete analysis of the other benefit components. Because the skin-cancer benefits alone vastly exceed the control costs, EPA limited its analysis of other components.

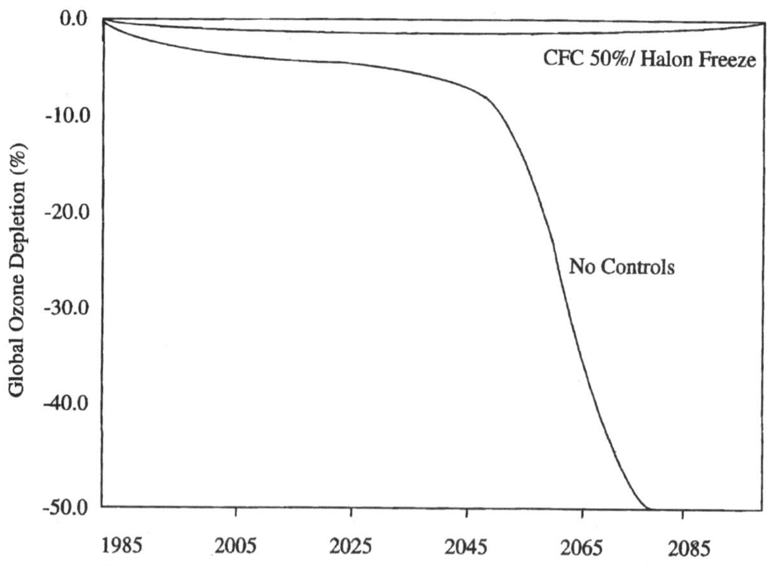

Figure 3. Global Ozone Depletion Estimates.

Note: These estimates are for the no-controls case and the CFC 50% reduction/ Halon freeze case.

Source: U.S. EPA 1987b: Exhibit 6-2.

The effect of ozone depletion on the amount of biologically damaging UV-B radiation reaching the earth's surface varies with frequency, the presence of ground-level ozone that can also absorb UV-B radiation, cloudiness, and other factors. EPA developed a model to represent the effect of decreased stratospheric ozone on UV-B flux by frequency, then weighted the frequencies in proportion to their biological activity and summed the changes into an effective change in a biological action spectrum.[6] The result is that, for ozone depletion of no more than 20% to 30%, the biologically effective UV-B flux was about double the ozone depletion, with increasing proportional response of UV-B radiation to ozone depletion for higher depletion levels (U.S. EPA 1987b; Appendix E).

Skin Cancers. The evidence linking UV-B flux to nonmelanoma (basal- and squamous-cell) skin cancer is fairly strong: these cancers are most often located on parts of the body that are routinely exposed to sunlight, and their incidence (primarily to Caucasians) is higher among outdoor workers and populations living nearer the equator. Based on epidemiological evidence, EPA developed an exposure-response model in which the incidence of nonmelanoma skin cancer increases roughly in proportion to the increase in UV-B flux. Because this skin cancer is quite common (without ozone depletion, the average lifetime risk for Caucasian Americans is about 30%; U.S. EPA 1987b; Appendix E), projected increases quickly lead to millions of additional cases: EPA estimated 153 million additional cases to its reference population under the medium ozone depletion scenario. Fortunately, these cancers are readily treatable, so

Table 4. Summary of Estimated Benefits of Regulations Implementing Montreal Protocol.

	Present value ($billions)		
Impact	*Base*	*Low*	*High*
Skin cancer mortality	6,350	17.4	342,000
Skin cancer morbidity	61.3	1.1	205.0
Cataracts	2.6	0.07	7.8
Crop damage from UV-B radiation	23.4	2.3	46.0
Crop damage from ground-level ozone	12.4	0.001	24.9
Reduction in commercial fish harvest	5.5	3.0	11.4
Damage to plastics	3.1	0.22	6.3
Sea-level rise (global warming)	4.3	na	na
Total	6,300	29	342,000

Notes: Totals as reported in original. Discount rate = 2%, 6%, 1% for Base, Low, and High cases, respectively. na: not reported

Source: U.S. EPA 1987a.

even with no assumption of improvements in medical technology, the projected increase in mortality is only three million.

The evidence linking melanoma cancer incidence to UV-B is much weaker than that for nonmelanoma cancers, but the risk appears to increase with acute episodes of sunburn, especially in childhood. EPA estimated an increased incidence of 782,000 cases and, because melanoma is much more frequently lethal, 187,000 additional deaths for its middle depletion scenario, yielding a total of about 3.2 million skin-cancer deaths.

The monetary value of averted skin-cancer mortality dominates the estimated benefits, accounting for 98% of the quantified benefits of regulation. Because of the long time horizon over which these deaths are averted, this value is extremely sensitive to the discount rate used and to assumptions about the rate at which the value of averted skin-cancer mortality changes over time. Possible changes in treatment technology that might reduce skin-cancer mortality were not quantified.

The value per averted cancer death was assumed to be $3 million in 1985, a value consistent with the standard range for value of statistical life of $1.6 to $8.4 million (Fisher and others 1989). In addition, the value of averted cancer deaths was assumed to grow at the per capita gross domestic product (GDP) growth rate, assumed to be 1.7% per year. Combined with the 2% annual discount rate, this yields an effective discount rate on skin-cancer mortality of about 0.3% per year in the base case, and a present value of averted skin-cancer deaths of $6.35 trillion. In an alternative, low-value case, the 1985 value of life is $2 million, increases at only half the rate of per capita GDP growth (0.85% per year), and benefits are discounted at 6% per year, yielding a present value of only $17.4 billion, 0.3% of the value in the base case. In the high-value case, the 1985 value of life is $4 million, grows at twice the per capita GDP rate, 3.4% per year, and benefits are discounted at only 1% per year. In this high case, the present value of averting a skin-cancer death is larger for future generations than for the present: it grows about 2.4% per year. The resulting present value is $342 trillion, about fifty times the value in the base case. Although these discount rates are far below the 10% discount rate advocated by the Office of Management and Budget, OMB apparently did not object to their use for this long-horizon analysis (personal communication, S. DeCanio).

The present value of averting the morbidity associated with increased skin-cancer incidence, $61.3 billion for the middle case, is based on the cost-of-illness method, by which the loss associated with disease is quantified as the value of the resources consumed in treating it (including lost work time). The estimated morbidity costs are about $15,000 per case for melanoma and $5,000 per case for nonmelanoma skin cancer. The cost-of-illness method would be expected to underestimate the value of averted morbidity because it does not account for the utility loss to the individual.

Increases in cataracts are quite plausibly linked to increases in UV-B radiation. EPA estimated 18.2 million additional cataracts in its reference population. Using cost-of-illness and individual willingness-to-pay measures, the value per cataract case is estimated as $15,000, the same as for a case of (nonfatal) melanoma. The possibility that increased UV-B would also lead to suppression of the human immune response and so increase the incidence of many diseases was recognized, but not quantified due to inadequate data.

Nonhealth Effects. In addition to human health effects, the benefits of reducing a number of other UV-related damages were quantified. These included damage to crops from UV-B and from higher concentrations of ground-level ozone (produced by the interaction of UV radiation, hydrocarbons, and nitrogen oxides), damages to commercial fishing because of increased larval mortality from UV-B exposure, increased weathering of outdoor plastics, and damage to seaports because of sea-level rise induced by global warming. Estimates of these benefits were based on more limited data, often case studies. For example, estimates of aquatic effects are based on studies of anchovies and UV effects on crops are based on studies of soybeans. As summarized in Table 4, these nonhealth effects did not contribute significantly to the overall benefit estimate. The nonhealth benefits of CFC control are probably underestimated because not all endpoints were included and assumptions made for the included endpoints are generally conservative. For example, the only damages from climate change included in the assessment are the costs of sea-level rise to port cities. One caveat is that the effects of UV increase on crops and other plants seem to be highly variable across species (Makhijani and Gurney 1995) so that if the species considered in the RIA are unusually sensitive to UV increases, these benefits may have been overestimated.

Comparison of Benefits and Costs

Estimated benefits and costs for alternative control strategies are summarized in Table 5. The base, low, and high cases differ only in the discount rate (applied to benefits and costs) and the valuation of skin-cancer mortality, as described earlier. For all of the proposed control policies, the net benefits are positive in all three cases, despite varying about four orders of magnitude. The control stringency and coverage required by the Montreal Protocol come close to maximizing net benefits for all three cases. For the base and high cases, net benefits are slightly larger if the United States were to unilaterally adopt the 80% CFC cut in 2003, and for the low

Table 5. Comparison of Estimated Global and U.S. Benefits and Costs for Alternative Control Levels. (billions of $1985)

Control	Base case			Low case			High case		
	Benefits	*Costs*	*Net*	*Benefits*	*Costs*	*Net*	*Benefits*	*Costs*	*Net*
CFC freeze	5,995	7	5988.00	16	0.7	15	324,000	12	323,988
CFC 20%	6,132	12	6,120	17	2	15	330,000	21	299,979
CFC 50%	6,299	24	6,275	18	5	13	339,000	41	338,959
CFC 80%	6,400	31	6,369	19	7	12	341,000	51	340,949
CFC 50%/Halon freeze	6,463	27	6,436	19	5	14	345,000	46	344,954
CFC 50%/Halon freeze/U.S. 80%	6,506	34	6,472	19	7	12	346,000	56	345,944
U.S. only CFC 50%	2,852	27	2,825	8	5	3	135,000	46	134,954

Note: Base, low, and high case reflect alternative assumptions about value of averted skin-cancer mortality and discount rate (2%, 6%, and 1%, respectively)

Source: U.S. EPA 1987a, Table 8.

case, net benefits would be increased by shifting to somewhat less stringent control measures.

Uncertainties about CFC emission growth in the absence of the proposed regulations, the resulting magnitudes of ozone depletion, UV-B flux, and responsiveness of the health and other endpoints to UV-B are not formally incorporated, nor are uncertainties about the cost and speed with which alternative technologies for reducing CFC use would become available and adopted. These omissions suggest that realistic uncertainty ranges for the benefits and costs of CFC control were much broader than those reported in the RIA. For example, a retrospective analysis of projected ozone depletion in the absence of additional regulation, incorporating uncertainties about CFC emissions and atmospheric processes, suggests an uncertainty range for the magnitude of ozone depletion in the year 2040 of at least zero to 50% (Hammitt 1995). Alternative assumptions about baseline growth in CFC emissions might have relatively modest effects on the estimated balance of benefits and costs, because larger baseline emissions would increase both the benefits and costs of control, and smaller baseline emissions would reduce both benefits and costs. Conditional on CFC emissions, uncertainties about the magnitude of ozone depletion and the response of health and other endpoints to ozone depletion (accounting for possible changes in medicine and other factors) could have substantial impacts on the estimated net benefits, however, as such uncertainties would alter the benefits but would presumably not alter the costs of control.

THE ROLE OF BENEFIT-COST ANALYSIS IN DECISIONMAKING

A number of factors were important in determining U.S. policy toward stratospheric-ozone depletion, including the decisions to support international negotiation and to accede to the Montreal Protocol. Although the formal RIA obviously could not have been influential in the decisions leading up to the signing of the Montreal Protocol (because it was not published until afterward), the benefit-cost framing of the problem and the preliminary estimates of the monetary values of benefits and costs appear to have contributed to government policymaking. It appears, however, that even a qualitative benefit-cost comparison was sufficient to support regulation; many participants in the debate viewed the choice between regulating and not as risking the possibility of global-scale catastrophic damages in exchange for improvements in consumer goods ranging from the important but mundane (refrigeration) to the frivolous (aerosol dispensers for personal care products). The precise level of control appears to have been determined by a combination of achievable compromise in international negotiations (splitting the difference between the U.S. proposal for 95% reductions and the European Commission's proposal to freeze production; Parson 1993: 60) and scientific projections showing that a 50% reduction would lead to minimal ozone depletion (Figure 3).

The draft RIA was published in December 1987, two months after the signing of the Montreal Protocol. The final RIA was published in August 1988. The RIA provided an analytic case supporting ratification of the protocol, but the important decisions about the compounds to regulate and the extent of control were made as part of the international negotiations leading to the protocol.

Although the RIA itself may have had only limited significance, benefit-cost analysis was clearly important in decisionmaking about this issue. Over the period from the initial discovery that CFCs could deplete stratospheric ozone to the time the Montreal Protocol was signed, numerous national and international assessments of the problem and of response options were conducted. The U.S. National Academy of Sciences published four studies evaluating the science of ozone depletion and its consequences, notably human skin cancer (NAS 1976, 1979, 1982, 1984). The World Meteorological Organization and other organizations coordinated a three-volume international assessment of the state of the atmospheric science (WMO 1985). The impacts of ozone depletion were subjected to less analysis, but the United Nations Environment Programme and EPA hosted a series of workshops on impacts and economic factors in 1986 (Titus 1986).

As noted earlier, the Council of Economic Advisers (CEA) conducted its own benefit-cost assessment of control options. This analysis was reported to be influential in decisionmaking within the Reagan administration since it countered the perspective offered by Interior Secretary Hodel and other administration members who suggested that adaptation was preferred to emission control. The CEA analysis may have been particularly influential within the administration, since it confirmed the EPA view that regulation was appropriate and CEA was viewed as a more central and trusted part of the administration team than was EPA (personal communications, S. DeCanio, J. Hoffman).

Observations of EPA Administrator Lee Thomas can be read to support the view that regulation was based primarily on the credible threat of global-scale damages, and not on quantification of the marginal benefits and costs of control options. Alternatively, his comments can be interpreted as confirming that the benefits exceeded the costs under any reasonable set of assumptions about the value of averting skin cancers and other damages (personal communication, S. DeCanio). Upon succeeding William Ruckelshaus as administrator in January 1985, Thomas was briefed on the CFC issue. He recalls "I just took a black-and-white view when I saw the data. I knew we had to get [CFCs] out of process. It didn't appear that even a little bit of them was going to be safe" (Litfin 1994: 73). This decision was not based solely on formal assessment: few scientists offered policy prescriptions, and most who did thought that 50% emission reductions were sufficient. In particular, Thomas's top science advisers did not advocate a near-total phase out of CFCs: Robert Watson of NASA testified to Congress that "the science doesn't justify a 95% cut" (Litfin 1994: 103–104). Thomas's position can be viewed as a risk management decision based on what would subsequently become known as the "precautionary principle"; in this, he differed from others in the Reagan administration. "[Presidential science adviser William] Graham looked at it from a purely scientific perspective, whereas I looked at it from a policy perspective. Where there was uncertainty, he thought we needed more research, and I thought we needed to be cautious. We just looked at the same thing and came to two different conclusions" (Litfin 1994: 104).

The U.S. CFC producers' decision to support global regulations also contributed to the decision. As described earlier, after discovery of the Antarctic ozone hole, Du Pont and other producers announced their support for international controls. Although they initially supported only a limit on the growth of CFC production, their acceptance of regulations removed a major source of opposition. The factors underlying the producers' change in position are not entirely clear, but it appears likely that producers were unwilling to risk costly public-relations battles to defend products that contributed only a few percent to overall profits. In addition, the sophisticated chemical producers may have perceived greater commercial opportunity in developing new compounds and processes to serve their customers' applications than in continuing to produce what had become commodity chemicals. At the time, there was suspicion that the U.S. producers shifted their position to support international controls because they believed they had a comparative advantage over foreign producers in developing and marketing substitute compounds. Indeed, the European Community suspected that the only reason the antiregulatory Reagan administration supported CFC controls was that U.S. producers had secretly developed substitutes (Litfin 1994: 108).

Whether or not benefit-cost analysis was central to the decision to impose additional controls on CFCs, the economic perspective clearly contributed to the design of the regulations. The Montreal Protocol and its implementing regulations represent one of the first major applications of market-based incentives for environmental regulation (Barthold 1994; Hoerner 1995; Stavins 1988). The protocol incorporates a tradable-permit approach: it does not limit use of individual compounds but attempts to limit total contribution to ozone depletion by restricting national use of multiple compounds weighted by their ozone depletion efficiencies (CFCs and Halons are subject to different limitations; trade-offs are allowed within but not between these classes). The protocol also allows national trading of production shares; this provision was initially intended to benefit producers within the European Community, by allowing consolidation of production shares to support efficient-scale plants. In addition, Dow Chemical used the provision to consolidate its U.S. and Canadian methyl chloroform production in the United States (Cook 1996). The U.S. implementing regulations for the protocol also incorporate tradable production permits because they were perceived as much more efficient than command and control regulations.

Part of the EPA regulatory strategy was to reduce the control costs, facilitating a decision in favor of regulation whether it was to be based on formal on informal comparison of the benefits and costs. To this end, EPA worked with CFC users and their customers to reduce the barriers to adopting non-CFC alternatives. For example, EPA was influential in revising the federal military specifications that required use of CFC-113 solvent in numerous products, worked with the Food Packaging Institute to promote substitution of HCFC-22 for CFC-12 in products like "clamshell" packages used by fast-food retailers, and advocated relaxing the purity standards to allow use of recycled CFC-12 in automobile air conditioners (Cook 1996; personal communication, J. Hoffman).

EX POST ANALYSIS OF *EX ANTE* ESTIMATES

Estimating the prospective benefits and costs of environmental regulations is always challenging. Estimates of the benefits of CFC control were made in a context of rapidly evolving scientific understanding of the factors determining stratospheric ozone concentrations. Estimates of control costs relied in part on assumptions about the availability of substitute compounds that had never been produced at commercial scale, nor introduced into industrial applications or consumer products. A decade has passed since the benefit and cost estimates were prepared; this section provides the beginnings of an *ex post* assessment of their accuracy.[8]

The Benefits. On the benefits side, there is strong observational evidence of global decreases in stratospheric ozone that are attributable to CFCs. Over the period 1979–94, cumulative ozone depletion is estimated as about 7% in the midlatitudes of the Northern and Southern Hemispheres, and about 2% in the tropics; the rate of ozone decrease appears to have accelerated in the later part of the period (WMO 1994: 1.1). The

Antarctic ozone hole has reappeared each spring, with the most widespread and deepest holes occurring in 1992 and 1993 (WMO 1994: 1.3). This depletion is substantially larger than forecast by models of the mid-1980s (see Figure 3).

Changes in UV-B flux at the earth's surface and in human health and other endpoints have been more difficult to measure and attribute to CFCs. Large increases in UV-B flux have been measured inside the ozone hole, with UV-B in Antarctica sometimes exceeding peak values in Southern California (WMO 1994: 9.1). Evidence of the effect of the increased UV-B radiation on Antarctic phytoplankton was reported in 1992 (Smith and others 1992).

Significant long-term trends in UV-B have not been reliably detected, in part because UV-B flux is sensitive to clouds and air pollution and monitoring instruments are not sufficiently accurate and stable for detection of small trends. However, ground-level UV-B flux was determined to exceed average levels in the midlatitudes in 1992 and 1993, years of anomalously low global ozone (WMO 1994: 9.1). Melanoma and nonmelanoma skin-cancer rates have increased substantially over the past few decades in the United States, Australia, and elsewhere, but these trends cannot be attributed to ozone depletion, as changes in fashion and leisure activities appear to have been more important (Makhijani and Gurney 1995; UNEP 1991; WHO 1994).

The Costs. No comprehensive assessments of the actual costs of controlling CFCs have been conducted, but there is a perception that the costs have been lower than those estimated at the time regulations were adopted. For example, the UNEP Economic Options Committee (an advisory committee established to help with implementation of the Montreal Protocol) states in its 1994 report that "ODS [ozone-depleting substance] replacement has been more rapid, less expensive and more innovative than had been anticipated at the beginning of the substitution process. The alternative technologies already adopted have been effective and inexpensive enough that consumers have not yet felt any noticeable impacts (except for an increase in automobile air conditioning service costs)" (UNEP 1994: 1–3).

Cook (1996) and Hoerner (1995) also assert that costs have been smaller than expected. Hoerner attributes this to unanticipated technical progress, stimulated by adoption of the excise tax that has held U.S. CFC consumption below the internationally determined limits. Cook attributes the favorable cost experience to the use of market-based regulatory instruments (providing desirable incentives for innovation), entrepreneurial government initiatives to facilitate interindustry cooperation and to remove regulatory barriers (such as military specifications requiring use of CFC-113 for cleaning electronic parts and purity standards for use of recycled CFC-12 in mobile air conditioners), the emergence of corporate competition for leadership in controlling CFCs (the voluntary phase out of CFC-12 in food-packaging foams), and the continued and enhanced government pressure for control in the form of subsequent amendments that strengthened the initial control provisions.

Comparison between marginal control costs estimated prior to adoption of the Montreal Protocol and subsequent CFC prices provides some evidence that control costs were overestimated for CFC-11 and -12, but not for CFC-113. The RIA does not present the marginal control-cost function used to estimate control costs: it reports the CFC prices at which alternative control options would be cost-effective, but not the associated reduction in CFC use. However, one of the reports prepared as input to the RIA (Camm and others 1986) does present estimated reductions in demand for CFC-11, -12, and -113 as a function of increases in price for these compounds. These demands can be compared with the realized prices and quantities of CFCs consumed following the Montreal Protocol.

The Camm and others (1986) estimates represent the equilibrium fractional reduction in CFC consumption as a function of price. For each compound, Camm and others report low and high estimates[9] of the reduction. The difference between the two cases reflects uncertainty about the extent to which certain low-cost control measures had already been adopted by industry, uncertainty about the allocation of total production of each compound to alternative applications, and different assumptions about the potential for control options about which the authors did not obtain quantitative data. The low estimate includes only control options for which Camm and others obtained quantitative estimates of cost and effectiveness and incorporates conservative estimates of the potential for these options; it clearly underestimates the reductions thought to be achievable at the time. The high estimate includes optimistic estimates of the potential for quantified options and some application of other control measures for which quantitative data were not obtained.

Realized prices (including the federal excise tax) represent the marginal cost of reducing CFC consumption. To the extent that user industries do not maintain substantial inventories of CFCs, these prices represent transient marginal costs; if transition costs are substantial, transient marginal costs would be expected to exceed the equilibrium marginal costs that Camm and others estimated.

Consumption and Costs. To compare the *ex ante* estimates with *ex post* experience, it is necessary to convert post–Montreal Protocol CFC consumption to fractional reductions from a baseline consumption scenario. Reductions may be estimated by comparing actual and forecast CFC consumption. Actual consumption for the years 1986 (the baseline) and 1989–94 is reported by EPA as part of its monitoring of compliance with the Montreal Protocol (U.S. EPA 1994; personal communication, M. James, U.S. EPA). Forecast consumption in the absence of the Montreal Protocol, from which reductions are determined, is estimated by applying the RIA middle-case growth rates (Table 2) to 1986 consumption. Prices and the excise-tax rates are reported by UNEP (1994) and are converted to constant (1986) dollars using the GDP deflator (President 1995).

Figures 4, 5, and 6 compare the estimates of Camm and others (1986) with actual consumption and prices for CFC-11, -12, and -113, respectively. The results differ substantially among compounds. For CFC-11 (Figure 4), the reduction in consumption is substantially larger than

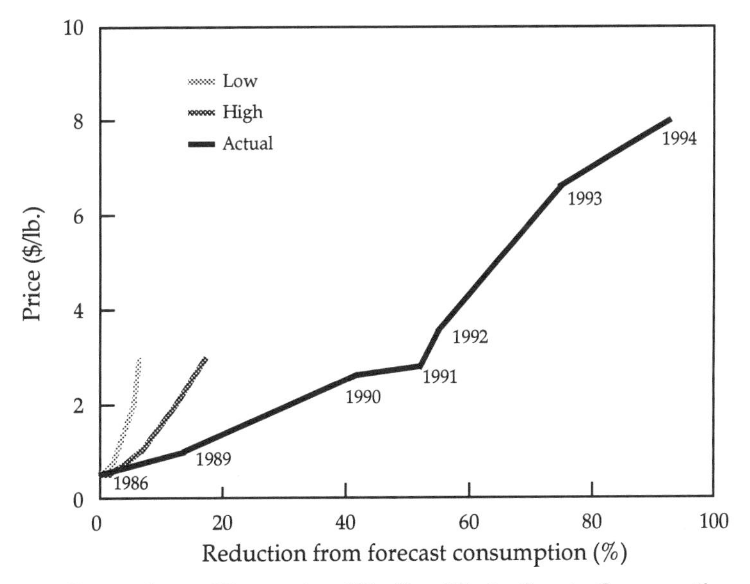

Figure 4. Comparison of Forecast and Realized Reduction in Consumption of CFC-11.

Note: The low estimate incorporates conservative assumptions about future adoption of technical options for which Camm and others were able to obtain quantitative estimates of cost and effectiveness. The high estimate includes more optimistic assumptions about adoption of technologies for which quantitative cost and effectiveness data were unavailable; Camm and others label this a "total" reduction estimate. The "actual" reduction is calculated as the fractional reduction from the RIA middle-case growth rates in Table 2.

Sources: Camm and others 1986 (low and high estimates); the actual reductions are from UNEP 1994 (prices and excise taxes) and EPA 1994 and personal communication, M. James (production).

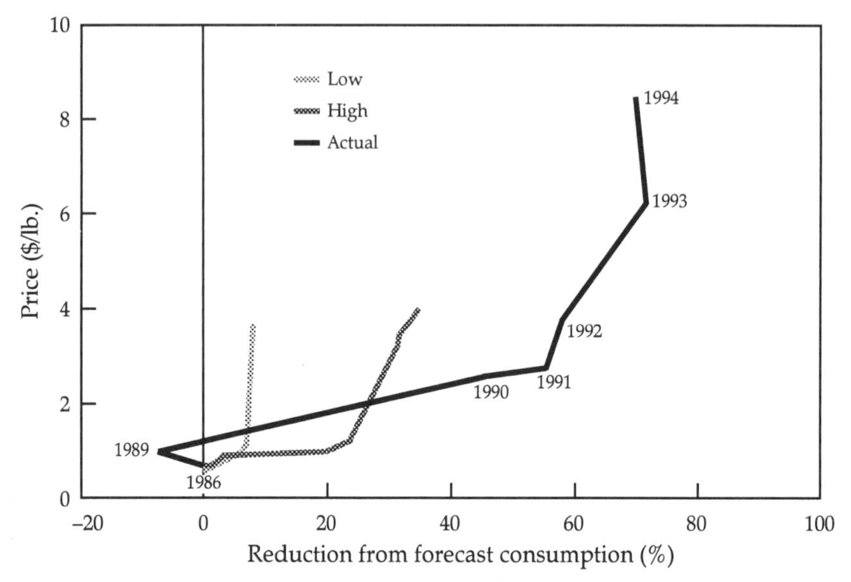

Figure 5. Comparison of Forecast and Realized Reduction in Consumption of CFC-12.

Notes and Sources: See Figure 4.

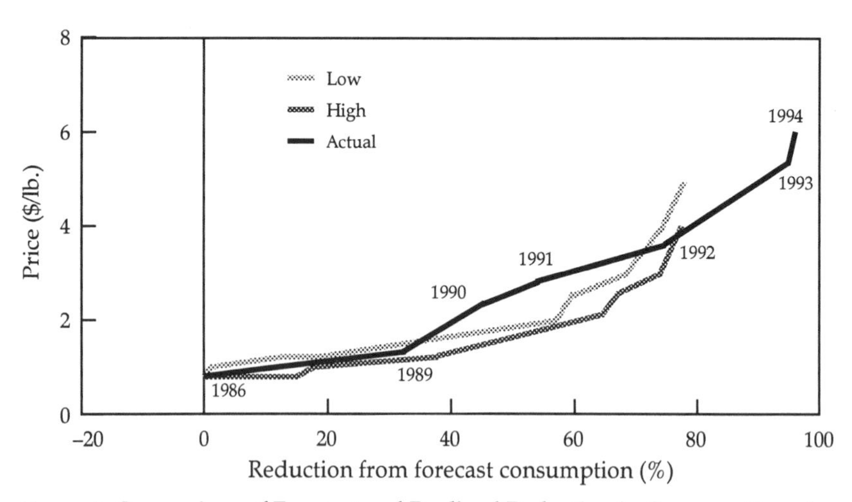

Figure 6. Comparison of Forecast and Realized Reduction in Consumption of CFC-113.

Notes and Sources: See Figure 4.

Camm and others estimated. At a price of $2 to $3 per pound, achieved in 1990–92, consumption was 40% to 60% below the forecast baseline, much larger than the 15% to 20% estimated reduction. The comparison for CFC-12 (Figure 5) suggests that control costs were also overestimated for this compound, although not as badly as for CFC-11. The realized prices suggest that consumption reductions in excess of about 60% are very costly, as the price increased from less than $3 to more than $8 with only modest additional reduction.

In contrast, control costs for CFC-113 appear to have been underestimated. As shown in Figure 6, the estimates by Camm and others suggest reductions approaching 80% at sixfold price increases. The 95% reductions achieved in 1993 and 1994 appear to be somewhat larger than one would have extrapolated from the 1986 estimates of Camm and others, but the reductions in prior years were consistent with or smaller than the forecast reductions for those prices.

The 1988 RIA forecasts the increase in aggregate CFC prices expected under the regulations. Comparing this forecast with the mandated consumption reduction provides an estimate of marginal control cost that is substantially smaller than the 1986 estimate by Camm and others that may even underestimate realized marginal control costs.[7]

Summary. Anecdotal evidence suggests that control costs have been substantially smaller than forecast at the time of the Montreal Protocol, but comparison of *ex ante* estimates of the marginal control costs and *ex post* realizations of reductions in CFC consumption and prices suggests a more tempered conclusion. The 1986 estimates by Camm and others appear to have overestimated the costs of controlling CFCs-11 and 12, but not 113. Later estimates by EPA appear more accurate and may have underestimated aggregate marginal control costs.

On the benefits side, global ozone depletion has been detected and the rates of depletion are larger than forecast. Although trends in UV-B radiation and adverse health effects that are attributable to CFCs have not been reliably detected, there is no evidence to contradict the magnitude of projected effects.

CONCLUSION

The Montreal Protocol and subsequent amendments that strengthened the requirements on ozone-depleting compounds appear to be achieving their primary objective of limiting ozone depletion and allowing for eventual restoration of natural stratospheric-ozone concentrations. Atmospheric measurements suggest that the quantity of chlorine introduced into the troposphere by ozone-depleting substances peaked in early 1994 and have begun to decline. Atmospheric modeling suggests that, if the Montreal Protocol and its amendments are adhered to, ozone depletion will peak between 1997 and 1999, and begin to recover early in the next century (Montzka and others 1996).

Regulation of CFCs and other ozone-depleting substances appears to have been driven primarily by extensive scientific assessment of the relationship between CFC emissions and stratospheric ozone, combined with comparatively informal and qualitative comparison of the benefits and costs of regulation. In the early stages of the transition, the comparison between potentially catastrophic global changes in ozone and the seemingly trivial benefits of CFCs as aerosol propellants was sufficient in the United States and a few other countries to lead to substantial reductions in CFC use. In later stages, more formal benefit-cost assessments were conducted, but they appear to have had modest effects on the policy process. Indeed, despite the quantification of health effects from ozone depletion, persistent uncertainty about the relationship between CFC emissions and ozone levels has subsequently led the policy discussion to focus its analysis on an earlier stage in the causal chain linking CFC production to ozone depletion and damages. After the discovery of the Antarctic ozone hole and the Ozone Trends Panel report (NASA 1988) showed that ozone depletion exceeded model forecasts, atmospheric chlorine concentration, rather than ozone depletion, has often been used as the analytic endpoint (Montzka and others 1996; WMO 1994). Policy judgments are grounded not in the estimated monetary value of damages associated with alternative levels of atmospheric chlorine or ozone, but rather in the goal of returning atmospheric chlorine concentrations to levels below those at which the ozone hole first appeared.

The relative unimportance of quantified health and environmental risks appears to reflect the global, uncertain, and potentially catastrophic consequences of CFC emissions. As was emphasized early in the debate, terrestrial life developed only after early photosynthetic organisms had produced oxygen, and with it, atmospheric ozone in sufficient quantity to screen out much of the ultraviolet radiation; humans and other terrestrial organisms could not be expected to survive without this shield. (Ironically, it was the ultraviolet radiation that initially provided the energy necessary to synthesize the initial organic compounds that produced life on Earth; Dotto and Schiff 1978: 27.) Thus, compelling evidence supporting the CFC-ozone link was decisive in leading to government regulation. The attitude prevailing at the 1989 Helsinki meeting of the parties to the Montreal Protocol was the attitude that led to the negotiation of the protocol itself: "Don't fool with Mother Nature" (Litfin 1994: 139).

ACKNOWLEDGMENTS

This chapter is based in part on "Protecting the Ozone Layer," by James K. Hammitt and Kimberly M. Thompson in *The Greening of Industry: A Risk Management Approach* (J. D. Graham and J. K. Hartwell, editors), 1997. The author thanks Dr. Thompson and Susan Swedish for their contributions to the earlier paper; Stephen DeCanio, Roger Dower, Arthur Fraas, John Hoffman, John Horowitz, Archie McCulloch, Mack McFarland, Pauline Midgley, Alan Miller, Mario Molina, Richard Morgenstern, Edward Parson, Tony Vogelsberg, Jonathan Wiener, and Donald Wuebbles for helpful comments; and the Harvard Center for Risk Analysis for support.

ENDNOTES

[1]During the early 1980s, the commercially important CFCs were CFC-11, CFC-12, and CFC-113. CFC-11 was used primarily for plastic foam production and CFC-12 for refrigeration and automobile air-conditioning. Before the aerosol ban, about half the U.S. consumption of CFC-11 and CFC-12 was as a propellant in dispensers for personal care and other products. CFC-113 was used primarily as a solvent for cleaning electronics and metal parts and for dry-cleaning clothing (Hammitt and others 1986).

[2]It is widely reported (for example, in Somerville 1996) that NASA failed to discover the hole because its data-processing algorithms discarded readings beyond a specified range, on the assumption that these would be erroneous. Pukelsheim (1990) reports that NASA had investigated the anomalously low springtime Antarctic results but concluded they were incorrect because they conflicted with much higher readings from the station at the South Pole, which itself was subsequently shown to have been reading too high.

[3]Economic analysis, provided by the technology and economics panel, has been influential in negotiating protocol revisions (Parson and Greene 1995). The panel provides the parties with periodic reports (such as UNEP 1994) on the feasibility of additional control measures, the effectiveness of implementation measures within signatory countries, and other technical and economic matters.

[4]Perhaps the clearest example of technological lock-in is the adoption of the standard QWERTY keyboard layout. This layout was adopted by early typewriter manufacturers because it slowed typists and thereby reduced the frequency of jammed keys; although acknowledged to be nonoptimal in a world of electric typewriters and word processors, the costs of converting from a standard that millions of typists know to some other keyboard appears to block a transition (David 1985).

[5]Engineering economic analysis is often used to estimate the costs of adopting new technologies for which historical data are unavailable. Estimates are

derived by adding up the estimated development, production, operating, and installation costs of new equipment.

[6]Two such spectra, representing effects on DNA and on erythmea, were developed. Their quantitative responses to ozone depletion are similar.

[7]The 1988 RIA forecast the increase in aggregated CFC prices for two scenarios. The marginal-cost curve for the high-cost scenario is comparable to the realized marginal cost for reductions up to about 50%. The 1987 RIA forecast costs one-third smaller than did the 1988 RIA, and so probably underestimated realized control costs. If consumption reductions had not been accelerated by the 1990 and 1992 amendments to the Montreal Protocol, the EPA estimates might have been quite accurate (Hammitt 1997).

[8]Prather and others (1996) analyzed the likely trends in CFC production and ozone loss that would have resulted if the possibility of CFC-induced ozone destruction was not recognized until the 1985 discovery of the Antarctic ozone hole.

[9]Labeled "quantitative low" and "total" in Camm and others (1986).

REFERENCES

AFEAS (Alternative Fluorocarbons Environmental Acceptability Study). 1995. *Production, Sales and Atmospheric Release of Fluorocarbons through 1993*. Washington, D.C.: AFEAS.

Barthold, T.A. 1994. Issues in the Design of Environmental Excise Taxes, *Journal of Economic Perspectives* 8: 133–51.

Benedick, R.E. 1991. *Ozone Diplomacy: New Directions in Safeguarding the Planet*. Cambridge: Harvard University Press.

Camm, F., T.H. Quinn, A. Bamezai, J.K. Hammitt, M. Meltzer, W.E. Mooz, and K. Wolf. 1986. *Social Cost of Technical Control Options to Reduce Emissions of Potential Ozone Depleters in the United States: An Update*. RAND N-2440-EPA. Santa Monica: RAND.

CMA (Chemical Manufacturers Association). 1986. *Production, Sales, and Calculated Release of CFC-11 and CFC-12 through 1985*. Washington, D.C.: CMA.

Cogan, D.G. 1988. *Stones in a Glass House,* Washington, D.C.: Investor Responsibility Research Center.

Connell, P.S. 1986. A Parameterized Numerical Fit to Total Column Ozone Changes Calculated by the LLNL 1-D Model of the Troposphere and Stratosphere. UCID-20762. Livermore, California: Lawrence Livermore National Laboratory.

Cook, E. 1996. Marking a Milestone in Ozone Protection: Learning from the CFC Phase-Out. *WRI Issues and Ideas* (January). Washington, D.C.: World Resources Institute.

Crutzen, P.J. 1970. The Influence of Nitrogen Oxides on the Atmospheric Ozone Content. *Quarterly Journal of the Royal Meteorological Society* 96: 320–25.

David, P.A. 1985. Clio and the Economics of QWERTY. *American Economic Review* 75: 332–37.

Dotto, L., and H. Schiff. 1978. *The Ozone War.* Garden City, New York: Doubleday.

Farman, J.C., B.G. Gardiner, and J.D. Shanklin. 1985. Large Losses of Total Ozone in Antarctica Reveal Seasonal ClO_x/NO_x Interaction, *Nature* 315: 207-10.

Felder, S. and T. F. Rutherford. 1993. Unilateral CO_2 Reductions and Carbon Leakage: The Consequences of International Trade in Oil and Basic Materials. *Journal of Environmental Economics and Management* 25: 162–76.

Fisher, A., L.G. Chestnut, and D.M. Violette. 1989. The Value of Reducing Risks of Death: A Note on New Evidence, *Journal of Policy Analysis and Management* 8: 88–100.

Gibbs, M.J. 1986. *Scenarios of CFC Use 1985–2075.* Washington, D.C.: ICF, Inc.

Hammitt, J.K. 1986. *The Timing of Regulations to Prevent Stratospheric-Ozone Depletion.* United Nations Environment Programme (UNEP) Workshop on the Control of Chlorofluorocarbons. September 1986, Leesburg, Virginia. Paper 5, Annex I, UNEP/WG.148/3. Nairobi: UNEP.

———. 1987. *Timing Regulations to Prevent Stratospheric-Ozone Depletion,* RAND R-3495-JMO/RC. Santa Monica: RAND.

———. 1995. Outcome and Value Uncertainties in Global-Change Policy. *Climatic Change* 30: 125–45.

———. 1997. *Are the Costs of Proposed Environmental Regulation Overestimated? Evidence from the CFC Phaseout.* Cambridge, Massachusetts: Harvard School of Public Health.

Hammitt, J.K., K. Wolf, F. Camm, W.E. Mooz, T.H. Quinn, and A. Bamezai. 1986. *Product Uses and Market Trends for Potential Ozone-Depleting Substances, 1985–2000.* RAND R-3386-EPA. Santa Monica: RAND.

Hammitt, J.K., F. Camm, P.S. Connell, W.E. Mooz, K. Wolf, D.J. Wuebbles, and A. Bamezai. 1987. Future Emission Scenarios for Chemicals that May Deplete Stratospheric Ozone. *Nature* 330: 711–16.

Hoerner, J.A. 1995. Tax Tools for Protecting the Atmosphere: The U.S. Ozone-Depleting Chemicals Tax. In R. Gale, S. Barg, and A. Gillies (eds.) *Green Budget Reform.* London: EarthScan Publications, Ltd.

Litfin, K.T. 1994. *Ozone Discourses: Science and Politics in Global Environmental Cooperation.* New York: Columbia University Press.

Lovelock, J.E., R.S. Maggs, and R.J. Wade. 1973. Halogenated Hydrocarbons In and Over the Atlantic. *Nature* 241: 194–96.

Makhijani, A., and K.R. Gurney. 1995. *Mending the Ozone Hole: Science, Technology, and Policy.* Cambridge, Massachusetts: MIT Press.

Maxwell, J.H., and S.T. Weiner. 1993. Green Consciousness or Dollar Diplomacy? The British Response to the Threat of Ozone Depletion. *International Environmental Affairs* 5: 19–41.

Molina, M.J., L-L Tso, L.T. Molina, and F.C.-Y. Wang. 1987. Antarctic Stratospheric Chemistry of Chlorine Nitrate, Hydrogen Chloride and Ice: Release of Active Chlorine. *Science* 238: 1253–58.

Molina, M.J., and R.F. Rowland. 1974. Stratospheric Ozone Sink for Chlorofluoromethanes: Chlorine Atom Catalyzed Destruction of Ozone. *Nature* 249: 810–11.

Montzka, S.A., J.H. Butler, R.C. Myers, T.M. Thompson, T.H. Swanson, A.D. Clarke, L.T. Lock, and J.W. Elkins. 1996. Decline in the Tropospheric Abundance of Halogen from Halocarbons: Implications for Stratospheric Ozone Depletion, *Science* 272: 1318–22.

NAS (National Academy of Sciences). 1976. *Halocarbons: Effects on Stratospheric Ozone*, Washington, D.C.: National Academy Press.

———. 1979. *Protection Against Depletion of Stratospheric Ozone by Chlorofluorocarbons*. Washington, D.C.: National Academy Press.

———. 1982. *Causes and Effects of Stratospheric Ozone Depletion: An Update*. Washington, D.C.: National Academy Press.

———. 1984. *Causes and Effects of Stratospheric Ozone Depletion: Update 1983*. Washington, D.C.: National Academy Press.

NASA (National Aeronautics and Space Administration). 1988. *Executive Summary, Ozone Trends Panel*. Washington, D.C.: NASA.

Nordhaus, W.D., and G. Yohe. 1986. *Probabilistic Projections of Chlorofluorocarbon Consumption: Stage One*. Prepared for U.S. EPA. Princeton: Yale and Wesleyan Universities.

Parson, E.A. and O. Greene. 1995. The Complex Chemistry of the International Ozone Agreements. *Environment* 37: 16–23, 35–43.

Parson, E.A. 1993. Protecting the Ozone Layer.. In P.M. Haas, R.O. Keohane, and M.A. Levy (eds.), *Institutions for the Earth: Sources of Effective International Environmental Protection*. Cambridge, Massachusetts: MIT Press.

Prather, M., P. Midgley, F.S. Rowland, and R. Stolarski. 1996. The Ozone Layer: The Road Not Taken. *Nature* 381: 551–54.

President. 1995. *Economic Report of the President 1995*. Washington, D.C.: U.S. Government Printing Office.

Pukelsheim, F. 1990. Robustness of Statistical Gossip and the Antarctic Ozone Hole. *The IMS Bulletin* 19: 540–42

Rowlands, I.H. 1995. *The Politics of Global Atmospheric Change*. Manchester, England: Manchester University Press.

Smith, R.C., B.B. Prezelin, K.S. Baker, R.R. Bidigare, N.P. Boucher, T. Coley, D. Karentz, S. MacIntyeare, H.A. Matlick, D. Menzies, M. Ondrusex, Z. Wan, and K.J. Waters. 1992. Ozone Depletion: Ultraviolet Radiation and Phytoplankton Biology in Antarctic Waters. *Science* 255: 952–59.

Solomon, S., R.R. Garcia, F.S. Rowland, and D.J. Wuebbles. 1986. On the Depletion of Antarctic Ozone. *Nature* 321: 755–58.

Somerville, R.C.J. 1996. *The Forgiving Air: Understanding Environmental Change.* Berkeley: University of California Press.

Stavins, R. (ed.). 1988. *Project 88—Harnessing Market Forces to Protect Our Environment: Initiatives for the New President.* Sponsored by Senators T.E. Wirth and J. Heinz. Washington, D.C.: U.S. Government Printing Office.

Stolarski, Richard S., and Ralph J. Cicerone. 1974. Stratospheric Chlorine: A Possible Sink for Ozone. *Canadian Journal of Chemistry* 52: 1610–15.

Titus, J.G. (ed.). 1986. *Effects of Changes in Stratospheric Ozone and Global Climate* (3 vols.). Washington, D.C.: U.S. EPA and UNEP.

UNEP (United Nations Environment Programme). 1991. *Environmental Effects of Ozone Depletion: Update 1991.* Nairobi: UNEP.

———. 1994. *1994 Report of the Economics Options Committee for the 1995 Assessment of the Montreal Protocol on Substances that Deplete the Ozone Layer.* Nairobi: UNEP.

U.S. EPA (Environmental Protection Agency). 1980. Ozone-Depleting Chlorofluorocarbons: Proposed Production Restriction. *Federal Register* 45: 66726, October 7, 1980.

———. 1987a. Protection of Stratospheric Ozone. *Federal Register* 52: 47489–523, December 14, 1987.

———. 1987b. *Regulatory Impact Analysis: Protection of Stratospheric Ozone.* December. Stratospheric Protection Program, Office of Program Development. Washington, D.C.: U.S. EPA, Office of Air and Radiation.

———. 1994. *Production, Use, and Consumption of Class I and Class II Ozone-Depleting Substances: Report to Congress,* Washington, D.C.: U.S. EPA.

WHO (World Health Organization). 1994. *Ultraviolet Radiation: An Authoritative Scientific Review of Environmental and Health Effects of UV, with Reference to Global Ozone Layer Depletion.* Environmental Health Criteria 160. Geneva: WHO.

WMO (World Meteorological Organization). 1985. *Atmospheric Ozone 1985.* Global Ozone Research and Monitoring Project. Publication #16. Geneva: WMO.

———. 1994. *Scientific Assessment of Ozone Depletion: 1994.* Global Ozone Research and Monitoring Project. Publication #37. Geneva: WMO.

Wuebbles, D.J. 1981. *The Relative Efficiency of a Number of Halocarbons for Destroying Stratospheric Ozone.* UCID-18924. Livermore, California: Lawrence Livermore National Laboratory.

———. 1983. Chlorocarbon Emission Scenarios: Potential Impact on Stratospheric Ozone. *Journal of Geophysical Research* 88 (C2): 1433–43.

Wuebbles, D.J., F.M. Luther, and J.E. Penner. 1983. Effect of Coupled Anthropogenic Perturbations on Stratospheric Ozone. *Journal of Geophysical Research* 88 (C2): 1444–56.

7

Asbestos

Christine M. Augustyniak

The Toxic Substances Control Act of 1976 (TSCA) authorizes EPA to control the content of products to eliminate "unreasonable risks." Over the period 1979-89 EPA dedicated an estimated seven million dollars to conducting sophisticated analyses of the costs and benefits of banning asbestos from more than thirty product categories. The analyses showed that asbestos could be banned from a number of products at low to moderate costs.

In promulgating the final asbestos rule, EPA management perceived a number of changes in the marketplace, such as the elimination of asbestos from new car brakes without noticeable increases in cost. These changes led EPA management to believe that the actual cost of substitutes would be lower than the estimates of record. In the end, EPA decided to phase out all asbestos use in covered product categories over a nine-year period and added a waiver procedure under which manufacturers could petition the agency to extend the deadlines if substitutes were not available.

The evidence shows that the economic analyses were used to categorize products containing asbestos, decide which product groupings to regulate, and divide the covered products into three groups for phased regulation based primarily on the availability of known substitutes. However, for a variety of reasons, management did not give great weight to the estimates of cost per life saved for the individual products. Here, Christine Augustyniak describes the economic analysis and shows how and to what degree the study influenced policy.

CHRISTINE M. AUGUSTYNIAK is the Associate Director of EPA's Environmental Assistance Division. While a member of the Regulatory Impacts Branch, she was responsible for development of the RIA that supported the final asbestos ban/phasedown rule. The opinions expressed in this chapter are those of the author.

INTRODUCTION

Asbestos is a naturally occurring fibrous mineral with a long history of usefulness. The first recorded uses occurred in the first millennium B.C.E.; Charlemagne had a tablecloth made from asbestos fibers that was burned rather than laundered after banquets. The dangers resulting from exposure to asbestos have been known for nearly as long as the fiber has been used; Romans noticed a high mortality among the slaves who worked with asbestos and required these slaves to wear pig bladders to cover their faces—an early attempt to use personal protective equipment to mitigate the dangers posed by exposure to asbestos. During the nineteenth century, high mortality was observed among the textile workers who produced the asbestos cloth and cloth items.

Since 1900, with the first autopsy report of pulmonary fibrosis in an asbestos worker, asbestos has been linked with a variety of serious illnesses including asbestosis, mesothelioma, and cancers of the lung and gastrointestinal tracts.[1] This evidence of the disease-causing properties of asbestos has led to two distinct responses from the U.S. government: attempts to mitigate the effect of asbestos exposure on workers through the inclusion of asbestos-related illnesses in worker compensation laws, and promulgation of regulations designed to reduce the exposure of workers to asbestos fibers. Regulations have been promulgated by the Consumer Product Safety Commission, the Department of Transportation, the Environmental Protection Agency (EPA), the Food and Drug Administration, the Mine Safety and Health Administration, and the Occupational Safety and Health Administration (OSHA). This paper will focus on one such regulation: EPA's ban of the manufacture (and importation) of most asbestos-containing products in 1989. Although the regulation was eventually overturned by the United States Court of Appeals in 1991, the development of the regulation and its supporting documents offers some insights into the usefulness of benefit-cost analyses for regulatory decisionmaking.

Regulatory History

The Toxic Substances Control Act (TSCA) of 1976 (Public Law 94-469, 90 Sta. 203, October 11, 1976) requires that a finding of unreasonable risk be made before EPA can take action on a chemical. Although "unreasonable risk" is not defined in the statute, a reading of the legislative history makes clear that the underlying notion is one of balancing costs and benefits: that small (or low-level) risks that are inexpensive to mitigate should be reduced while larger risks whose mitigation is costly may not merit action. Consequently, benefit-cost analysis has played a prominent role in the regulatory process of EPA's Office of Toxic Substances (OTS, now known as the Office of Pollution Prevention and Toxics) since its inception in 1976.

Section 6 of TSCA, which governs the regulation of (existing) hazardous chemical substances and mixtures, states that:

> If the Administrator finds that there is a reasonable basis to conclude that the manufacture, processing, distribution in commerce, use or disposal of a chemical substance or mixture, or that any combination of such activities presents or will present an unreasonable risk of injury to health or the environment, the administrator shall by rule apply one or more of the following requirements to such mixture to the extent necessary to protect adequately against such a risk using the least burdensome requirements....

The listed requirements include prohibiting or limiting the manufacturing, processing, distribution in commerce, use, or disposal of a chemical substance or mixture, either absolutely or in particular uses or concentrations; requiring that the substance, mixture, or any article containing the substance or mixture be accompanied by clear and adequate warnings; records retention; and otherwise regulating use and disposal of the substance or mixture.

The regulation to reduce exposure to asbestos fiber was to be the first major agency action to use Section 6 of TSCA. As a consequence, considerable staff and extramural resources were devoted to this project. During the ten year course of the regulation development, some $5 to $10 million were spent on consultants to help EPA develop the economic model and to locate the supporting studies of health effects, exposure, costs, and benefits. These extramural resources were directed by a large cadre of economists, engineers, health scientists, and chemists. There was broad workgroup participation from across the agency.

During the course of regulation development, a number of options for the control of asbestos fibers were considered. These options ranged from requiring that asbestos-containing products be labeled to requiring that engineering controls be used in workplaces, to a complete prohibition on the use of asbestos fiber. Even after analysis had led the workgroup to conclude that remedies that addressed only some portion of the product life cycle[2] would still result in an unreasonable risk of exposure to asbestos, a variety of different options to effect a ban were considered. These options ranged from using then standard command-and-control regulations with different timings of product category bans depending on availability of substitutes, to using market mechanisms such as permits to raise fiber price while allowing asbestos to be used in the most commercially important uses. Within the "permit option" both quantity-denominated and price-denominated permits[3] were evaluated. As promulgated, the rule required both product bans and permits for fiber use in unbanned product markets and was known as "The Asbestos Ban and Phasedown Rule."As finally promulgated, it required that the manufacture of most asbestos products be banned between 1990 and 1996, with all distribution in commerce to stop by 1997. It was the first important use of Section 6 of TSCA for a chemical other than polychlorinated biphenyls (PCBs).[4]

Problem Definition

In the case of asbestos, the questions posed by TSCA and therefore addressed by EPA were: "Does exposure to asbestos pose an unreasonable risk?" "If so, how should this risk be addressed?" TSCA did not allow regulators to immediately consider the ban of a substance, even if the risk had already been determined to be "unreasonable." In order to proceed with a regulatory action on asbestos under Section 6, the agency was required to address the costs and benefits of various regulatory options. Of the many possible actions that could have been taken to mitigate the unreasonable risks, the agency had to determine which of the alternatives was "least burdensome." (TSCA requires that whatever action is taken be the least burdensome one that addresses the unreasonable risk.)

Because both the costs of the action and the risk mitigation (benefits) would continue into the future, many analytic issues had to be addressed. Of major importance was the issue of the baseline: in the absence of regulation what would be the level of asbestos use? The use of asbestos had been declining prior to EPA's initiation of the asbestos ban rulemaking, but agency authorities did not have reason to believe that the risks had declined to a level that was no longer unreasonable.

The growth and decay of product use over time is not an important issue in situations where the costs and benefits occur at the same time. For example, in the control of a substance whose risk is posed at the time of manufacture but not during other stages of the product life cycle such as during product use, repair, and so . More accurately, if the costs (control of the substance) and the benefits (risk reduction) occur at the same point in time—for instance, when the exposure is controlled—being inaccurate about the level of use results in being wrong about the absolute level of benefits and costs, but not wrong in the conclusion as to whether the regulation is sound from a benefit-cost standpoint.[5] In the case of asbestos, however, where the imposition of the costs and the realization of some of the benefits[6] occur at different points in time, choice of a discount rate can affect conclusions both about the absolute level of costs and/or benefits, and about the overall conclusion as to whether the regulation should be promulgated.

The question of determining baseline use is more complicated than it first appears. Asbestos is a "producer good;" that is, it is not valued as a product in and of itself, it is valued because it is used in valuable products. These products, many and diverse, might be expected to have different growth rates in the future, both because products are used in different industrial sectors and because some products can be more easily substituted for than others. Moreover, use of some of these products is dependent on volatile markets like automobiles—asbestos is used in friction products like brakes and clutch facings, for example—while the use of others is not. The demand pattern of asbestos through time is the summation of the demands in individual markets.

The benefits portion of the analysis is also complicated. Asbestos causes both mesothelioma and other cancers including lung cancer. Although asbestos is virtually the only cause of mesothelioma in the United States, the same cannot be said for lung cancer. Lung cancer is caused by many factors including exposure to tobacco smoke. Asbestos and tobacco smoke interact synergistically[7] to produce lung cancer so that if the smoking rate changes, the number of lung cancers among those exposed to asbestos will change independently of a change in the rate of exposure to asbestos. Overlaying these baseline issues and general problems of quantification of costs and benefits are questions of how to deal with general uncertainty, which becomes more acute as the time horizon is lengthened.

Economic Sectors Affected

To make the asbestos-containing products, asbestos fiber from foreign[8] and domestic producers (miners and millers) is combined with other substances by a "primary processor." The material is generally processed further by a "secondary processor" who actually manufactures the product that is sold to domestic and foreign consumers. The consumers of the asbestos-containing products are usually producers of other products (cars, for example) for which the asbestos product is an input. For example, "asbestos diaphragms" are inputs into the production of chlorine and caustic soda; asbestos brakes are inputs into car production.

In addition to affecting these groups, other groups may be affected by a ban or severe restriction of use of asbestos. If asbestos were a major import from some country, regulations governing asbestos could have an impact on the balance of payments. If asbestos-related activities were both large employers and concentrated in some geographic area, regulations could affect levels of unemployment at least locally. Although these and other impacts were considered for further analysis, the generally small dollar volume (in relative terms) of asbestos imports and the distribution of asbestos-related industries across the United States suggested that the most important impacts would be to miners, millers, and processors, and that the secondary effects could be neglected without harm to the usefulness of the analysis.

Regulatory Approaches Considered

Under TSCA, EPA has wide latitude in the actions it can take once it has made the finding that a chemical substance presents or will present an unreasonable risk of injury to health or the environment. The agency can restrict or prohibit the manufacture, processing, or distribution in commerce; it can require labels or other forms of notification to the public, and record keeping; it can prohibit or otherwise regulate commercial use and disposal. In short, TSCA is a very powerful statute. Before any of these actions can be taken, however, the agency must determine not only that there is an unreasonable risk, but that the measure chosen is the least burdensome one that will provide adequate protection to human health and the environment. As part of the analysis performed for what ultimately became the asbestos ban regulation, a number of alternatives were considered.

One regulatory approach often used in an occupational setting is to require that workers use personal protective equipment. Examples of personal protective equipment include such items as goggles, gloves, and respirators. When the risks posed by a substance are mainly to workers, in particular when workers perform their tasks in a centralized location, such a requirement is a useful approach. In the context of asbestos, however, exposure and consequently risk occur not only during manufacture, but also during use and repair. Many of the occupational exposures will, therefore, occur in settings where it would be difficult to enforce a requirement to use personal protective equipment. The risks posed by asbestos are from exposure through inhalation that is mitigated through use of a respirator, which is uncomfortable to wear.

A further characteristic of asbestos is that it is very long lived, and that the aerodynamic properties of the fibers cause them to remain suspended, and consequently respirable, for long periods of time. Once they have settled they can become resuspended easily. It was concluded that personal protective equipment did not mitigate a significant portion of the risk.

Another approach to mitigate the risks might be to make people aware of the substance to which they are being exposed, to require labeling of the product so that people in the vicinity of the product could take appropriate precautions. Although this approach might work to some extent for products in cases where the label could be attached outside of the space in which the asbestos was contained, this too was judged to mitigate an insufficient portion of the risk.[9] For example, much of the occupational exposure to asbestos in brakes occurs when auto mechanics replace the drums or discs. When the housing is removed, the degraded brake material, including asbestos, falls out of the contained assembly and into the workplace where that material can become entrained (suspended in the air). Unless the mechanic can be made aware of the asbestos before the housing is opened, a label does no good. In addition, a label cannot be used if it impedes the action of the brake. Finally, there are also risks of exposure to asbestos in brakes to bystanders. In a study done of the Connecticut Turnpike, levels of airborne asbestos were measurably higher in the immediate vicinity of the toll plazas, presumably due to the application of brakes, than in other portions of the highway. There is no label that would mitigate this risk.

Similarly, a change in the manufacturing process could mitigate some, but not all of the occupational risks posed by asbestos. None of the risks that occur during other portions of the life cycle (such as during use, repair, or disposal) and none of the risks posed by ambient exposure, however, could be mitigated by a change in the manufacturing process.

After examining some of the other regulatory options available to the agency, the consensus of the workgroup was that only a ban on virtually all uses of asbestos would mitigate the unreasonable risks, and that no other option or combination of options could mitigate the risks in a cost-effective manner because no other option addressed a significant portion of the risk; that is, the risk associated with the full life cycle of the asbestos product: manufacturing, use, repair, and disposal. In arguing that the only appropriate policy for dealing with the risks posed by asbestos aside from doing nothing was a ban,[10] the agency did not assume that products would be banned immediately. The analysis examined banning products according to some schedule, or, alteratively, allowing less and less asbestos to be used in products over a schedule of years (phasing down the use of asbestos). It should be noted, however, that the possibility of neither using some market-based mechanism (such as permits) to ban individual products, nor severely restricting the amount of fiber that could be used was addressed at this point.

As will be clear from the cost tables, EPA conducted its analysis on a product-by-product basis and could have taken different regulatory action (including no regulation) for each product category.

In fact, EPA had considered alternatives to banning the manufacture and distribution in commerce of asbestos and asbestos-containing products. Specifically, the agency evaluated whether labeling requirements and/or engineering controls could be used to mitigate a sufficient percentage of the risk posed by exposure to asbestos such that the residual risk would not be "unreasonable." EPA found that neither alone nor in combination could these alternatives mitigate meaningful amounts of risk at reasonable cost.

EPA also assessed whether some kind of market-based incentives, such as marketable permits, could be used to restrict asbestos use to the higher-valued uses, causing asbestos to be replaced in lower-valued uses.

At the time of the analysis, EPA had no experience with an incentive-based market system for achieving regulatory ends. Although the theoretical construct was attractive, the practical difficulties proved daunting. Ninety percent of the asbestos used in the United States was imported, not only as fiber, but also embedded in products. In order for a permit system to work, the amounts of fiber imported as a portion of thousands of different products would have to be tracked against permits. The practical difficulties soon overwhelmed the theoretical advantages.

In the end, EPA banned the manufacture and distribution in commerce of asbestos and asbestos-containing products in three stages:

- Stage 1—Manufacture, importation, and processing banned as of August 27, 1990, and distribution in commerce banned as of August 25, 1992, for flooring felt, roofing felt, pipeline wrap, A-C (asbestos-cement) flat sheet, A-C corrugated sheet, V-A floor tile, and clothing.
- Stage 2—Manufacture, importation, and processing banned as of August 25, 1993, distribution in commerce banned August 25, 1994, for beater-add and sheet gaskets (except specialty industrial gaskets), clutch facings, automatic transmission components, commercial and industrial friction products, drum brake lining (OEM), and LMV disc brake pads (OEM).
- Stage 3—Manufacture, importation, and processing banned as of August 26, 1996; distribution in commerce banned as of August 25, 1997, for A-C pipe, commercial paper, corrugated paper, rollboard, millboard, A-C shingle, specialty paper, roof coatings, nonroof coatings, brake blocks, drum brake linings (AM), LMV disc brake pads (AM), and HV disc brake pads (AM).
- In addition, any new products would be banned along with Stage 1 products.

Products were grouped together based on the availability of substitutes; products for which there were good substitutes were scheduled for banning in Stage 1, those for which substitutes were not as advanced were scheduled for later bans. EPA also instituted an exemption procedure so that if substitutes for some products did not develop as expected there would not be undue hardship on the regulated community.

THE BENEFIT-COST ANALYSIS

As indicated earlier, the asbestos ban rulemaking, and the analyses that supported it, occurred over ten years or more. The model described in this chapter reflects the analyses that occurred after promulgation of the proposed rule, which occurred in 1986. This model shares many features, including the product-by-product analyses, with the earlier work. Having the earlier work allowed the agency to avoid many of the less fruitful approaches and to refine the analyses. Additionally, it allowed for a retrospective look at the assumptions made in the earlier analyses and improved the selection of the assumptions for the final model.

As part of the earlier analyses, the agency had examined the issues of personal protective equipment and labeling as methods of reducing exposure to the population. As indicated above, these options were rejected because each dealt with only a small portion of the risk, leaving a large residual unreasonable risk that required mitigation. The reanalysis concentrated on the effects of banning specific asbestos-containing products and reducing the amount of asbestos fiber that could be used in products.

Important to estimating both benefits and costs was the development of a microeconomic model of the market for asbestos fiber and for the markets for asbestos-containing products (the downstream markets). The fiber market and product markets are obviously linked. Demand for asbestos fiber arises from demand for asbestos-containing products; decreases in the effective amount of available asbestos, through a cap on fiber use, are translated into higher prices and thus into a decreased quantity demanded of the asbestos-containing products caused by the higher prices of those products. The magnitude of the change in quantity demanded depends on the shape and location of the product-specific supply and demand curves. (See the figures on the next two pages.) Specifically, two factors are important: the shape of the demand curve in the relevant range and the magnitude of the shift in the supply curve. The shape of the demand curve is determined by the prices and market shares of the substitute products. The effect on the supply curve is determined by the increase in the asbestos fiber price caused by the cap, and the way in which the increased fiber price increases the cost of producing the asbestos-containing product.

A more complicated situation can arise in the case of product bans. If all asbestos-containing products are banned, the analysis is trivial. If,

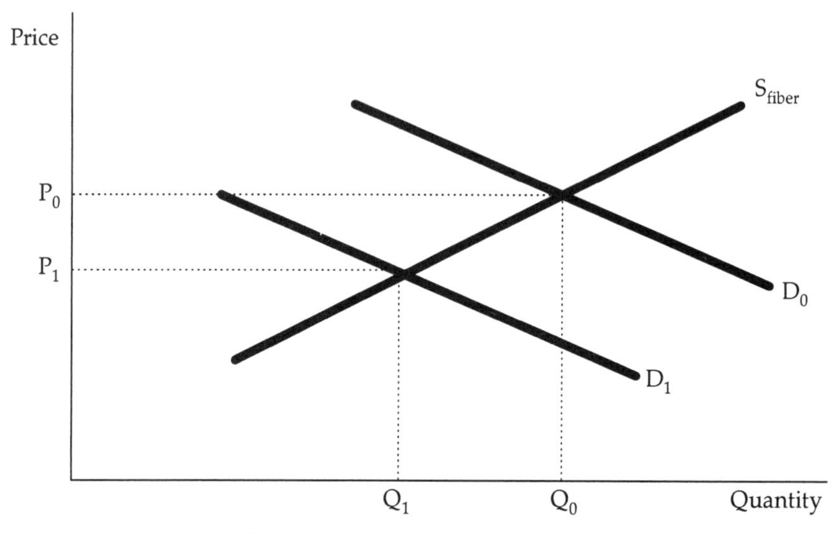

Figure 1. Fiber Market After Banning Some Products.

however, bans are applied to some products but not to others and caps are not applied to maintain the price of asbestos fiber, the resulting decreased demand for asbestos lowers the market price of fiber. (See Figures 1 and 2.) A lower fiber price increases the supply of the unbanned products, increasing the equilibrium quantity of those goods (assuming that the demand is not completely inelastic; see Figure 3). Although the total amount of asbestos fiber used decreases, exposure to asbestos fibers need not decrease. Different products release differing amounts of fiber during manufacture, processing, and so forth, resulting in differing amounts of exposure.

The uses of asbestos were divided into approximately thirty-five categories,[11] categories that were thought to be homogeneous with respect to the set of substances that could be used to substitute for the asbestos-containing product. Extensive data were collected from a variety of sources including surveys and studies conducted by contractors for the purpose of promulgating this regulation, studies performed by other agencies such as OSHA and the National Institute for Occupational Safety and Health (NIOSH) for a variety of purposes, and information collected under TSCA rulemaking authority.

Beyond the fundamental imperative of determining whether the benefits of the regulation exceeded the costs—or in other words, if the risk posed by the asbestos was "unreasonable"—it was also important to keep in mind a number of issues. First, under different forms of the regulation, who were the relative winners and losers? TSCA was concerned with the

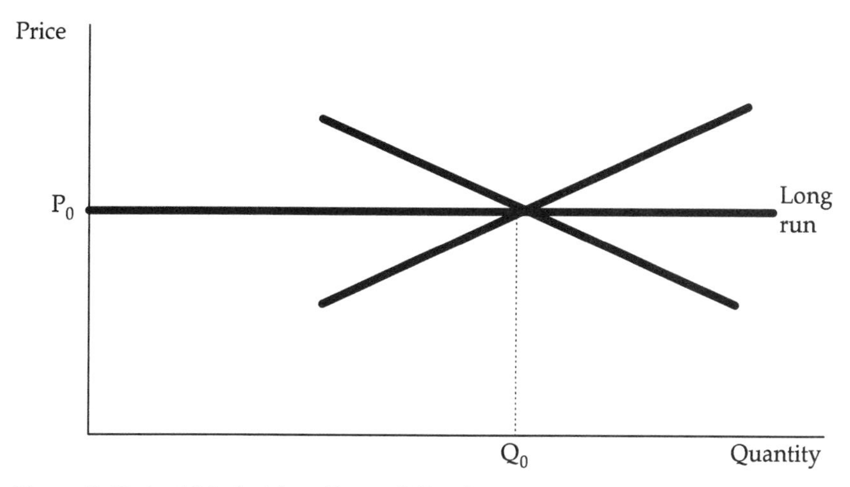

Figure 2. Output Market for a Banned Good.

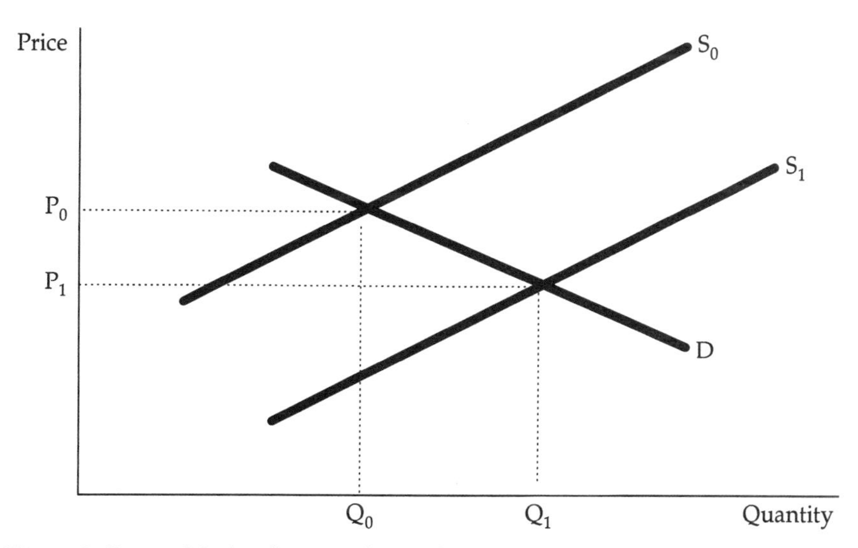

Figure 3. Output Market for an Unbanned Good.

domestic benefit-cost ratio.[12] Because foreign entities played a large role in the domestic market for asbestos-containing products, it was possible that the overall benefit-cost ratio differed from the domestic relationship. In the case of asbestos, because the substance was imported, it was likely that some of the costs were borne by foreigners, making the regulation look more attractive. In cases where the exposure to a toxic substance occurred over a border, a regulation that made sense from a global benefit-cost perspective would not necessarily be justified under TSCA.

Another consideration was to build a model that was sufficiently flexible so that a wide variety of options could be examined. At the time of model development the agency had already amassed large amounts of information about asbestos. It was known that asbestos was used in a wide variety of situations. The uses could be grouped into a number of fairly homogeneous product categories. Within each category there was a common set of substitutes for the asbestos uses. The difference in price between the asbestos products and their substitutes differed from product to product. In some cases substitutes were available at a very small price difference. In other cases the substitutes were far more costly than the asbestos product. In a similar manner, exposure and the potential for exposure differed from product category to product category. The large differences among asbestos products led naturally to analyses that examined issues on a product by product basis as opposed to consideration simply of the asbestos fiber market. All analyses were conducted on this individual product basis.

Cost Analysis

In order to develop a cost model the following information was needed: the amount of asbestos fiber used in each product category; the market price and performance characteristics of each asbestos product; and, the identity, price, and performance characteristics of each substitute for each asbestos product.

Characterizing the substitutes was necessary since the cost of banning asbestos is estimated by the cost of using alternatives, adjusted for the difference in performance characteristics of the alternatives. Characterizing the substitutes required determining the viable substitutes for each asbestos product, determining the price of the substitute product, identifying and pricing the difference between the useful characteristics of the substitute and the asbestos product,[13] and estimating the share of the asbestos product market that that substitute could be expected to capture if asbestos could no longer be used. This information was used to adjust the price of the substitute product so that it could be compared to

Table 1. Asbestos Cement Shingles and Substitute Products.

Product	*Price ($/unit)*	*Useful life (years)*	*Market share (%)*	*Adjusted price ($/unit)*
A-C shingles	113	40	n/a	113
Wood siding	162	30	32	216
Vinyl siding	106	50	27	84.4
Asphalt shingles	49	20	20	98
Aluminum siding	128	50	19	102.4
Tile roofing	173	50	2	138.4

*Source:*U.S. EPA 1989.

the asbestos product. Table 1 presents such information for asbestos-cement (A-C) shingle.

A-C shingle is a product used on the exterior of buildings. It is selected as a building material because it is inexpensive but also durable and fireproof. In addition, A-C shingle can be molded to resemble other materials so that it is used in restorations of historical structures. The substitutes for A-C shingle include other shingles, such as those made from cedar, but also building materials such as aluminum siding, brick, and so forth. An engineering estimate done for the regulatory impact analysis estimated that A-C shingle has a useful life of forty years, while cedar shingle (wood siding) has an effective life span of only thirty years before requiring large-scale replacement. In order to properly compare the two alternative building materials, the analysis would need to account not only for the difference in unit prices, but also for the fact that the cedar shingles must be replaced more often. Replacement involves additional costs. Not only must one use one-third more shingle-equivalents over the forty-year life span of the asbestos shingles, there is the labor cost to install them, and there may be other associated costs such as disposal. All of these costs must be factored in to adjust the "price" of the cedar shingle to have the same meaning as the "price" of the asbestos shingle.

The height of the "step" represents the relative price difference of the substitute in question compared to the next cheapest product. The length of the "step" is the market share that product is expected to capture. Using this construction implies that if the market were to expand, the amount of each product bought and sold in the market also expands, the market share remaining constant. Figure 4 is an example of the "stepped" demand function.

Ideally, other characteristics of the product would be accounted for as well. In the market we currently observe that some consumers choose to buy cedar shingle even though it is more expensive than asbestos-cement shingle. One possible explanation is aesthetic—cedar shingle looks better

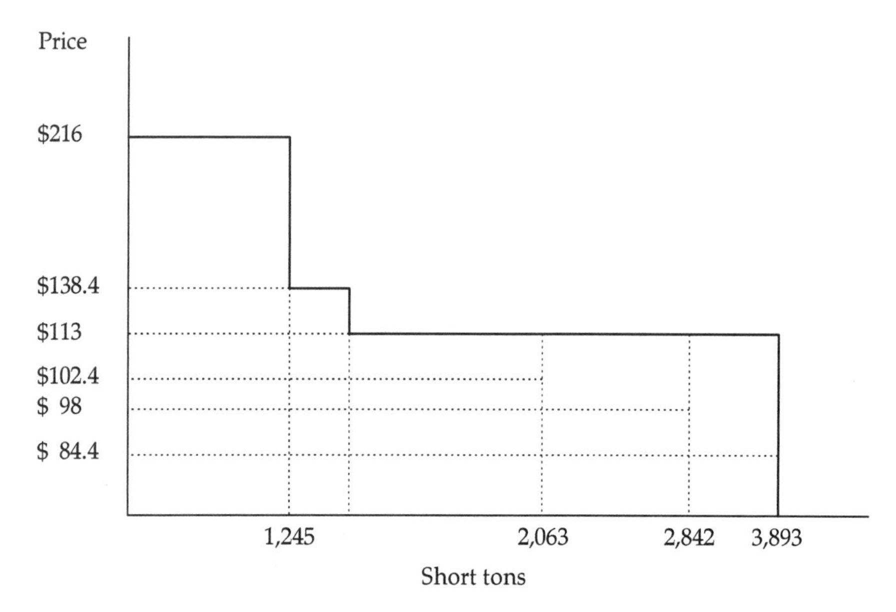

Figure 4. Stepped Demand Function for Asbestos-Cement Shingles (in $1987).
Source: U.S. EPA 1989.

than A-C shingle. If cedar shingle is somehow a "better" product than A-C shingle, consumers who are "forced" to buy cedar shingle if regulators ban asbestos will not suffer the entire extra cost that this model would attribute because part of the cost increase they experience is the result of the consumer acquiring a "better" (more aesthetic) product. Unfortunately it is not possible to assign a value to the improvement.[14]

In some cases, even after the adjustments described above, the price of the substitute remained below the price of the asbestos product. For an economist, this raises a perplexing theoretical issue: Why hasn't the substitute replaced the asbestos product? There are two possibilities (assuming that accurate information is available to consumers, and further assuming that the information developed is correct): either this is a market not in equilibrium, and once all adjustments have taken place the asbestos product will no longer be used; or there is some important element that has not been captured by the analysis. In considering why aluminum siding has not replaced A-C shingle, the analysis might have neglected to consider aesthetics—some people don't like the way aluminum siding looks, and even those who like the look of aluminum might not like the look of a building with some A-C shingle and some siding. Judging that the second explanation was more likely than the first, in each case where the price of the substitute product, after adjustment, remained or became less than the price of the asbestos product, the price of the substitute product was set equal to the asbestos product price.

In cases where more than one substitute was identified for an asbestos product, the relative share of the market that each substitute would capture was determined by surveying knowledgeable individuals in each field. From the price and market share information on each product a market demand curve was calculated. This method of modeling demand is "conservative" in the sense that it uses only information generated by the study—the prices and anticipated market shares of substitute products—rather than estimating the demand curve econometrically. This procedure avoids the need to assign an intersection between the demand curve and the price axis; however, it does result in a likely underestimation of the consumer surplus.[15]

One of the concerns of the analysis was to calculate consumer surplus[16] and changes in consumer surplus. In order to make this calculation it is necessary to locate the demand curve, including the intercept with the price axis. The typical methods of estimating demand curves do not provide sufficient information to accomplish this task. For this analysis, demand curves for each product category were constructed from the information developed on the substitute products, rather than econometrically from market information.

In the long run, the supply function in the asbestos product markets is assumed to be perfectly elastic, and hence horizontal. In the short run, however, the supply curves are not assumed to be horizontal. In each product market there may be "quasi-rents,"[17] payments to factors that are immobile in the short run and that can, therefore, be forgone in the absence of alternative uses for those factors. Over the longer time horizon these factors are mobile and must be compensated in order to remain in the asbestos product industry. For some product markets, these quasi-rents are quite small; for instance when the equipment is not highly specialized and can be used with little modification to make another product. In situations where the equipment cannot be easily converted the quasi-rents can be quite large. For this analysis the total quasi-rents were converted to an amount of rent per unit of product to construct a linear, upward sloping supply curve.

The demand for asbestos fiber is the horizontal summation of asbestos fiber demanded in each asbestos product market. The fiber supply curve is estimated econometrically from time series data on asbestos production and importation. Unlike the long-run supply curves in the product markets, the long-run supply curve in the fiber market is not perfectly elastic. It is positively sloped.

Baseline Issue. In the analysis, all costs are generated relative to the baseline; that is, based on what amount of fiber would be used, and for which products, in the absence of additional regulations. The issue of the baseline is of particular concern to this rulemaking. In some environmental regulations the costs and benefits of the regulation occur at the same point in time. For example, if the exposure to a chemical occurs only during the manufacture of that chemical, the costs imposed to control that exposure, assuming that they are variable costs such as process changes or materials changes, will occur at the same point in time as the reductions in risks that are attributable to that regulation—the reduction in risk is the benefit brought about through reduced exposure. Under this circumstance, if the prediction of the future levels of chemical production is wrong, the absolute levels of costs and benefits will be wrong, but the benefit-cost ratio will be unaffected. As a consequence, the decision as to whether to undertake the regulation will also be unaffected.

With asbestos, however, exposure to the substance occurs at all points of the life-cycle: during mining, primary and secondary processing (manufacturing), use, repair , and disposal. The costs of the regulation occur at the beginning of the process- during mining, milling, and manufacture, when substitute products are used. Although some of the benefits[18] are concurrent—reduced occupational exposure—some of the benefits are estimated to occur fifty years after the costs are incurred.[19] Unless a discount rate of zero is used, making inaccurate assumptions about the growth path of asbestos product markets can lead not only to inaccurate absolute levels of costs and benefits, but also to an inaccurate benefit-cost ratio and therefore a potentially incorrect policy choice.

Several baselines were estimated. For most of the products, the baseline estimation involved use of information on asbestos product markets collected at two points in time, in 1980 and 1986. The 1980 information was collected using a rule issued under TSCA authority that required manufacturers (including importers) and processors of asbestos fiber and products to report the amount of asbestos fiber used and product produced. In 1986, all firms that had reported the 1980 information were contacted to update the information.[20] Using the two data points, a number of alternative baselines on a product by product basis were estimated. The most "optimistic" baseline assumption—in the sense that it resulted in the lowest costs being attributed to the regulation—was the one that projected past rates of growth (largely negative) into the future. In the analysis this is referred to as the "high decline" baseline. The most "pessimistic" baseline—in the sense that it resulted in the highest costs being attributed to the regulation—assumed that there would be no further decline in asbestos use. In other words, any effect of lawsuits and forthcoming additional regulation had already been incorporated into production-use decisions. In the analysis this is referred to as the " low decline" baseline. It is this "low decline" baseline that was chosen as the base case for the analysis.

For brakes, the consumption of which is determined by automotive production and miles driven, estimation of the baseline was more sophisticated. An additional complication in estimating baseline asbestos use is that during the period of analysis disc brakes were coming into more general use.[21] The analysis requires a distinction between disc and drum brakes because different amounts of asbestos are used in the different brakes, and they have different effective life spans. In terms of replacement brakes (the aftermarket), although there is a debate as to whether nonasbestos can be used to replace asbestos brakes, there is no debate that disc brakes can only be replaced by disc brakes, and drums by drums.

The baseline for the brakes categories was estimated from models developed to estimate future automotive demand, projecting the then prevailing rates of switching from drums to discs in new cars and an aftermarket situation in which cars that had disc brakes as original equipment were assumed to be fitted with replacement disc brakes. Likewise, cars originally equipped with drum brakes were assumed to use replacement drum brakes.

In addition to the question of the future level of production of asbestos products, there is another "baseline" issue: what will be the price of the substitute products in the future? Although some of the products that are substitutes for the asbestos-containing products are established ones whose prices already reflect price reductions achieved through experience and economies of scale, for other products this is not the case. At the time of the analysis there were a number of relatively new fibers and other products that were being used to produce replacement products. Kevlar® was a new man-made fiber that could replace asbestos in some uses, while PTFE (Teflon®) was another relatively new product that could replace asbestos in some uses. For products like these, which are new (at the time of the analysis) and have probably not yet achieved the price reductions that come form manufacturing experience and economies of scale, the market price is likely to drop, in real terms, through time. If the price of the substitutes decreases, the cost of the regulation will also decrease. The question, then: how to account for this?

In the main analysis, the costs of the regulation were estimated, assuming that prices of the substitutes remained constant through the period of analysis. In order to determine the effect of moving along a learning curve, the costs were re-estimated assuming that the prices of all substitute products decreased 1% per year.[22]

In Table 2, the consequence of the different assumptions regarding the future prices of substitutes is demonstrated. Using otherwise identical

Table 2. Costs of Product Bans Using Different Assumptions about the Prices of Substitutes (millions of $1987).

Product description	*Gross domestic loss (given substitute prices)*	*Gross domestic loss (declining substitute prices)*
1. Commercial paper	.00	.00
2. Rollboard	.00	.00
3. Millboard	–.43	–.43
4. Pipeline wrap	1.97	1.07
5. Beater-add gaskets	207.42	168.67
6. Electrical paper	–.73	–.73
7. Roofing felt	8.90	7.31
8. Acetylene cylinders	–.45	–.45
9. Flooring felt	.00	.00
10. Corrugated paper	.00	.00
11. Specialty paper	–.09	–.09
12. V/A floor tile	.00	.00
13. Asbestos diaphragms	–.97	–.97
14. A-C pipe	225.01	125.72
15. A-C flat sheet	2.73	2.37
16. A-C corrugated sheet	.62	.29
17. A-C shingles	71.42	54.87
18. Drum brake linings (OEM)	14.43	6.89
19. Disc brake pads LMV (OEM)	3.54	3.54
20. Disc brake pads HV	.33	.33
21. Brake blocks	19.27	3.24
22. Clutch facings	25.48	12.83
23. Automatic trans. comp	.30	.22
24. Friction materials	2.06	2.06
25. Protective clothing	.00	.00
26. Thread, etc.	–.83	–.83
27. Sheet gaskets	164.60	133.05
28. Packing	.00	.00
29. Roof coatings	–21.5	–21.5
30. Nonroofing coatings	–2.35	–2.35
31. Reinforced plastics	–.73	–.73
32. Missile liner	–.69	–.69
33. Sealant tape	–1.63	–1.63
34. Battery separators	.00	.00
35. Arc chutes	.00	.00
36. Drum brake linings (AM)	19.87	7.45
37. Disc brake pads LMV	3.59	3.59
38. Mining and milling	7.31	7.31
Total (U.S. net welfare loss)	748.43	510.40

Note: The costs presented in this table are based on the assumption that products 1, 2, 4, 7, 9, 10, 12, 15, 16, 17, and 25 are banned in 1987, that products 5, 18, 19, 20, 21, 22, 23, 24, and 27 are banned in 1991, that products 14, 36, and 37 are banned in 1994, and that products 3, 6, 8, 11, 13, 26, 28, 29, 30, 31, 32, 33, 34, and 35 are not banned.

Source: U.S. EPA 1989.

assumptions about the products to be banned, the timing of those bans, the growth (or decline) in the market for the asbestos product, the appropriate discount rates to be used,[23] assuming that the prices of substitutes will decline 1% per year results in a cost of regulation that is one-third less than assuming no change in prices of substitutes.

Table 2 presents costs of the rule based on the assumption that each of three separate groups of products is banned in 1987,[24] 1991, and 1994, and that the remaining products are not banned.[25] Assignment of products to a particular group was determined by the availability of substitutes, so that where there was uncertainty that the substitutes had the necessary performance characteristics the product ban was pushed into the future to allow longer time for the development of substitute products. The analysis is conducted assuming that the bans remain in effect through the year 2000. The associated benefits are all those resulting from the change in exposure to asbestos that results from the product bans. Given that exposures result from disposal, which occurs at the end of the product's life, and given the latency period, some benefits occur well into the next century.

Negative losses represent gains in consumer surplus as a consequence of the regulation. These gains arise in cases where a product is banned at the end of the scenario (or is not banned). The consumer surplus gains result from the drop in the asbestos fiber price brought about by reduced fiber demand in the market as a whole consequent to the bans of other asbestos-containing products.[26] (See Figure 3, above.) Using the assumption of declining prices of the substitute products lowers the total (domestic) cost of the asbestos ban by over 30%. In this, as in other estimates of costs, a discount rate of 3% is used as the social rate of discount.[27]

Cost estimates were generated assuming a "low decline" baseline.[28] A number of sensitivity analyses were performed, using differing assumptions about baselines, different discount rates, different timing of bans and fiber caps, and different combinations of products banned. For example, banning all of the products at the beginning of the analysis period (1987 in the convention) resulted in costs of nearly $7 billion. Removing just two products from the list to be banned (asbestos diaphragms and missile liner) reduced costs by two-thirds to $2.3 billion.

Benefits Analysis

As discussed earlier, asbestos has been linked with a variety of serious illnesses, including asbestosis, mesothelioma, and cancers of the lung and gastrointestinal tract. Although it is possible that exposure to asbestos fibers produces other losses (for example, it could produce cancers in economically important species of animals or plants, or other damage to the environment), evidence of other effects is not suggested by the literature. It follows that the benefits of controlling exposure to asbestos are primarily[29] to be found in the reduction in humans of these serious health effects.

Among the health effects considered, asbestosis generally occurs with the high levels of exposure that were formerly found in the occupational setting prior to federal involvement in asbestos control. At levels of exposure that could be found in workplaces under the *permissible exposure limit* (PEL)[30] current at the time when work began on this analysis, asbestosis occurred rarely if ever. Consequently, reductions in the incidence of asbestosis were not included among the benefits of more stringent asbestos controls.

Lung cancer, which is caused by exposure to asbestos, is also caused by a variety of other factors, most importantly, smoking habits. It also demonstrates an association with demographic characteristics including age, sex, and race. In order to isolate the effect of reductions in exposure to asbestos from other factors that affect the probability of developing lung cancer it is necessary to account for the underlying rates.

Smoking and exposure to asbestos have a synergistic relationship in producing lung cancer—a smoker who is exposed to asbestos experiences an increased risk of lung cancer that is more than the sum of the increases in risk attributable to asbestos and smoking: it is the product of the increases in risk.[31] A consequence of this relationship between smoking and asbestos exposure is that it becomes necessary to forecast smoking rates in order to appropriately capture benefits due to changes in asbestos exposure.

Mesothelioma, a cancer of the lining of the lung (pleural mesothelioma) or abdominal (peritoneal mesothelioma) cavities, on the other hands is a disease that is associated only with exposure to asbestos[32] and has not been shown to be affected by age, sex, race, or smoking habits. Changes in the incidence of this disease in the United States can be completely attributed to changes in exposure to asbestos.

Cancers of the gastrointestinal tract (cancers of the throat, stomach, and so forth), like cancers of the lung, have many causes. Although individual cancers of the gastrointestinal system demonstrate no association with exposure to asbestos, taken as a whole complex they do exhibit a dose-response relationship with asbestos exposure. In this analysis the rate of gastrointestinal cancers is assumed to be 10% of the lung cancer rate.

Fortunately for the purposes of analyzing benefits, all of the health effects associated with asbestos exposure have a dose-response effect that allows a probability of mortality to be calculated from exposure information. There is not yet a model that can determine which effects are associated with which exposure level when the harm caused by exposure to the substance is something other than cancer. Therefore, these benefits must be described in qualitative rather than quantitative terms.

In the initial analyses of benefits of the regulation, an effort was made not only to determine the number of cancer cases avoided by the regulation, but to quantify other benefits that result from reductions in asbestos exposure. For example, costs of hospitalization during end stages of cancer were calculated. Having conducted the exercise, it became clear that given both the latency period between time of exposure and hospitalization, and the short duration of hospitalization,[33] the added precision in that aspect of the analysis did not appear to improve the precision of the final regulation to a level that would be commensurate with the cost of obtaining the information. For the final rule the calculated benefits of the regulation were reductions in the risk of developing this set of cancers, and the associated reduction in mortality. Other benefits were not analyzed.

In order to analyze the benefits, a wide variety of data were required. First of all exposure data were needed. Exposure occurs during several stages in the life cycle of an asbestos product: during manufacturing (both primary and secondary), installation, use, repair, and disposal. Furthermore, the exposure is not only to those with an occupational involvement, such as those who do the manufacturing, but also to those who might live or work near the site of manufacture (ambient exposure) and to consumers who might use, repair, or dispose of an asbestos-containing product.

Occupational exposure estimates were developed from a variety of sources: data developed by NIOSH, as well as studies conducted by academics and industry groups were used. As a component of updating the data, firms were asked for exposure estimates. OSHA compliance data were used. In general the level of aggregation for the exposure data was greater than for the economic data; consequently the exposure estimates were developed for each group of asbestos products. The data were applied to product-specific worker populations. In situations where the jobs are intermittent—in installation and repair jobs, for example—full time equivalent employees were used.

Two approaches were used for estimating nonoccupational exposures, depending on whether the exposure occurred during "do it yourself" installation or repair; or whether ambient emissions associated with production activities were involved.

Not every product caused exposure in every category. There were, however, many instances where there was good reason to believe that exposure occurred, but there were no data as to the level of exposure. In some cases a level could be inferred from the level observed in analogous situations, those involving a similar process or product; in other cases, generally for the the nonoccupational settings, there was no information.[34] Although benefits were estimated assuming zero exposure where no data were available, unless the exposures to asbestos from thoses products were in fact zero, this procedure would clearly bias the benefit-cost information in the direction of not supporting a regulation when regulation is warranted.

Additional estimates were prepared assuming that, for products that are subject to wear, 1% of the asbestos content of the product would be released per year of product life, due to weathering and maintenence (cutting, sanding, scraping, polishing) that occur during the use of a product. The model results presented in Table 3 demonstrate the difference in the number of cancer cases that the regulation is predicted to avoid when only those exposures that can be quantified are used, and compares that to the number that is predicted when the assumptions about additional exposure are also used. The assumptions of additional exposure (both occupational and nonoccupational) result in nearly a 70% increase in the number of cancer cases avoided by the regulation. The number of cancer cases avoided is not discounted in this table.[35]

Dose-response data, combined with demographic data[36] and cancer cure rates, were used to translate the exposure reduction into cases of cancer avoided by the regulation. Mortality information was also used to avoid having the same statistical individual die more than once.

Because lung cancer has many causes, baseline lung cancer rates were taken from the most recent (1977) U.S. data, and effects of changed smoking rates were projected. The complex of gastrointestinal cancers were assumed to be 10% of the rate of excess lung cancers. This is the lower end of the estimates found in the literature. For all of the cancers, EPA assumed a minimum latency period of ten years.

As was the case for estimating the costs of the regulation, there were methodological issues in estimating the benefits.

At the time that these analyses were performed, OSHA was in the process of lowering the permissible level of exposure in the workplace. All of the occupational exposure data at hand were collected when the PEL was an order of magnitude higher than the level expected to be in effect when the rule became operational.[37] In some cases the exposures in the workplace were already below the new PEL and it seemed reasonable to assume that no further reductions would be undertaken.[38] In other cases, however, workplace exposures were above the new PEL. What exposures should be assumed in calculating the benefits of the regulation? It was assumed that all exposures above the new PEL would drop to 0.2 fibers per cubic centimeter (f/cc), but no further. It is possible to argue that workplaces would be on average "cleaner" than the requirement in order to be certain of meeting the new standard in case of an inspection.

Table 3. Cancer Cases Avoided by Reduction in Exposure.

Additional exposures per product	*Fiber exposure quantified*	*Nonquantified assumptions*
Commercial paper	.0000	.0000
Rollboard	.0000	.0000
Millboard	.0000	.0000
Pipeline wrap	1.7416	5.4137
Beater-add gaskets	7.7573	50.2777
Electrical paper	.0000	.0000
Roofing felt	1.5116	1.5116
Acetylene cylinders	.0000	.0000
Flooring felt	.0000	.0000
Corrugated paper	.0000	.0000
Specialty paper	.0000	.0000
V/A floor tile	.0000	.0000
Asbestos diaphragms	.0000	.0000
A-C pipe	3.1110	35.5412
A-C flat sheet	1.0504	2.1813
A-C corrugated sheet	.1435	.7727
A-C shingles	.6395	8.0623
Drum brake linings (OEM)	8.3800	8.3800
Disc brake pads LMV (OEM)	.9927	.9927
Disc brake pads HV	.2165	.5368
Brake blocks	14.4204	14.4204
Clutch facings	.6049	1.9553
Automatic trans. comp	.0005	.0005
Friction materials	.5244	4.8945
Protective clothing	.0000	.0000
Thread, etc.	.0000	.0000
Sheet gaskets	2.4658	23.1871
Packing	.0000	.0000
Roof coatings	.0000	.0000
Nonroofing coatings	.0000	.0000
Reinforced plastics	.0000	.0000
Missile liner	.0000	.0000
Sealant tape	.0000	.0000
Battery separators	.0000	.0000
Arc chutes	.0000	.0000
Drum brake linings (AM)	106.2551	106.2551
Disc brake pads LMV	15.8541	15.8541
Mining and milling	1.1258	1.1258
Total	166.7950	281.3628

Note: The numbers in these columns are statistical cancer cases.

Source: U.S. EPA 1989.

The issue of how the new standard would be met was not addressed. One way to meet the new standard would be increased venting of the fibers to the outside of the facility, which would increase the non-occupational exposure. It was assumed that there was no change in the non-occupational exposure.

Once the changes in the PEL were accounted for, it was assumed that there would be no further reductions in the level of asbestos exposure apart from those resulting from lower production levels. This decision was justified on the basis that we were dealing with a dying industry. It was evident that there were many firms that had ceased using asbestos between the 1981 reporting rule and the 1986–87 survey, and it seemed unlikely that additional investments in new technology (such as would be needed to produce lower exposures) would be made. This choice however, did result in a higher level of benefits than would have been the case if continued lowering of fiber levels had been assumed.

There was no attempt to account for asbestos exposure to families of those occupationally exposed. Here again there was evidence that some workers might be carrying fibers from the worksite to home, increasing risk to household occupants, but with no way to determine the extent of the exposure, this effect was ignored in the analysis.

One of the most complicated issues involved the projection of smoking rates. During the lifetimes of the exposed cohort, smoking rates had increased, then decreased with males and females exhibiting different timing and magnitudes for the change. What would happen in the future? The synergistic effect of smoking and asbestos exposure on lung cancer made the benefits associated with the regulation sensitive to the choices made about appropriate assumptions of smoking behavior. A separate model was developed that predicted changes in smoking rates. Rates among males were predicted to continue to decline, smoking rates for females were predicted to peak, and then to decline.

Other Analyses

A number of analyses were conducted as inputs to the benefits model or the cost model. Most of these have been discussed already. On the cost side they include a model of automobile use and replacement in order to estimate brake use. On the benefits side they include projecting changes in smoking rates as the exposed cohort of the population changes.

In the original regulatory impact analysis (RIA), completed in 1985, a number of additional studies were conducted. These include a projection of unemployment rates and lost wage effects on displaced workers, and hospital costs to those affected by asbestos-related disease. Consideration was also given to examining the balance of payments effects that might result given that virtually all of the asbestos used in the United States is imported. Balance of payments effects were not noted, despite the high proportion of imported material, presumably because the total volume of asbestos imports is small and the price of asbestos is also low. In preparing the 1989 RIA, these studies were not updated, and, consequently, not included.

Asbestos use had been declining prior to 1981. Certainly over the period 1981 to 1985 asbestos use continued to decline sharply. In updating the production information EPA learned that many companies manufactured both an asbestos-containing product and a non asbestos product, often in the same facility. Given the small number of persons occupationally involved with asbestos, and the apparent ease with which many of the workers could be employed in producing substitute products, employment effects were not expected to be large enough to justify the effort needed to accurately capture their effect, particularly given that few if any employment effects were identified in the earlier study.

For similar reasons, updating information about costs of treatment and wages lost to workers who are stricken with diseases caused by exposure to asbestos was determined to be unnecessary: more accurate information was deemed unlikely to improve the quality of the decision.

Upon further examination, most of the possible effects of exposure to asbestos apart from the cancers were held to be sufficiently small that they could be ignored without damage to the quality of the analysis or the decisions that would result. Distributional analyses were conducted as part of the cost analysis, to determine who the gainers and losers from actions would be in each time period. Losses to miners and millers of asbestos fiber are foreign to the United States, while those in the other categories are a mixture of foreign and domestic actors.

There is one additional analysis that was important in achieving promulgation of the final rule. At earlier stages of the analysis, one option under consideration was that of using market mechanisms to reduce or eliminate the use of asbestos in the United States. At the time that the initial work was done, EPA did not have much experience with the use of a permit system to achieve environmental regulation. The proposal had included discussion of a permit system, one that would issue permits for the use of asbestos in a product to firms that produced asbestos-containing product as of 1980, regardless of whether firms were currently producing these goods at the time of permit distribution. The allocation of permits would be in proportion to the firm's asbestos use as reported to EPA in 1980. These permits could be bought and sold freely.

The notion behind this plan to distribute permits to people who no longer used asbestos was not to give firms a reason to continue production using the asbestos. If firms were given permits only if they produced asbestos-containing products at the time the ban took effect, firms would continue to use asbestos rather than switch to a substitute product, in order to have the "access" to the valuable permit. Permits would have value even to firms that could fully substitute for the asbestos fiber they used because the permits could be sold to other firms which needed them. By "grandmothering" the asbestos use to 1980, firms continue the process of substituting away from asbestos fiber without the disincentive of the future loss of a valuable asset—the "permit," which could be sold to other firms.

Permits were also to be "bankable"; that is, they could be saved for use in the future. Asbestos is neither difficult nor expensive to store, It does not require protection from sun or rain or heat or cold. Rather than have firms purchase unneeded fiber to store for future use, given the risk that fibers would be released to the environment, it was thought preferable from an environmental standpoint to allow "storage" of the permit instead of the fiber.

Through the course of the rulemaking activity the notion of using a market-based mechanism to reduce asbestos use was abandoned. This appears not to have been the result of an explicit decision as much as a lack of experience with permit programs and no strong interest in pushing the idea from the ranks of senior EPA management. Certainly implementation of a permit program for asbestos would have been difficult given the necessity of enlisting the assistance of the Department of Treasury's Customs Service to track the amount of asbestos imported in items, for example.

No further work was done on permit or other market-based mechanisms for about two years after the proposal. In mid 1987 the Office of Management and Budget, Office of Information and Regulatory Affairs began once again to urge consideration of permit mechanisms. Although further work was done, a number of issues remained intractable, such as whether EPA had authority to auction or sell permits, or whether such activity was legally a "tax" and consequently something that must be enacted by Congress. Additional difficult issues revolved around how imported fiber embodied in products would be treated.

In the end a permit scheme was not made part of the final rule; however, the work EPA commissioned (McCauley, Bowes, and Palmer 1992) did advance the debate.

After the Fifth Circuit Court overturned the asbestos ban/phase down rule in 1991, the agency returned to consideration of market-based mechanisms to control the use of asbestos. A consulting firm—ICF Inc. of Vienna, Virginia—was commissioned to develop an asbestos product tax that would internalize the cost of some of the damages caused by asbestos.

The analysis served to point out a number of difficulties that arise when attempting to translate this method, well examined in the academic arena, into decisions in the public arena. Setting a tax to internalize cost requires a well characterized notion of the value of the costs associated with asbestos exposure. Given that the benefits of this regulation are largely a reduction in certain cancers, to implement this tax would require explicit valuation of human life.[39]

REFLECTIONS ON THE ANALYSIS

The proposed ban on asbestos was to be the first major ban promulgated under TSCA. Unlike previous legislation giving the executive branch regulatory authority, TSCA explicitly requires balancing the risk reduction achieved by the regulation with the costs imposed by it. As such, TSCA was the ideal mechanism to incorporate economic considerations into decisionmaking. Economists were involved early in the process and had major responsibility during the ten-year development of the regulation.

The RIA clearly determined the "shape" of the asbestos regulation. The surprisingly high cost of labeling asbestos-containing products, coupled with an inability to demonstrate any value gained by a labeling requirement, caused the agency to drop consideration of that option. For too many products it was clear that the labels in the absense of other regulatory requirements would not allow the behavior change needed to reduce the risks posed by exposure to asbestos. The analysis of engineering controls also clarified the need for a more comprehensive solution to the problem of asbestos exposure than could be provided by merely workplace controls.

Economic information about the availability and costs of substitute products helped to determine the timing of the proposed product bans, which products could be banned immediately, and which product bans could be postponed to allow time for the substitutes markets to mature. Economic information made the case that some products should not be banned at all, such as diaphragms and missile liner. In other cases, although whole product categories were not exempt from the ban, the economic information caused some markets to be further subdivided. Packings, for example, were divided into cases involving high temperature and pressure and all other uses, based on the economic analysis.

Acknowledging the above, it still must be said that not as much use was made of the information developed as part of the regulatory impact analysis as some might have wished. For example, there was a strong desire on the part of decisionmakers to treat elements of a class of products together, even when it could be argued from a benefit-cost perspective that differences should be recognized. Treating automotive components as a class led to including automotive transmission components in the ban along with other automotive items even though the analysis indicated that the cost per case avoided exceed a half billion dollars.[40] Although the decision to ban automatic transmission components makes sense from a regulatory perspective—it leads to a rule that is clear to the regulated community (no asbestos in automobiles) and one that is relatively easy to enforce from the agency's perspective—the decision also appears to violate the "no unreasonable risk" basis of rulemaking under TSCA. This decision and ones like it led to problems when the rule was challenged.

There were other problems with the way in which the analysis was carried out. In many cases excessively conservative assumptions were used when data were lacking. In the short term these decisions led to easy agreements between EPA and OMB regarding the costs and benefits of the rule, and avoided charges that the agency might be overstating risk or understating costs. In the longer term, however, the clear underestimation of benefits made it difficult to justify the scope of the rule.

For example, a method of assigning surrogate exposures was not developed until after the proposed rule was issued: missing exposure information resulted in treating the case as if exposure, for that product in that circumstance, was zero. This omission led to a clear underestimate of the benefits of the rule. Failure to allow the public to comment on the solution that was developed later allowed the overturn of the rule on procedural rather than substantive grounds.

How could such an extreme oversight occur?

Part of the problem may have been a lack of experience on the part of the agency with a regulatory framework in which economic analysis is used to justify a regulation. The far more common experience, which arises from the form of most environmental legislation, is that economic information is developed after the parameters of the proposed regulation are set, and serves merely to indicate the economic consequences of a regulation. The impact of this lack of experience meant that no one anticipated the controversy that would arise, for example, from incorporating assumptions about occupational and nonoccupational exposures that would serve as surrogates in cases where no direct measurements were available. In hindsight maybe the rule should have been reproposed to allow public examination of the methods introduced into the analysis after the original proposal.

Discussion long after the fact with some EPA decisionmakers suggests that, in many important ways, they did not "believe" the information developed by the workgroup. Not that the analysts were thought to be lying; rather, that EPA has had some experience in developing regulations, and that experience suggests, in many cases, the estimates of the costs of rules developed as part of the analytic excercise tended to be higher than subsequent experience demonstrated them to be. Another possibility was that, in the minds of the decisionmakers, the staging of the bans allowed time to make adjustments in the rule. If substitute products failed to be developed on schedule or to perform properly, the timing of the bans could be adjusted to ensure the costs of the rule would be lower than estimated.

In some respects, the analysts themselves did not "believe" the analysis. They were aware of the underestimate of benefits attributable to the proposed rule. This low estimate occurred at many points in the analysis, including the result of the choice of the stepped demand function in which the consumer surpluses are underestimated to equating unknown exposures with zero exposures.

Did the Fifth Circuit overstep its authority in criticizing EPA policy for not discounting benefits when expressed as lives saved? Certainly there are those who have made that argument. The federal government, however, did not choose to appeal the Fifth Circuit ruling.

No other important rules using the authority of Section 6 of TSCA (apart for rules dealing with PCBs) have been promulgated. Given the promise that TSCA holds for developing a regulatory scheme that demonstrably improves social welfare, it would be a shame if this remained the case.

ENDNOTES

[1]The health consequences arising from exposure to asbestos are due almost entirely to respiration of the asbestos fibers, rather than ingestion of the fibers. The cancers which arise in the gastrointestinal tract, and the mesothelioma of the peritoneum are thought to arise from the migration of inhaled fibers from the lung through body tissues.

[2]For example, engineering controls might reduce exposure to workers in the manufacturing sector but would not reduce exposure to workers in the repair sector or to consumers.

[3]The "price-denominated permit" option would be a situation in which the number of permits to be offered for sale is unknown but the price at which the permits would be offered is known. Ideally the permit price is equal to the marginal damage imposed by the asbestos. Thus a price-denominated permit option would be analogous to a pigovian tax on this product.

[4]Congress had specifically banned the manufacture and use of PCBs in the original legislation. The work of EPA has been largely in the area of promulgating regulations to allow exceptions to the ban.

[5]The absolute magnitude of the social surplus achieved by regulation is dependent on the magnitude of the exposure reduction, but the ratio of benefits

to costs need not be affected by an incorrect determination of the number of units affected. If, for example, both the costs per unit and benefits per unit are constant—and this is often the way in which they are estimated given our limited data—only the absolute magnitude of the social surplus is affected by incorrectly estimating the number of units.

[6]Risk reduction that occurs through reduced exposure during use, repair, and disposal of the asbestos-containing product, for example, produces benefits that are delayed relative to the costs of the reduced exposure that occurs at the time of manufacture.

[7]The probability of developing lung cancer if one is both exposed to asbestos and to tobacco smoke is the product rather than the sum of the relative risks of developing lung cancer caused by exposure to each.

[8]In 1985 over 90% of the asbestos used in the United States was imported from Canada.

[9]Many asbestos products use a small amount of fiber in each piece, but many individual pieces are produced. Thus, a labeling requirement would prove surprisingly costly even though each label is inexpensive.

[10]EPA still had to determine whether the "ban" option passed the benefit-cost test.

[11]The exact number of product markets modeled changed as the analysis became more sophisticated. At the time the regulation was first proposed, asbestos-containing products were assigned to thirty-two product categories, one of which was "other." After analyzing the comments on the proposal, and after the public hearings and the first round of cross-examination hearings, the number of categories was increased to ensure that categories were homogeneous. As the need arose to refine the analysis, more categories were added.

[12]Although TSCA does not explicitly state that only domestic costs and benefits are to be considered, one of the findings upon which the law is based is the effective regulation of interstate commerce necessitates regulation (in this case) of intrastate commerce. Furthermore, in Section 12, the ability of the Administrator to regulate the export of chemical substances or mixtures is dependant on whether "...the Administrator finds that the substance, mixture, or article will present an unreasonable risk of injury to health within the United States or to the environment of the United States." The interpretation that only domestic costs and benefits should be considered was upheld by the Fifth Circuit Court in *Corrosion Proof Fittings v. EPA and W. K. Reilly*, which found that TSCA "speaks of the necessity of cleaning up the national environment and protecting United States workers but largely is silent concerning the international effects of agency action."

[13]The differences between the asbestos product and the substitute included not only the difference in useful life, but the different costs for installation and disposal. Costs such as these, which are imposed on private parties, were annualized using a discount rate of 7% to equate them to the asbestos product prices.

[14]To the extent that the substitute products are "improvements" relative to the the asbestos products in ways that are not captured in our adjustments, the costs of the regulation are overstated by this model.

[15]Standard consumer theory assumes that the marginal utility or marginal benefit of each additional unit of consumption declines as an extra unit is consumed. The demand curves generated by the method described above are "stepped." They consist of flat portions whose length is determined by the market share that particular substitute is presumed to capture, and whose height is the adjusted price of the substitute. No rational consumer would purchase some unit of good unless the value were at least as large as the cost, so that the marginal value of the last unit of the substitute good is at least equal to the adjusted price. It follows that consumer surplus is underestimated.

[16]*Consumer surplus* is the term used by economists to recognize that for most of the transactions that are freely made the value of the good to the individual who acquires it exceeds the price s/he pays. Only for the transaction that occurs at the margin is the price paid merely equal to the value of the good. The value of consumer surplus in any market is the difference between the value placed on the good by the demander (purchaser) and the price which is paid (the market price) for each unit which is purchased.

[17]This terminology was used in the regulatory impact analysis.

[18]Where benefits of the regulation are assumed to be the reduction in risks of developing any of the cancers which have been identified as the main focus of regulatory concern described previously.

[19]Asbestos-cement pipe, used in water and sewer systems, is estimated to have an effective life of fifty years. The assumption made for this analysis is that after the end of the useful life, a product must be removed from underground and disposed of, leading to potential exposures.

[20]In choosing to update the information by recontacting previous submitters rather than by promulgating a new information collection rule, it was implicitly assumed that an insignificant, if any, amount of fiber would be used by firms who had not responded to the earlier information collection rule. Leaving aside questions of compliance with the earlier rule, it seemed plausible to assume that no new firms would be entering the asbestos product industry. This seemed a reasonable assumption given that firms were shifting out of production of asbestos products, there were ongoing lawsuits involving asbestos exposure, and both EPA and OSHA were engaged in rulemaking activity.

[21]When first introduced in the United States, disc brakes were used for the front wheels of some relatively expensive cars. In a fairly short period of time discs replaced drums on the front wheels of most new cars. Through time, disc brakes have come to be used on all four wheels.

[22]No literature was found that described how prices of products changed within a period of time; however there was one paper that reported that for new products, the prices declined 1% with each doubling of the production volume. This led to the choice of 1% per year price decline for comparison purposes.

[23]Throughout the analysis a two-stage discounting procedure was used. Costs involving private expenditure, such as control equipment or reformulation costs, were annualized using a 7% real rate of interest. Present values were calculated using a 3% rate as the social rate of discount.

[24]The model was designed to accept information for any year as input. Because all data for the model were updated to 1985, we refer to 1985 as the baseline conditions and describe results relative to 1985, so that a ban two years later is referred to as a ban which occurs in 1987.

[25]The costs of the rule as presented in Table 2 do not correspond to the final rule, in that more products were banned in the final rule. The scenario reflected in Table 2 was chosen to illustrate the effect on the benefits and costs of changing some of the assumptions.

[26]From an economic perspective these costs are transfers. The lower prices which consumers experience come at the "expense" of lower prices received by suppliers of asbestos fiber. Approximately 90% of the asbestos fiber is supplied by foreign firms; TSCA, however, considers only domestic costs.

[27]A review of the literature suggested a long term real social discount rate of 2%–4%.

[28]The low decline baseline assumed that all market declines, except for cases of automotive-related products for which a separate baseline estimation was performed, had already occurred. The alternative, high decline baseline, projected product growth rates which had occurred between 1981 and 1985 to the end of the period of analysis. Because the 1981 to 1985 rates were largely negative, using a high-decline baseline resulted in the lowest estimated costs of the regulation.

[29]Asbestos, being highly corrosive, damages some equipment.

[30]PELs are set by OSHA of the Department of Labor. At the time that analytic work for the asbestos ban rule was being performed, OSHA was investigating whether to reduce the PEL from 2 fibers per cubic centimeter (f/cc) of air to some lower value, either 0.5 or 0.2 f/cc of air. Because of the uncertainty about whether the PEL would be changed, the analysis was performed using several different levels. For the base case analysis it was assumed that the new OSHA PEL of 0.2 f/cc would become effective prior to promulgation of the final asbestos rule.

[31]If smoking one pack per day increases the risk of lung cancer by a factor of ten, and exposure to a given level of asbestos fiber increases one's risk of developing lung cancer by a factor of five, a pack a day smoker who is exposed to a given level of asbestos fibers faces a risk of lung cancer that is fifty times that of a non-smoker who is not exposed.

[32]There is a fiber, in addition to asbestos which is a potent inducer of mesothelioma. This fiber, erionite, is found in Turkey. In the United States, however, mesothelioma is a marker for exposure to asbestos.

[33]Hospitalizations for the kinds of cancers induced by asbestos are relatively short before death intervenes.

[34]For example, how would one determine the ambient exposure from asbestos-cement products used on the outside of a building where the product was subject to weathering? It was known that a study had detected fibers in rainwater in the vicinity of buildings which were constructed of asbestos-cement, strongly suggesting that weathering of the A-C product and release of fibers occurs but providing no way to determine an air concentration.

[35]When benefits of a regulation are expressed as lives saved (or cancer cases avoided), EPA's practice was not to discount. The internal workings of the model would allow cases to be discounted since the cases are generated by product by year based on the specific exposures avoided and persons affected as modeled.

[36]Excess lung cancer and gastrointestinal cancer rates are affected by age, sex, race, and smoking habits. For the purpose of this analysis, the nonoccupational population is assumed to have the same characteristics as the general U.S. population. For occupational exposure, the population profile was estimated from 1983 industry (blue collar) data, except for smoking habits which are assumed to be those of the general population.

[37]At the time EPA was developing the regulation the PEL for asbestos was 2.0 f/cc. OSHA was proposing to lower the PEL to 0.2 f/cc, but was also considering as an alternative 0.5 f/cc.

[38]In its analysis OSHA assumed that if the PEL dropped by an order of magnitude, exposures in each workplace would also drop by an order of magnitude, regardless of the level at which they started.

[39]Using the approach of "cost per case avoided" does not entail the same explicit determination of the value of a life. One can divide the total costs incurred by the number of cases avoided and determine whether the value of a case avoided is at least as great as the average cost. This seems to be a task which is politically palatable as opposed to assigning an explicit value to a life.

[40]The total costs of a ban of this product were also estimated to be quite small, $300,000 or less depending on other assumptions.

REFERENCES

McCauley, Molly K., Michael D. Bowes, and Karen L. Palmer. 1992. *Using Economic Incentives to Regulate Toxic Substances*. Prepared for U.S. EPA. Washington, D.C.: Resources for the Future.

U.S. EPA (Environmental Protection Agency). 1989. *Regulatory Impact Analysis of Controls on Asbestos and Asbestos Products*. January 19. Washington, D.C.: U.S. EPA.

8

Lead in Drinking Water

Ronnie Levin

On June 7, 1991, EPA published a rule to control lead contamination of drinking water. Initiated on the heels of the successful lead-in-gasoline phase down, the lead-in-drinking-water rule represented a major shift in drinking water policy. Water utilities, once responsible only for controlling source-water quality, were required by this rule to account for the quality of the water reaching the customer instead of that leaving the water treatment plant. That quality could be achieved either by controlling corrosion chemically or, if necessary, by replacing service pipes. Adoption of this tough new rule was due largely to the economic analysis reflected in the 1991 lead-in-drinking-water RIA, which projected annual net benefits of more than $2.2 billion. In this case study, Ronnie Levin (principal author of the economic analysis that formed the basis of the 1991 RIA) reviews the history and economic analysis that led to the rulemaking, reflects on the policymaking process, and concludes, among other things, that robust economic analysis performed in a timely manner can play a critical role in environmental rulemaking.

INTRODUCTION

To better understand this case study, it is helpful to be familiar with the basics of lead contamination, the regulatory and statutory background to the rulemaking, and the scope of the benefit analysis underlying the 1991 rulemaking. These topics are discussed briefly below.

RONNIE LEVIN is a senior scientist at EPA where she has worked since 1980.

Lead Contamination of Drinking Water

Lead contamination of drinking water in the United States occurs primarily as a by-product of corrosion in public water systems. Water leaving a central treatment plant is generally lead-free, but lead leaches into the water as it passes through the pipes of the public distribution system, in particular the pipes and other plumbing components within the home.

Several factors are associated with the risk of elevated lead levels in drinking water. One is how corrosive the water is: corrosivity is related to pH, alkalinity and hardness levels, and other water qualities. Another is how long the water has been sitting in the pipes: "first-flush water"—water that has been sitting for several hours or more, such as overnight or all day while the residents are at work or school—will have higher lead levels than water that has been in the pipes for only a few minutes. The amount of lead-containing material that is in public and/or private plumbing is also a factor: lead solder used to join copper pipes and lead service lines contribute the most lead, while brass faucets containing lead can also contaminate drinking water. The water's temperature is yet another factor: lead levels are higher in warmer water, whether the result of hot weather or simply of water being within the home or in the water heater. The age of the plumbing can also contribute to lead contamination: new plumbing that contains lead and that has not had time for any internal buildup of insoluble layers is most likely to have high lead levels.

These factors complicate regulatory efforts to control lead exposure from drinking water because its contamination generally occurs within the home and is related to the actions of both the water utility and its residential customers.

Statutory Requirements

The Safe Drinking Water Act (SDWA), passed by Congress in 1974, requires EPA to establish maximum contaminant level goals (MCLGs) and national primary drinking water regulations (NPDWRs) for drinking water contaminants. These two requirements, sometimes confused with one another, are distinguishable in two ways: MCLGs are nonenforceable health goals, while NPDWRs set the enforceable standards for contaminants.

MCLGs neither constitute regulatory requirements nor impose obligations on public water supplies. They are strictly health-based goals, to be set at a level where "no known or anticipated adverse effects on the health of persons occur and which allows an adequate margin of safety" (Section 1412[b][4]). The MCLG should be set at zero where "there is no safe threshold for a contaminant."

NPDWRs include either numerical levels—maximum contaminant levels (MCLs)—or treatment requirements, as well as compliance monitoring requirements. The MCL must be set as close to the MCLG as "feasible" (Section 1412[b][4]).

The 1991 Rule and Its Regulatory Background

In 1975, EPA set the MCL for lead in drinking water at 0.05 milligrams per liter (mg/l),[1] as an interim NPDWR, based on the 1962 Public Health Service standard (40 CFR 141.11[b]). (At this time, the MCLG for lead was also 0.05 mg/l.) Samples were to be taken to be representative of water in the distribution system. In 1985, EPA proposed to reduce the MCLGs for lead and for copper[2] (50 *Federal Register* 46936).

In 1986, the SDWA Amendments banned the use of lead solder or flux containing more than 0.2% lead, and pipes or fittings containing more than 8% lead, among many other actions. These regulations also required public notification where there was any risk of lead contamination of drinking water. Other changes in the SDWA in 1986 (such as a mandate that EPA propose the MCLG and MCL treatment technique simultaneously) required EPA to repropose the lead and copper MCLGs.

1988 Reproposal. In 1988, EPA proposed an MCLG for lead of zero, and proposed an MCL for lead in source water of 0.005 mg/l, to be measured at the entrance to the water distribution system (53 *Federal Register* 31516). EPA also proposed a treatment technique requiring "optimal" corrosion control to minimize lead and copper corrosion and proposed that water systems implement a public education program concerning the risk of exposure to lead and copper in drinking water.

By this proposed standard, corrosion control was to be triggered if any of the following measurements were found:

- The average lead levels in first-flush samples from kitchen taps in targeted houses exceeded 0.01 mg/l;
- The copper levels exceeded 1.3 mg/l (both the proposed copper MCL and MCLG) in more than 5% of the targeted samples; or
- pH was less than 8.0 in more than 5% of the targeted samples.

Furthermore, the proposed public education requirements would have been triggered if the lead levels exceeded an average of 0.01 mg/l or if more than 5% of the targeted tap samples exceeded 0.02 mg/l.

1991 Final Rule. In 1991, EPA established an MCLG of zero for lead and an NPDWR for lead and copper consisting of a *treatment technique* that included corrosion control treatment and public education and, if necessary, source-water treatment and/or lead service line replacement (56 *Federal Register* 26460). No MCL for lead was set, and a single action level of 0.015 mg/l at the 90th percentile replaced the multiple action levels in the 1988 proposal. Exceeding the action level triggered a requirement to implement the treatment technique, especially corrosion control. To so exceed, however, was not itself a violation of the drinking water standard; as such, the action level was "unenforceable."[3]

Tap water samples—samples taken at kitchen taps in private residences, as opposed to distribution or source-water samples—were required for all water systems and were targeted to risk factors related to lead contamination of drinking water. (Risk factors for elevated lead levels in drinking water include water that has been sitting unused in pipes for over six hours, newly installed plumbing that uses lead solder, and the presence of lead pipes.) Requirements included first-flush samples (defined as water that had been sitting in the pipe for over six hours) from up to 100 residences known to have lead pipes, relatively recently installed lead solder with copper pipes (that is, since 1982), or both.

All systems were required to implement corrosion control treatment that minimized lead and copper corrosion, termed *"optimal" corrosion control* in the 1991 rule. The specifications for treatment, however, were to be determined by the state and by system size. Large systems (serving more than 50,000 people) were to maintain a minimal difference—less than 0.005 mg/l—between the 90th percentile tap water samples and the highest source-water lead concentration entering the water system. Also, large systems were required to install "optimal" corrosion control or to demonstrate that their water was minimally corrosive. Small systems (serving 3,300 or fewer people) and medium systems (serving from 3,301 to 50,000 people) were required to meet a "lead action level" of 0.015 mg/l at the 90th percentile of targeted tap samples or were required to install "optimal" corrosion control treatment, to be determined by the state. ("Action level" was a term created by this rule.)

A water system exceeding the action level after installing corrosion control was to evaluate its source water and check for the presence of lead pipes. If *source water* was identified as the problem, the water utility was to remove the lead from the water entering the distribution system. If *lead pipes* were identified, the utility had to develop a plan for remediating that problem, which would probably include removing and replacing the pipes. Utilities had up to fifteen years to remove all lead pipes contributing excessive lead to the drinking water, with funding arrangements, timetables, and other such requirements left to the discretion of the water system and state.

Scope of the 1991 Rulemaking

From the regulatory history, one can see that the 1991 final rule significantly extended the scope of previous drinking water regulations and also established several important precedents.

- Tap water samples from within private homes, as opposed to samples of source water or water in the distribution system, were the basis of the rule.
- System size determined some important regulatory requirements, including implementation schedules and how "optimal" corrosion treatment was defined.
- More monitoring samples were required than in previous rules.
- The *best available technology* (BAT) was determined by each state on a system-by-system basis. (In other regulations, a single BAT is established nationally for all systems in all states.)
- The MCLG for a carcinogen was set at zero because EPA asserted that there may be *no threshold for some of its noncarcinogenic health effects.*[4]
- Public education to inform consumers of potential risks of exposure to lead (with the tacit acknowledgment that unaddressed risks may remain) was required for the first time.
- Use of a 90th percentile in monitoring and for determining compliance was established, as opposed to arithmetic averages or maximum values, which had been used previously.
- The concept of the action level—an unenforceable level that triggers action—was developed; exceeding the action level triggered implementation of the treatment technique, but did not itself constitute a drinking water standard violation.

Scope of the 1986 Benefit Analysis

The benefit analysis underlying the 1991 rule, as well as the initial cost assessment, was undertaken to support the 1985 proposed revision to the MCLG and was completed in 1986 (referred to in this document as "the 1986 benefit analysis"). It was conducted within EPA and reviewed extensively within the agency. Methodologically, it refined and extended the health valuation methods developed in the benefit-cost analysis to support reducing lead in gasoline. It was conducted by EPA's Office of Policy Analysis (OPA),[5] not by the Office of Drinking Water. (OPA had a tradition of doing independent economic analyses and had also conducted the lead-in-gasoline benefit-cost analysis.) A new category of benefits was monetized for the drinking water analysis: materials damages reduced as a result of corrosion control treatment. Resources required to complete this analysis were 0.5 full-time equivalent (or work years), computer support costing about $5,000.

The 1986 benefit analysis was extensive and presented several unexpected findings.

- Forty-two million people were estimated as the population exposed annually to significant lead levels in their drinking water, a higher number than had been anticipated.
- About $1 billion ($1985) was estimated as the quantified annual health and other economic benefits, dwarfing the estimated costs.
- Large numbers of people were estimated to be at risk of serious health effects every year due to exposure to lead in drinking water.
- Unnecessarily corrosive water was estimated to cause about $500 million ($1985) in preventable corrosion damage to public and private plumbing. (This amounted to about half of the quantified annual benefits.)
- EPA's approach to regulating lead in drinking water, by implication, should be revised.

Scope of the Two Regulatory Impact Analyses. Separate RIAs were completed for the 1988 proposal and 1991 final rule. Both built on the methodologies used in the 1986 drinking water benefit analysis, which contained three categories of monetized benefits: health benefits to children, health benefits to adults, and materials damages avoided by control of corrosive water. The categories of monetized benefits varied, however, between the 1988 and 1991 RIAs. A key component of the 1986 benefit analysis and the only monetized benefit in the RIA for the 1988 proposal—materials damages related to control of corrosive water—was dropped from the 1991 RIA. On the other hand, health benefits were monetized in the 1991 RIA, but not in the 1988 RIA.

Scope of This Case Study

This assessment reviews the final 1991 RIA as it was released, with two exceptions. First, results are rounded to two or three significant digits. (The RIA itself presents results in up to nine significant digits.)[6] Second, where there are discrepancies between the rule and the RIA, both are included.

THE RIA: AN OVERVIEW

The Office of Policy Analysis offered support to the Office of Drinking Water (ODW) following the 1985 proposal to reduce the lead MCLG. In agreement with ODW, OPA assessed benefits and ODW evaluated the costs of the proposal. The intent was to build on the success of the lead-in-gasoline benefit-cost analysis.

The 1986 OPA drinking water benefit analysis suggested that a fundamentally different approach to regulating lead in drinking water would yield enormous public health and other economic benefits. Neither the policy implications of such a change nor the details of a regulatory framework, however, were included in the 1986 benefit analysis.

Key Policy Issues

Several key policy decisions faced EPA managers once the 1986 benefit analysis was conducted. One related to agency agreement on the importance of certain lead health effects, especially those occurring at low exposure levels. Because drinking water usually contributes only small amounts of lead and because U.S. blood-lead levels had already decreased significantly, the health implications for the drinking water rule involved typically low levels of both exposure and health effects. Consensus on the importance of these effects had developed during the lead-in-gasoline rulemaking and included the key EPA offices—the Office of Research and Development (ORD) and the Office of Policy Planning and Evaluation—as well as the Office of Air and Radiation. A 1986–87 EPA review of the health evidence resulted in a strengthened EPA position on the significance of these health effects. At the same time, concurrent assessment activities within ORD culminated in the incorporation (including quantification) of two new categories of health effects in EPA analysis: cardiovascular effects in adult males and fetal growth effects.

Another policy issue was related to the tendency of lead to contaminate drinking water *within* the house. The policy question was whether to address *exposure* to lead in drinking water within private homes or the *water* distributed by the utility. This decision would be a significant departure from past regulatory approaches, with far-reaching political and programmatic implications. EPA decided on an exposure-based approach because the corrosivity of water determines how much lead will leach into the water and how quickly lead levels will rise. No matter what the source of the lead (private or public plumbing), the quality of the water entering the home is the principal determinant of lead levels within the home, and that quality is controlled directly by the water utility.

Economic arguments also existed for a water quality rule: it was the *corrosivity* of the water that resulted in lead contamination. Reducing the corrosivity of the water would reduce damage to all plumbing components (pipes, faucets, radiators, and so forth), representing savings for both the water utility and consumers.

Finally, the lead-in-drinking-water regulation raised issues related to multimedia contamination; that is, contamination from multiple sources within all environmental media: air, water and soil. Following this regulation, EPA initiated *regulatory clusters*: combining rules that fell under different statutory authorities and were regulated by different EPA offices but that affected the same industry or contaminant.

Major Scientific and Analytical Problems

Regulating lead in drinking water is neither simple nor straightforward, with complexities at all stages, including the following:

- There were *no baseline data* on lead levels in tap water.
- There is significant *variability* in levels of lead in drinking water, even at the same faucet under similar conditions, because lead levels are a function of the combination and interactions of many factors affecting the microenvironment at each tap.
- Corrosion control treatment is *system-specific* and depends on the exact characteristics (such as the chemical composition) of the water, the conditions and layout of the distribution system, the use patterns within the system, the water temperature, and many other such factors. Even after the system has been evaluated and treatment determined, the treatment may require a period of trial and adjustment.
- Numerous uncertainties existed about the *reliability of corrosion control* treatment to consistently meet the lead action level.
- Corrosion control treatment may require changes in water quality with implications for treatment or compliance with other drinking water requirements. For instance, raising the pH to reduce corrosivity may require changes in disinfection procedures.

Major Stakeholders

Drinking water regulations apply to public water systems; that is, water utilities serving at least twenty-five customers, or fifteen connections, for sixty days or more annually. Of the approximately 60,000 community water systems in the United States, about 37,000 systems (almost two-thirds) are classified as very small, serving 25–500 people. Altogether, these systems serve only 2%–3% of the U.S. population served by community water systems. On the other hand, there are only 266 very large systems (serving more than 100,000 people), but together they serve about 100 million people, which is about 44% of the population served by community water systems.

Institutional History

The drinking water program is among EPA's oldest programs and has been characterized by its stability. There has always been a close relation between EPA's professional drinking water staff and professionals in the water supply industry.

From a regulatory perspective as well, the drinking water program has been quite stable. The fifteen or so MCLs that the program administered until 1985 were not new to EPA; in general, EPA codified standards that had been set by the U.S. Public Health Service in the 1950s and 1960s.

Regulatory Approaches Considered

While the 1986 benefit analysis showed that there were potentially large health and other economic benefits to regulating lead in drinking water, a regulation significantly different from previous drinking water standards was required to achieve these benefits. By necessity, such regulation was complicated for EPA, burdensome to the states, and vigorously opposed by the water utilities. It was also likely to face legal challenges.

EPA senior management decided to proceed with an innovative drinking water rule. Having made that policy decision, there were then two key decisions related to an approach to this regulation. First, EPA had to decide whether to establish an MCL for lead or design an alternative regulatory framework. Then, EPA had to decide whether to target the regulation (that is, monitoring and sampling) to high risks or average exposures.

MCL versus Treatment Technique

An MCL is a single enforceable water quality standard to which a water utility is held; EPA must identify a BAT to achieve an MCL.

An MCL for lead could be set either in the distribution system or at the tap. Setting one in the distribution system, either alone or in combination with a regulation to control corrosion, had the advantage that it would limit the lead contribution from source water and would provide an easily enforceable standard. It would not, however, restrict the corrosion within the house that is the principal source of lead contamination of drinking water.

An MCL at the tap, on the other hand, had different disadvantages. Unless this standard was set very high (and thus was not protective of health), it would be vulnerable to legal challenges that could claim that meeting the standard was beyond the control of the water system because the materials containing lead were within the private home. In addition, there was no single BAT for corrosion control; treatment needed to be set on a system-by-system basis.

An alternative approach—establishing a corrosion treatment requirement triggered by a non-enforceable standard—resolved some difficulties. Elevated lead levels in the home (that is, lead levels higher than in the distribution system) under standardized conditions indicated excessive corrosivity and triggered a treatment requirement: to treat the water to achieve "optimal" corrosion control. The specific treatment could be set on a system-by-system basis. The use of a non-enforceable trigger also addressed concerns regarding the technical feasibility of corrosion treatment to produce very low lead levels (that is, greater than or equal to 0.01 mg/l).

A treatment technique triggered by a non-enforceable action level at the 90th percentile, however, also had some disadvantages. Because exceeding the action level did not itself constitute a violation, one problem was enforceability: it is more difficult to establish compliance with a complicated corrosion treatment plan than with a single numerical limit. Another problem was the tacit decision to permit some fraction of the population (that is, the upper 10% of the population) to receive drinking water with relatively high lead levels. No guidance was issued limiting that exposure to protect against an unreasonable risk to health.

EPA chose to establish an NPDWR for lead that was a treatment technique triggered by exceeding an "action level" at the consumer's tap. It was set at 0.015 mg/l to minimize questions of technical feasibility associated with corrosion control treatment.

Sampling Issues

In general, lead levels in U.S. drinking water are low (<0.01 mg/l). However, there are many risk factors that result in occasional high levels. The more critical of these risk factors are discussed.

High versus Average Risks. EPA decided to target sampling to routine high exposures: first-flush samples in private homes with relatively newly installed lead solder, lead pipes, or both.

First, virtually everyone refrains from using water for most of the night, so morning first-flush samples are common—at each faucet in each home every morning. In addition, most of the U.S. adult population works outside the home and most children attend school during the day, so first-flush conditions also occur routinely again in the evening when the family members return home.

Second, newly installed lead solder can contribute high levels of lead to drinking water, and until the ban on the use of lead solder was implemented in the late 1980s, it was a risk factor common to many people. In

1986, EPA estimated that about 10 million people lived in housing built within the last two years.

Finally, the 1991 rule estimated that, as a high bound, about 15,000 water systems had some lead pipes somewhere within the distribution system. EPA estimated that about 5.6 million housing units, occupied by over 15 million people, receive water with excessive lead levels due to corrosion of their lead pipes.

Sample Volume. EPA decided to require a one-liter sample, which would adequately assess the effectiveness of corrosion control efforts but would indicate lower lead levels than a smaller sample size.

Variability in Lead Levels. Lead levels vary significantly even at the same tap under seemingly identical conditions. To address this concern, EPA greatly increased the number and frequency of sampling required in order to more confidently characterize water quality (and exposure). The number of samples to be collected was based on the size of the service population, and is presented in Table 4. Followup sampling could be reduced if the water system exhibited consistently low lead levels.

Also to address the issue of variability, EPA decided to use a 90th percentile limit, assuaging utilities' fears that a single high value would mean a drinking water violation.

Best Available Technology

Corrosion control cannot be addressed with a single, off-the-shelf technology. While several general approaches are applicable to most systems, the exact measures to be implemented must be tailored to the specifications of the water quality, other water treatment, climate, and the physical configuration of each system. EPA left the determination of what constitutes compliance and implementation of BAT to the states.

Alternatives Not Considered

Alternative regulatory approaches, such as market mechanisms, were not considered for several reasons. First, the regulation was driven by health concerns and the standards were set as low as possible, considering costs and technical feasibility. In addition, the health effects are direct and accrue to each person exposed; therefore, there was no alternative to having each system meet the same standard. Second, each water utility is a local monopoly, so no trading or permit system would have worked.

Other options, such as requiring or encouraging residents to flush their faucets several times a day, were considered neither environmentally sound nor enforceable. Options such as requiring replacement of the plumbing in housing that presented the highest lead levels were unacceptable due to financial or statutory constraints and because high lead levels in specific homes were understood to indicate water quality issues for the entire system.

Finally, alternative multimedia approaches, such as targeting lead exposures likely to result in lead *poisoning*, were not considered then, although some were proposed subsequently. The 1986 SDWA Amendments explicitly required EPA to address lead in drinking water. Also, at that same time, EPA was developing an agency lead strategy that called for each program to undertake to reduce both general population exposures and high exposures as much as possible. Because drinking water represented the highest uncontrolled lead exposure at the time, it would not have been acceptable within EPA to leave it unregulated—especially as there was an economic argument for regulation.

COST ANALYSIS

Three sources of lead contamination of drinking water were addressed by the 1991 regulation: source-water contamination, leaching from lead solder (and other interior plumbing materials), and leaching from lead pipes. Strategies for controlling these different sources of contamination exhibit differences in cost-effectiveness. A treatment technique requirement allows the specification of different performance targets for each of the different control methods.

The 1991 RIA examined the comparative costs and benefits of three alternative regulatory treatment technique formulations (Table 1). In the RIA, Alternative B represented the estimated costs of the final rule as published; Alternative A was somewhat more stringent and Alternative C was somewhat less stringent.

The 1991 RIA estimated costs for six components of the rule: source-water treatment, corrosion control treatment, lead pipe replacement, public education, state implementation, and monitoring. Each is discussed separately. Total annualized costs for the final regulation were estimated to range from $500 to $800 million (Table 2).

Methodology for Estimating Costs

A model used for all drinking water regulations provides the base data and/or estimates for the structure of the regulated drinking water industry, including the total number of systems by system size and water source (surface or ground), the average number of entry points per sys-

Table 1. Alternative Regulatory Treatment Technique Formulations for Evaluating the Costs and Benefits of the Final Rule.

	Action levels (mg/l)		
Regulatory alternatives[1]	*Source water contamination*[2]	*Leaching from pipe*[3]	*Leaching from solder*[4]
A (more stringent regulation)			
Small systems	.005–.015	.005	.015
Large systems	.005–.015	.005	.015
B (the rule as published)			
Small systems	.005–.015	.015	.015
Large systems	.005–.015	.005	.015
C (less stringent regulation)			
Small systems	.005–.015	.015	.015
Large systems	.005–.015	.015	.015

[1]Regulatory alternatives A, B, and C are distinguished by the stringency of their requirements.
[2]Measured at the water treatment plant; states to determine if treatment is required.
[3]Single sample for each pipe; measured after optimization of corrosion control treatment if lead service connection samples still exceed the action level of 0.15 mg/l.
[4]90th percentile in first draw samples from targeted taps.

Source: U.S. EPA 1991, Exhibit 4–1.

Table 2. Total Estimated Costs of the National Primary Drinking Water Regulation for Lead. ($1991 million)

Regulatory components	*Number of systems affected*	*Total capital cost*[1]	*Total annualized cost*[2]
Source water requirements[2]	880*	$450	$90
Corrosion control	40,000	$990*	$220
Lead pipe replacement[3]	8,300	$1,500*–6,250	$80*–370
Public education[4]	40,000	—	$30
State implementation	—	$50*[5]	$40
Monitoring			
Source water	40,000	—	<$1
Corrosion control	79,000[6]	—	$27
Lead pipe replacement	8,300	—	$12

Notes: Costs are rounded. $year not specified; assumed to be "current year" ($1991).
*Discrepancy between RIA (Exhibit 4-7, pp. 4-33; and Exhibit 4-11, p. 4-38) and cost summary in final rule (*Federal Register*, Table 20, p. 26539).
[1]Annualization is over a 18–20 year period at a discounting rate of 3%.
[2]Costs assume all systems with lead levels > 0.005 mg/l and/or copper levels > 1.3 mg/l in source water will be required to implement source water treatment.
[3]Range in costs reflects uncertainties in the number of pipes that will be removed and costs for replacement.
[4]Includes all systems that may ever require it.
[5]These are one-time, startup costs.
[6]Includes all community and nontransient noncommunity systems.

Source: Summary in 56 *Federal Register* 26539; June 7, 1991.

tem, treatments in place, system capacity, and so forth. Costs are calculated by system size for twelve size categories and then summed. All costs are annualized, but the dollar year was not specified; it is assumed to be "current" (that is, in $1991; however, health benefits are presented as $1988). Costs are annualized over an eighteen- to twenty-year period, at a 3% discount rate.

Estimates were also made, by system size, of current lead levels in source water and at the tap, presence and extent of lead pipes, extent of current corrosion control treatment (if any), contribution of lead solder to lead levels at the tap, contribution of lead pipes to lead levels at the tap, and lead levels after corrosion treatment. Specific data used for these assumptions are discussed in each section.

This regulation is to be implemented over a period extending more than fifteen years into the future for systems having to remove lead pipes. There are additional lags in implementation at the state level to permit the development of comparable state regulations. Further lags will occur in the process of monitoring and fine-tuning corrosion control treatment. For these reasons, many of the costs estimated in the RIA will not occur for many years. Uniform discounting of costs was used to standardize costs. Conforming to EPA practice, this was done in two steps: first, capital costs were amortized using a 10% interest rate over the economic life of the purchase; then all amortized capital and operating and maintenance costs were summed in the years they occur and then discounted to present value at 3%.

While EPA used the best data and judgments available in estimating national costs, the agency noted that at least some of the results may be overestimated. For instance, the unit costs for sample collection and analysis were probably high, as were the estimates for corrosion studies and for capital costs for corrosion control treatment. EPA could not estimate the magnitude of the overestimation.

The following sections describe the six regulatory components for which costs were estimated.

Source-Water Treatment. Systems with high lead levels in water entering the distribution system must undergo source-water treatment. About 880 water systems were assumed to incur costs related to source-water contamination by lead at an estimated annualized cost of about $90 million.

Two cost elements are used for addressing source water contaminated by lead: water treatment and waste disposal. Costs for treatment were projected by a decision tree based on the best engineering judgment and assumptions about water quality and compliance rates based on EPA occurrence data; all estimates were made by system size. Unit costs are contained in the costs and technologies documents (U.S. EPA 1989), which are prepared by the Office of Drinking Water for every regulation.

In the RIA, the total national cost of source-water treatment was estimated and presented as a range, because the final decisions about lead or copper levels and related required treatment were left to the states. The range was bounded by two extremes: one where no system exceeded a contribution of 0.005 mg/l from source water and the other where no system reduced lead levels in source water below 0.015 mg/l. A single point estimate was presented in the rule itself.

Corrosion Control Treatment. Three general approaches to corrosion control are available, depending on the specific parameters of the water quality, climate, and physical configuration of the particular water system: pH adjustment, water stabilization, and use of chemical corrosion inhibitors. These approaches can be used alone or in combination, and a host of options exists for each. EPA assumed that water utilities exceeding the action level would begin with the least expensive and difficult treatment and only progress if more desirable options were unsuccessful or inappropriate. While some off-the-shelf technologies that are best suited to small systems are available, for medium and large systems (that is, systems serving over 3,300 people), corrosion treatment must be tailored to the unique specifications of the system.

For estimating the costs of this component of the rule, only the six largest system size categories were analyzed, and systems were assigned to one of five assumed "treatment-in-place" categories, based on treatment and pH level. Data for this were taken from two sources: a 1986 survey of community water systems (RTI 1987) for systems serving communities of fewer than 10,000, and the American Water Works Association's *1984 Water Utility Operating Data* (as presented in U.S. EPA 1989) for systems serving communities with populations larger than 10,000. The data from those two sources were combined with data from the 1988 "Ciccone survey," *Corrosivity Monitoring Data from U.S. Public Water Systems* (as presented in U.S. EPA 1989). About 40,000 water systems were estimated to require some corrosion control treatment, at an estimated annualized cost of $220 million.

Data on the efficacy of corrosion treatment were taken from published and unpublished literature and from data from the American Water Works Service Company. This information was compiled by the Office of Drinking Water in U.S. EPA 1990. While none of the data were definitive—they were taken generally from monitoring results collected for reasons other than evaluating the efficacy of alternative corrosion treatments—they were useful to EPA as a guide for what could reasonably be achieved under a variety of conditions.

Another cost included in this section was the corrosion control studies required of large and medium systems as a means of documenting for the state that "optimal" corrosion control was in place.

Leaching from Lead Pipes. Projections on the extent of lead pipes in U.S. public water systems were based upon data from the 1988 American Water Works Association lead survey (AWWA 1989). That survey indicated that about 26% of public water systems had some lead pipes and estimated that more than half of those systems were likely to fail the lead action level due to those lead pipes. EPA estimated the number of systems in each of the twelve system size categories that were likely to have lead pipes, what treatment was in place, how many pipes were likely to cause the system to fail, and so forth. Useful information was also provided by two other documents: *Lead Control Strategies* (AWWA-RF 1990) and *Lead Service Line Replacement: A Benefit-to-Cost Analysis* (AWWA 1990). About 8,300 water systems were estimated to require some actions to address leaching from lead pipes.

A fifteen-year compliance schedule was promulgated for lead pipe replacement due to cost considerations.

Where lead pipes are found to contribute to excessive lead levels at the tap, the least expensive solution is to increase the level of corrosion treatment. The methods described above to estimate the costs of corrosion control for lead solder were also used to evaluate lead leaching from lead pipes. Only systems that were projected to exhaust all available corrosion treatments and still did not meet the action levels were identified as having to undertake public education programs or pipe replacement. The total replacement cost includes both physical replacement of pipes and required monitoring, discounted for implementation to occur over the fifteen-year period permitted in the rule.

Public Education. EPA estimated that about 40,000 water systems will incur costs initially to conduct public education activities to inform consumers about the risks of lead contamination of drinking water and what consumers can do to reduce those risks. The agency estimated that most of those systems will then install corrosion control treatment and successfully meet the lead action level, obviating their need to continue a public education program. Some water utilities—EPA estimated about 12,000 systems (8,300 with lead service lines)—will remain above the lead action level and be required to continue public education activities. EPA estimated the costs of public education at about $30 million per year, assuming that EPA will develop the materials and utilities will distribute them, and that while most systems will conduct programs for only a short time, some systems may continue their programs for up to twenty-five years.

State Implementation Costs. In 1988, EPA and the Association of State Drinking Water Administrators conducted a survey of state drinking water resource needs resulting from the 1986 SDWA Amendments, *State Costs of Implementing the 1986 Safe Drinking Water Act Amendments* (ASDWA 1990). Data from this survey were used to assess state implementation costs, which EPA estimated at $40 million per year.

Costs of Monitoring. Total national monitoring costs consist of two components: source-water monitoring costs and corrosion monitoring costs. Costs of monitoring for lead pipes were included in that section. Table 3 displays the estimated monitoring costs, by rule component and by category. Monitoring costs would occur in all 79,000 water systems regulated by this rule, at an estimated annual cost of about $40 million.

For source water monitoring, the cost estimates include one sample for each public water system per entry to the system initially, and then one sample every year (for surface water systems) or one sample every three years (for groundwater systems).

For corrosion monitoring, the cost estimates include the number of samples and frequency required in the rule, presented in Table 4. The beginning date for monitoring was phased by system size. All large systems (serving over 50,000 people) were required to conduct tap sampling for a variety of water quality parameters. Small and medium systems fail-

Table 3. Estimated Annualized Monitoring Costs by Rule Component. ($million)[1]

	Regulatory alternatives		
Rule component	*A*	*B*	*C*
Source Water	0.6	0.5	0.5
Corrosion Control			
Lead Solder	23	20	20
Initial compliance testing[1]	17	16	16
Water quality at the tap	3	1	1
Water quality at the plant	4	3	3
Lead Pipes	19	12	12
Initial compliance testing[1]	8	7	7
Water quality at the tap	2	1	1
Water quality at the plant	10	3	3
Training and implementation	7	7	7
Total	50	39	39

Note: Totals may differ from sum of subtotals due to rounding. The $year is not specified; assumed to be "current year" ($1991).

[1]Initial monitoring to determine compliance with lead and copper action levels.

Source: U.S. EPA 1991, Exhibit 4-1.

Table 4. Monitoring Requirements: Number of Samples and Frequency.

System size	*Monitoring requirements*	
(population served)	*Normal (samples/six months)*	*Reduced[1] (samples/year)*
>100,000	100	50
10,001–100,000	60	30
3,301–10,000	40	20
501–3,300	20	10
101–500	10	5
<100	5	5

Note: Sampling numbers are per water system, with one sample per site required.

[1]A system is permitted to reduce its monitoring if it has two successive six-month monitoring periods without exceeding the action level.

Source: U.S. EPA 1991, Exhibit 4-8.

ing the lead or copper action level must also monitor water quality during the monitoring period in which they exceed the action levels. The additional analyses of water quality are necessary for designing a corrosion control program. Reduced monitoring applies to systems that have not exceeded the lead or copper action level(s) for two successive six-month monitoring periods.

Costs Not Included

Costs of secondary effects and competing risks were not included.

BENEFIT ANALYSIS

The SDWA does not direct EPA to consider benefits in establishing drinking water standards. EPA established the requirements of the regulation based on the criteria contained in Section 1412 of the SDWA. However, EPA is directed by Executive Order 12291 to estimate both the benefits and the costs of its major rules.

Methodology for Estimating Benefits

Several categories of benefits were quantified and monetized in the benefit analyses conducted for this regulation. Quantified health benefits included avoided health damage for children, adult men, and pregnant women and fetuses. The methods used to estimate the health benefits of reduced lead exposure were, for the most part, extensions of methods developed to support EPA's 1985 lead-in-gasoline regulation. New to the 1986 analysis was an assessment of the number of pregnant women and fetuses at risk from exposure to lead in drinking water; those effects were not monetized. All the health effects quantified and the methods used to value the effects that were monetized were reviewed within and outside EPA.

The method used to estimate avoided materials damages—that is, reduced damage to plumbing components due to reduced corrosivity of the water—was developed specifically for this regulation. Materials benefits were quantified and monetized in the 1986 benefit analysis (U.S. EPA 1987) and included in the RIA for the 1988 proposal. While they were discussed qualitatively in the final rule, no quantitative estimate of these benefits was included in the 1991 RIA for the final rule.

As with the cost estimates, the final RIA examined the comparative benefits of three alternative treatment technique formulations; this examination is outlined in Table 1. In the RIA, Alternative B represented the estimated benefits of the final rule as published; Alternative A was somewhat more stringent and Alternative C was somewhat less stringent.

In each category of quantified benefits (child and adult health benefits, and materials damages avoided), the estimates were incomplete because of gaps in the data or difficulties in monetizing some types of benefits. Nonetheless, each category of estimated benefits was substantial and exceeded the estimated costs of the rule, with one exception: the estimated costs and benefits of lead pipe replacement in the 1991 RIA. Total monetized annual health benefits were estimated at $3–$4 billion, not including any estimated materials damage benefits. In the final rule, total benefits of corrosion control alone were estimated to exceed the total costs by a factor of more than ten to one.

The three categories of estimated benefits are described below.

Health Benefits. The health benefits of reduced exposure to lead were estimated for three populations: children (aged 0–6 years), pregnant women and fetuses, and adult men (aged 40–59 years). Benefits for children and adult men were partially monetized; those related to fetal exposures were not. Exposure estimates rely on actual data on the distribution of blood-lead levels in the U.S. population provided by the Second National Health and Nutrition Examination Survey (also known as NHANES II [U.S. EPA 1986]). (Blood-lead levels are expressed as micrograms of lead per deciliter [μg/dl] of blood.) Blood-lead distributions were adjusted to account for changes in exposure that occurred or were estimated to occur between 1980 and 1991.

Estimates of preregulation lead levels in U.S. public drinking water were based on data collected and evaluated for EPA's Office of Drinking Water on partially flushed (thirty-second flush), random daytime "grab" samples from kitchen taps (Patterson 1981). The postregulatory distribution of water-lead levels was estimated by making assumptions about the efficacy of corrosion control treatment based on data in U.S. EPA 1990. Different coefficients, based on age, were used to relate water-lead levels and blood-lead levels.[7]

Medical costs are presented as $1988 in the 1991 RIA. Note, however, that costs are presented as "current year," but the dollar year is not specified; it is assumed to be $1991.

Children's Health Benefits. Two categories of benefits were estimated monetarily: avoided medical costs to treat children with elevated blood-lead levels and averted costs to compensate for lead-induced cognitive damage.

Medical treatment cost estimates include only direct costs: testing, medical treatment, and followup, using established protocols for treatment and data on the likelihood of required medical interventions. The lead-in-gasoline benefit-cost analysis and the 1986 drinking water benefit analysis only estimated costs for medical treatment for children exceeding the then-current lead toxicity level set by the Centers for Disease Control (CDC) of 25 µg/dl. Because CDC lowered its "blood-lead level of concern" in 1989 to 10 µg/dl, the 1991 RIA estimated costs for children with blood-lead levels > 10 µg/dl. The 1991 RIA estimated benefits for reducing a child's blood-lead level ranging from $300–$350 per child to $3,200 per child depending on the blood level and other biochemical analyses as classified by the CDC.

Two alternative methods were used for valuing lead-induced cognitive damage and addressed two aspects of this damage: the cost of compensatory education (part-time, in class) to address cognitive damage affecting school performance, and the decrease of future lifetime earnings as a result of IQ decrements (discounted to age seven), covering short- and long-term effects, respectively. Compensatory education for low-level cognitive damage affecting school performance was valued at $5,800 per child, as was additional education for children with IQs below 70. The effect of IQ decrement on future earnings was valued at $4,600 per lost IQ point. Both aspects are assumed to be underestimates of actual benefits.

Adult Health Benefits. Estimated adult health benefits were also based on blood-lead data from NHANES II (U.S. EPA 1986). Studies associating increased blood-lead levels with increases in blood pressure in adult men led to calculating benefits of reduced hypertension and other cardiovascular effects, including deaths. Morbidity was monetized; the 1986 benefit analysis used specific, published cost-of-illness studies to value avoided cases of hypertension, nonfatal heart attacks, and strokes. The 1991 RIA used a uniform estimate of willingness to pay of $1 million per event for nonfatal heart attacks and strokes; reduced incidence of hypertension was valued at $628 per case for direct medical costs and lost productivity. Deaths were valued at $2.5 million each in the final RIA.

In addition, in the 1986 benefit analysis, risks to fetuses from lead exposure were assessed by estimating the number of pregnant women with blood-lead levels over 15 ug/dl, but no benefits were monetized.

Materials Damages Avoided

The corrosive action of water on plumbing materials containing lead results in lead contamination of drinking water; treatment here is defined as reducing the corrosivity of the water. Reduced corrosivity has other benefits as well, such as reduced damage to pipes and other plumbing components, radiators, hot water heaters, faucets, and so forth. Such benefits accrue to both the water utility and the private home or building owner.

The published literature, summarized in the 1986 benefit analysis and in AWWA-RF 1989, indicates that at least 20%–30% of total corrosion damage is avoidable through water treatment. These benefits were calculated monetarily in the 1986 benefit analysis and included in the RIA for the 1988 proposal, but they were discussed only qualitatively in the final 1991 rule itself and were omitted entirely from the 1991 RIA.

Key Uncertainties and Data Gaps

Many uncertainties and data gaps exist in the 1991 RIA. The lack of complete data necessitated assumptions and best engineering judgments about current lead levels in drinking water and the efficacy of corrosion treatment; about the likelihood of decreasing blood-lead levels or reversing—or preventing—adverse effects on health from lead; about the relationship between lead exposure and cognitive damage, fetal effects, or cardiovascular health; and about the effectiveness of corrosion treatment to reduce corrosion damage.

Sensitivity Analyses

The final 1991 RIA showed graphically the uncertainties believed to be associated with the point estimates of the *total* (not annualized) costs and benefits of corrosion control treatment to reduce lead leaching from solder (Figure 1) but did not provide quantitative data. Cost estimates were assumed to have an uncertainty of ±50%. Benefit estimates were assumed to be as much as two times greater than the point estimates but not less than 70% of them. While the RIA text is peppered with judgments by EPA that particular costs or benefits are likely to be high or low, no compilation of such is made, nor is documentation provided.

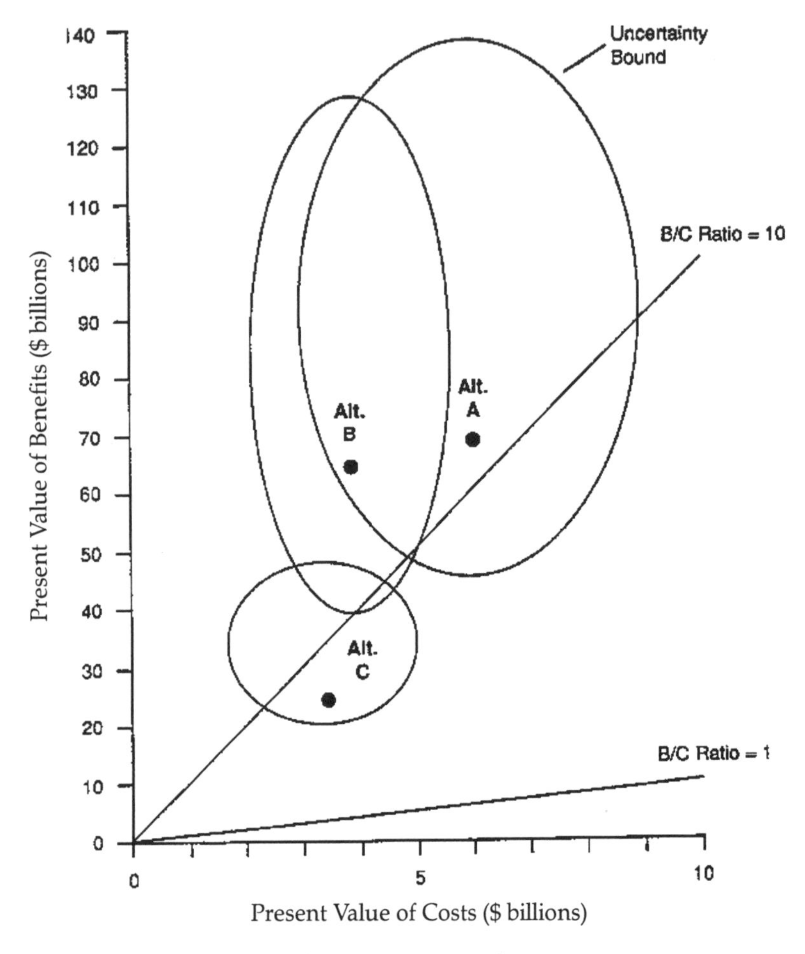

Figure 1. Benefits and Costs of Corrosion Control.

Source: U.S. EPA 1991, Exhibit 1-1

For lead pipe replacement, the RIA presented several scenarios using different specifications for high and low bounds for a base case and high case, for each of three alternative regulatory formulations, but no sensitivity analysis on any factors or assumptions was presented. No sensitivity analysis was done related to source-water contamination.

The 1986 benefit analysis did not include any sensitivity analysis.

Net Benefits

The 1991 RIA did not compile a section on the net benefits of the entire final regulation. Instead, the three components of the rule (source water treatment, corrosion control, and lead pipe replacement) were summarized separately.

The middle section—on corrosion control—is discussed most fully. Annual health benefits for the final rule were estimated at $4.3 billion, as outlined in Table 5; materials damage benefits from reduced corrosion were not included. The RIA presents a table of the costs and benefits of the rule for a twenty-year period, which estimates total present values of $63.8 billion for benefits and $4.2 billion for costs. In concluding this section, the RIA states that "the surplus of benefits over costs ... implies substantial room for error. In addition, the benefits estimates procedure employed in developing the analysis does not include a number of conceivably important categories of yet unquantifiable health effects as well as the potential benefits of extending the life of pipe materials." The diagram presenting the perceived uncertainty bounds on the benefit estimates (see Figure 1) indicates that even when all uncertainties are weighed, the net benefits of corrosion control under even the more stringent conditions evaluated exceed the costs by a factor of almost ten to one.

The net benefits of controlling lead in source water were inferred from the excess net benefits of corrosion control. The benefits of lead pipe

Table 5. Estimated Annual Health Benefits of Corrosion Control to Reduce Exposure to Lead in Drinking Water. ($1988)

	Cases avoided	*Benefits ($million)*
Adults		
Hypertension	635,199	$399
Heart attacks	818	$818[1]
Strokes	609	$609[1]
Deaths from heart disease	622	$1,555[2]
Children		
Medical treatment	NA[3]	$0.07
Compensatory education	NA[3]	$2
Lost IQ points	188,318[4]	$864[5]
Children with IQ < 70	738	$38[6]
Total		$4,300

[1]The unit cost estimate is based on qualitative assessments of willingness to pay, which is assumed to be $1 million per case.
[2]Unit cost estimate is $2.5 million/death.
[3]It was not possible to determine from the data presented in the RIA how many children were calculated to benefit.
[4]This is an estimate of lost IQ points, and not an estimate of the number of children suffering IQ decrements.
[5]Uses relationship of IQ to future earnings: each IQ point decrement is estimated to result in a 1.76% decrease in future earnings, including direct and indirect effects.
[6]Based on estimates of the costs of part-time, in-class compensatory education.

Source: U.S. EPA 1991, Exhibit 5-16.

replacement were not clearly positive; the RIA estimates annual benefits of pipe replacement ranging from $70 to $170 million, with costs estimated at $280–$370 million (the rule itself presents a range of $80–$370 million).

Discussions throughout the 1991 final RIA indicate that benefits were perceived to be underestimated and costs possibly overestimated.

In the 1986 benefit analysis, health and materials damage benefits were estimated at about $1 billion per year ($1985) and costs at about $230 million per year ($1985), for an estimated benefit-to-cost ratio of about four-to-one. In addition, a case study of potential benefits in the city of Boston was included in the 1986 analysis; the case study estimated a benefit-to-cost ratio of over ten-to-one for that city.

REFLECTIONS ON THE PROCESS AND THE ANALYSES

Role of Analysis in Decisionmaking for the Regulation

Quantitative analysis—in particular monetized costs and benefits—played an unusually prominent role in the 1991 regulation for lead in drinking water, as it did in the 1985 lead-in-gasoline rule. The gasoline regulation had left EPA sensitive to the adverse health impacts of lead exposure and receptive to restrictions on those exposures, especially if an economic argument for control was evident. Data on lead's adverse health effects continued to grow throughout the 1980s. The 1986 drinking water benefit analysis clarified the multimedia aspects of lead exposure, and EPA developed a multimedia approach to lead research and control, culminating in the development of the EPA multimedia lead strategy and a clustering of all potential lead regulations. Senior EPA officials viewed drinking water as a conspicuously large—and uncontrolled—exposure source.

The lead-in-drinking-water regulation published by EPA in 1991 was far more stringent than the Office of Drinking Water had originally intended and represented a major change in approach from previous drinking water regulations. How such a strict regulation became law most likely resulted from many factors, including the five described below.

Quality and Robustness of the Analysis. The scientific database on lead's health effects was extensive. In addition, the benefit methods had withstood both internal and external challenges and had been accepted by other federal agencies, such as CDC and the Department of Housing and Urban Development. Finally, the benefit analyses took a fairly cautious approach to quantifying benefits. Thus, although there were uncertainties around each of the estimates, the net benefits were substantial, even under pessimistic assumptions. Eliminating any one benefit category altogether, doubling costs, or halving all benefit estimates, for instance, still resulted in positive net benefits. Senior EPA officials had reasonable confidence in their decision to promulgate a stringent drinking water rule.

Widespread EPA Support. The 1985 lead-in-gasoline rule had left a cadre of EPA senior managers (administrator and deputy administrator, assistant administrators, office directors, and division directors) with an unusually strong background on lead and a commitment to reduce exposure through all vehicles available to EPA. This was imparted to incoming senior managers, so support remained high at the most senior levels. Support for an analytic approach and for a stringent multimedia approach to lead regulations also extended to lower levels of management and staff.

Analysis Preceding Decisions. The 1986 benefit analysis raised the possibility of significant public health and other economic benefits associated with a fundamentally different, stringent approach to regulation. Unlike most rules where the RIA is conducted at the end of the regulatory process to support (or defend) the finalized rule, this analysis was completed before key decisions were made and bureaucratic momentum inhibited changes.

In addition, strong analytical support remained available *within* EPA. As new information became available or new issues arose, additional analyses were performed quickly by staff known to senior managers. While the formal RIAs for the 1988 proposal and 1991 final rule were written by contractors, most of the analysis itself was done within EPA.

Widespread Public Support. During the period of these developments, there was public support for a stringent drinking water standard, as well as support from key environmental groups. There was also keen congressional interest, with relatively little opposition and one strong advocate. EPA's drinking water proposal immediately followed congressional efforts to address lead contamination of drinking water. External opposition was relatively localized and light.

Timing. The timing of the 1986 benefit analysis was fortuitous. It closely followed the 1985 lead-in-gasoline regulation, which was highly regarded within the agency. It even more closely followed the 1986 Amendments to the Safe Drinking Water Act, which had highlighted lead contamination of drinking water as a congressional concern. And finally, a window of opportunity had occurred in 1986 and 1987 when a new director was appointed to the Office of Drinking Water.

Lessons for the Future

In the case of this RIA, conditions were unusually favorable for analysis to influence policy, as was true for the lead-in-gasoline regulation. Nonetheless, there are some general lessons for the future.

First, a quantitative analysis is more likely to have a positive impact if it is done early in the process, even if the results suggest that a new approach is warranted. Analyses to defend decisions that have already occurred will have little impact. If the analysis is sufficiently robust and done early, it can engender significant programmatic change.

Second, a regulation that has a good public health and economic basis can rally support both inside and outside the agency.

Limitations of the Process and Recommendations

Current analytical tools, including benefit-cost analysis, are still in their infancy. Serious gaps in data and methodology constrain the utility of these tools. Typically, only a few potential health or other benefits can be quantified, and even fewer can be valued monetarily. Consequently, when the sum of the limited subset of benefits that can be quantified and monetized is shown to be less than the estimated costs, it is often impossible to conclude anything about the relative magnitude of the full benefits. *The two lead regulations discussed in this volume are anomalies.*

EPA should support the improvement of existing valuation methodologies, including cost-of-illness studies for noncarcinogenic health endpoints, as well as the development of new valuation methods for health and other endpoints.

In addition, current methods are clearly not yet capable of fully addressing many risk-risk trade-offs and other fundamental complications or competing needs. This is particularly relevant to the lead-in-drinking-water regulation, as altering water quality to attain "optimal" corrosion control for lead may adversely affect other treatments, such as drinking water disinfection. Some of these issues were considered at different points in the analysis (for instance, EPA's cost model considers existing water treatments), but neither comprehensively nor systematically. Such sophistication will develop as the tool is more extensively used.

Finally, to evaluate its estimation methods EPA should collect data on actual costs of regulations and, where possible, the anticipated benefits.

ENDNOTES

[1]1 mg/l = 1,000 micrograms per liter (μg/l) = 1 part per million; so 0.05 mg/l = 50 μg/l = 50 parts per billion.

[2]From 1985 on, the rules for lead and copper were linked because both are corrosion by-products, and monitoring, analysis, treatment, and other such activities will be similar or identical for both. Only the details relevant to the lead component are discussed in this chapter.

[3]This means, for instance, that a water system with a 90th percentile exceeding 0.015 mg/l had to implement corrosion control procedures but was not violating the standard. Only if the system failed to implement those procedures was it violating the standard. Compliance/violation is "enforceable" and has significant legal and public perception implications.

[4]EPA determined that although lead is a likely human carcinogen, its non-cancer health effects occur at much lower levels and, in fact, show no evidence of a threshold.

[5]OPA was within the Office of Policy, Planning, and Evaluation (OPPE) in 1986. In 1995, OPA functions were reorganized within OPPE.

[6]It should be noted that discrepancies exist, such as in cost estimates, between the RIA and data in the final rule itself. Such discrepancies, combined with the complexities of the presentation of data in the 1991 RIA and departures from accepted methods, have led many to continue using the 1986 benefit analysis in their studies. Two such studies are the benefit-cost analyses conducted to support the Lead Strategy of the Centers for Disease Control and the National Academy of Sciences Committee on Lead Exposure.

[7]The dose-response curve for children was derived from an unpublished study (Maes and others 1990–91) at the National Centers for Disease Control. The Maes study underwent several revisions. The 1991 RIA relied on a 1990 version of the study. Subsequent revisions by Maes and coworkers estimated dose-response curves that were considerably lower. In retrospect, it appears that the final rule (56 *Federal Register* 26542) erroneously referred to a 1991 version as the basis of the RIA. The review and revisions of the Maes manuscript continued through the period of the promulgation of this rule, but the paper was never published.

REFERENCES

ASDWA (Association of State Drinking Water Administrators). 1990. *State Costs of Implementing the 1986 Safe Drinking Water Act Amendments.* August. Washington, D.C.: ASDWA.

AWWA (American Water Works Association). 1989. *Lead Information Survey.* Denver: AWWA.

——. 1990. *Lead Service Line Replacement: A Benefit-to-Cost Analysis.* Prepared by Roy F. Weston, Inc. and Economic and Engineering Services, Inc.

AWWA-RF (AWWA Research Foundation). 1989. *Economics of Internal Corrosion Control.* Prepared by Economic and Engineering Services, Inc. and Kennedy/Jenks/Chilton, Inc. Denver: AWWA-RF.

———. 1990. *Lead Control Strategies.* Prepared by Economic and Engineering Services, Inc. in association with the Illinois State Water Survey. Denver: AWWA-RF.

Maes, E.S., L.A. Swygot, D.C., Pascal, and B.S. Anderson. 1990–91. *The Contribution of Lead in Drinking Water to Levels of Blood Lead: I. A Cross-sectional Study.* Atlanta: Centers for Disease Control. Unpublished.

Patterson, J.W. 1981. *Corrosion in Drinking Water Systems.* Report prepared for U.S. EPA Office of Drinking Water. Washington, D.C.

RTI (Research Triangle Institute). 1987. *Final Descriptive Summary: 1986 Survey of Community Water Systems.* Prepared for U.S. EPA, Office of Drinking Water. Research Triangle Park, North Carolina: RTI.

U.S. EPA (Environmental Protection Agency). 1986. *Air Quality Criteria Document for Lead.* (Volumes 1–4 and Addendum). June. Research Triangle Park, North Carolina: Environmental Criteria and Assessment Office, Office of Research and Development, U.S. EPA.

———. 1987. *Reducing Lead in Drinking Water: A Benefit Analysis.* Prepared by R. Levin, December 1986, revised Spring 1987. EPA 230-09-86-019. Washington, D.C.: Office of Policy, Planning, and Evaluation, U.S. EPA.

———. 1989. *Technologies and Costs for the Removal of Lead and Copper from Potable Water Supplies.* April. Washington, D.C.: Office of Drinking Water, U.S. EPA.

———. 1990. *Influence of Plumbing, Lead Service Lines, and Drinking Water Treatment of Lead Levels at the Tap.* August. Washington, D.C.: Office of Drinking Water, U.S. EPA.

———. 1991. *Final Regulatory Impact Analysis for National Primary Drinking Water Regulations for Lead and Copper.* Prepared by Wade Milks Associates and Abt Associates for U.S. EPA, Office of Drinking Water. April. EPA contract 68-C0-0069.

9

Municipal Landfill Management

Sara Rasmussen

In 1988, EPA concluded that existing criteria for municipal solid waste landfills (MSWLFs) were not sufficient to protect human health and the environment. As part of its process for formulating additional criteria, EPA conducted a regulatory impact analysis to assess the costs and benefits of various regulatory options. The scope of the RIA expanded over time, to accommodate increasing concerns about the costs of regulation. In 1991, after much debate, EPA promulgated the Subtitle D (Municipal) Landfill Criteria to establish design and management requirements for MSWLFs. A number of RIA-related issues were given significant attention during final rule development, including those associated with the true incremental costs of the rule (additional costs beyond those already imposed by state and local governments), the protectiveness of the statutory minimum option (that relied on corrective action to clean up groundwater contaminated by landfills), the technical feasibility and cost of groundwater cleanup at various types of landfills, the economic impact on small communities, the value of clean groundwater, and the level of benefits (including nonquantified, noncancer-related benefits) associated with various regulatory options. The extent to which the RIA influenced the final rule is still debated within the policy community, with Office of Management and Budget staff arguing that its role was minor

---

During the final years of promulgation of the Subtitle D (Municipal) Landfill Criteria, SARA RASMUSSEN was responsible for completion of the regulatory impact analysis. She was a member of the Nonhazardous Waste Section of the Regulatory Analysis Branch in EPA's Office of Solid Waste. The opinions expressed in this chapter are those of the author and do not reflect those of any organization.

(since the most cost-efficient option according to the RIA was not chosen) and EPA analysts arguing that it was significant (at least in identifying ways to keep the costs of the final rule down). At any rate, it is clear that political considerations dominated the decisionmaking process. In this case study, Sara Rasmussen explores the many issues surrounding the development of the rule, discusses the scope and use of the economic analysis, and critiques specific aspects of the MSWLF analysis.

INTRODUCTION

The Subtitle D (Municipal) Landfill Criteria, issued in 1991, were the culmination of seven years of study and work. In considering the final Subtitle D Criteria, it is helpful to review the driving forces that contributed significantly to their development.

In 1976, Congress passed the Resource Conservation and Recovery Act (RCRA). RCRA established a framework for managing both hazardous and solid (that is, nonhazardous) waste. While RCRA provided for extensive federal involvement in the management of hazardous waste (Subtitle C), RCRA called for a much more limited role concerning nonhazardous waste (Subtitle D). For Subtitle D, Congress banned "open dumping" (as defined in RCRA, Section 4004) nationally but left the remainder of management issues to the states. States were required, however, to develop and receive EPA approval for "state solid waste management plans." As a result, in 1979, EPA published the *Criteria for Classification of Solid Waste Disposal Facilities and Practices* (40 CFR Part 257; U.S. EPA 1979). These criteria set forth the requirements for defining "open dumps" and established the "guidelines" for state plans.

Legislative Requirements

In 1984, Congress passed the Hazardous and Solid Waste Amendments (HSWA). HSWA (Section 4010) directed EPA to prepare a report for Congress by November 1987, indicating whether the current criteria were adequate to protect human health and the environment from groundwater contamination and recommending whether additional enforcement authorities were needed.

Section 4010 also directed EPA to revise (by March 1988) the criteria promulgated for facilities that may receive *hazardous household waste or small-quantity generator* hazardous waste. Such facilities presumably included all landfills accepting municipal solid waste. The revised criteria were to protect human health and the environment and, at a minimum, to require groundwater monitoring, establish criteria for the acceptable location of new or existing *municipal solid waste landfills* (MSWLFs), and provide for corrective action as appropriate. Finally, in developing the criteria, EPA was allowed to take into account the "practicable capability" of MSWLF owners/operators.

Section 4005 of HSWA required each state to "adopt and implement a permit system or other program of prior approval and conditions" to ensure that units receiving household hazardous waste or small-quantity generator waste comply with the revised criteria. Each state's program was to be approved by the administrator of EPA.

To meet the HSWA requirement, EPA completed a study on the adequacy of the existing criteria for municipal solid waste landfills to protect human health and the environment. In 1988, the agency completed *A Report to Congress: Solid Waste Disposal in the United States.* The report concluded that existing criteria were not sufficient to protect human health and the environment, and that additional criteria were needed. The report concluded that "federal and some state solid waste regulations lack the following essential requirements ... location criteria, appropriate design criteria, groundwater monitoring, corrective action, closure, post-closure care and financial responsibility (U.S. EPA 1988), ES2)."

Old Landfills on Superfund List

As of May 1986, 184 of the 850 Superfund sites listed or proposed for listing on the National Priority List were MSWLFs. In reality, many of these old municipal landfills probably accepted hazardous waste as well, making it difficult to determine the hazards of municipal landfills versus hazardous waste landfills.

Public Concerns

Public concerns surrounding the management of waste and MSWLFs shifted significantly in the 1980s. Public sentiment grew from a semi-indifferent attitude to one of active public crusades against new landfills.

Potential Risk. Throughout the 1980s, public concern regarding the environmental and health impacts of landfills increased. Potential health effects, environmental impacts, and costs of cleaning up Love Canal, Times Beach, and other hazardous waste sites were highly publicized. In 1991, a Roper poll showed that over 50% of respondents cited solid waste as a very serious environmental problem.

Not In My Backyard. A general theme of "not in my backyard" emerged. Several communities took action to fight the siting of new landfills in their neighborhoods, making siting landfills extremely difficult in some areas. Some communities gave up trying to site a landfill and opted instead to transport their waste great distances to larger, regional landfills rather than confront the expense and political nightmare of siting new landfills within state. In the 1980s, the highly publicized "Garbage Barge" incident occurred, where the Marbor barge, containing garbage from New York, was turned away from numerous ports.

Concern about Landfill Capacity. In the late 1980s, articles began to flood the news about the garbage and landfill "crisis." According to the 1986 landfill survey, fewer new landfills were being opened each year. There were an estimated 18,500 municipal solid waste disposal units operating in the United States as of 1974 (Waste Age 1977, 26). This number had dropped to just over 6,000 by 1986. However, the size of new landfills was increasing. At the same time, the quantity of waste disposed was increasing. In 1988, EPA stated in a report to Congress (U.S. EPA 1988) that the nation generated nearly 180 million tons of municipal solid waste and that quantity would likely grow to 216 million tons by the year 2000.

Costs. Increasing MSWLF management costs also raised concern. The 1979 criteria that required "open dumps" to be closed and restricted some of the most questionable management activities, such as open burning, were having some effect. Many small "dumps" had closed rather than increase standards. In addition, several states established their own landfill criteria, requiring liners and groundwater monitoring. The cost of these additional controls was alarming to landfill owners. Tipping fees (the fee for dropping waste off at landfills) were rising, and there was concern that fees would rise to unaffordable heights in some areas of the country.

Changing State and Local Controls. The universe of MSWLFs was significantly changing. State and local landfill requirements expanded continuously during the late 1980s, leaving EPA with outdated information on actual controls and designs of landfills. By 1988, the draft regulatory impact analysis (RIA) reported that fifty of the states and territories had some form of standards. By the time of the final rule promulgation in 1991, many states had adopted more stringent controls.

Political Concerns

Federalism versus State Control. There was a strong sentiment in parts of Congress and the administration in the 1980s that federalism should be minimized where possible. At the same time, some states were looking to the federal government to set minimum standards to aid them in their MSWLF programs.

Antiregulation Sentiment. During the 1980s, there was an increasing clamor to restrict burden from all aspects of the federal system and to back away from "command and control" regulation. Reports such as "Project 88" and proposals calling for "economic incentives," flexible regulations, and voluntary industry efforts were prominent in federal policy debates. In this environment, it became clear that EPA would have a harder time justifying any type of restrictive regulation.

Impacts on Small Communities. Impacts of federal regulations on small local governments and tribes were a major concern in the late 1980s and early 1990s. Several small entities were being financially overwhelmed from a number of regulatory requirements, including safe drinking water criteria, sewage treatment controls, and other federal requirements. There was a strong concern that many local governments would go bankrupt if they were forced to meet all the current federal requirements. EPA initiated a project to look at cumulative economic impacts of regulations on small communities.

ANALYTICAL CHALLENGES

EPA faced many obstacles in trying to develop the RIA for the Subtitle D criteria. The largest obstacle, by far, was the general lack of data. Other obstacles included limited resources, the state of the art of risk assessment tools and methodology, quantification of the benefits, and the debate over the value of groundwater resources.

Lack of Data and Resources

The most prevalent and obvious limitation early on was the lack of data available for use in the analyses. Municipal waste management was primarily a local government service, and the federal government did not collect data on municipal landfills in any systematic way (nor does it now). To develop a site-specific database on landfills, EPA conducted a one-time survey of 1,250 landfills, the Subtitle D [municipal] landfill survey, in 1986.

While the final rule was promulgated several years after the project was initiated, there were several deadlines set early on that placed limitations on the analysis. For example, first runs of the risk and cost analyses were needed before all the survey results were received and analyzed. Thus, EPA ended up using the first complete 701 responses and extrapolating those up to the full universe estimated to be 6,034 landfills. EPA gathered currently available leachate data (time and expense prohibited an extensive testing program) from only forty-four landfills.

Issues that arose in the final stages of the rulemaking were not analyzed rigorously due to time constraints in getting the rule out and due to limitations in the analytical tools developed earlier. Such a lack of analytical rigor was true especially for the studies of small landfills. EPA did not know until after the 1986 survey results were in, how many very small landfills (1 ton per day [TPD] or less) were in the system. EPA designed its cost methodology early on, and the smallest landfill costed out by the analyses was 10 TPD. If there had been more time, a more rigorous analysis of costs for the very small landfills could have been conducted.

While significant resources were dedicated to this project, resource constraints still played a role in limiting the amount of data collected and the amount of outside analyses conducted. Particularly during the final phases of the rule development, questions were raised that were beyond the scope of the current analysis (such as the quantification of additional benefits of the rule), and a rigorous analysis was not conducted. Some of the recommended improvements to the analysis were not implemented because they would have required additional data collection, model runs, and analysis taking too much time and too many resources. The MSWLF rule was one of many being worked on at EPA, and the distribution of resources, both funding and staffing—in terms of full-time employees—was a continual juggle.

State-of-the-Art Risk Assessment Tools

In 1985, when the RIA was initiated, only a limited number of risk-modeling tools were available. In the 1980s, EPA was expanding the state of the art for risk assessments,[1] and much changed from 1985 to 1991. The risk analysis conducted for the RIA was a major effort at the time. EPA spent a considerable amount of time and number of resources to develop the Subtitle D Risk Model used in the RIA (a mainframe model), which adapted the liner/location model.

There continue, however, to be numerous uncertainties associated with risk assessments. Risk assessments, which include fate and transport through groundwater, rely upon countless numbers of assumptions to represent complex processes. Risk assessment was not (and still is not) developed to the point where science can provide exact, complete estimations of human health risks from all waste constituents, or risks at particular sites. While the science is continually emerging, the scientific community clearly was not in complete agreement regarding several aspects of risk assessments. For instance, there were several methods that could be used to develop carcinogenic dose-response curves. There were and continues to be incomplete data on the toxicity of numerous wastes and chemicals. Computer modeling of the volume, fate, and transport of releases from landfills through the ground to the groundwater system, and other aspects of risk assessments, are simplifications of nature and were still being developed. Few models in the Office of Solid Waste had been verified in the real world.

The uncertainties surrounding the "results" from all risk assessments can cloud issues during the decisionmaking process. When there is uncertainty regarding the toxicity of a chemical or waste, incomplete estimates or bounded estimates (providing a range of potential risk) are often given. Reactions to and interpretations of these uncertainties and risk ranges are divergent. The risks presented in the MSWLF RIA were based on several simplifying assumptions, for instance, an assumption that cleanup can be effective at all sites. (In reality, we know that cleanup is not possible at some sites.) It was clear that the perceptions of real risks associated with MSWLFs varied dramatically among those working on the MSWLF rule both within the agency and elsewhere.

Quantifying Benefits

The RIA initially quantified two types of benefits (that is, avoided impacts from landfills): reduced cancer cases caused by drinking groundwater contaminated from landfills and reduced costs associated with replacing water for drinking wells that were contaminated. There were several other benefits that people associated with the rule that were not quantified in the RIA. Because they were not quantified, they were not included in benefit-cost comparisons for the rule. The emphasis placed by management on the benefit-cost comparisons grew during the rule development period. EPA is continually challenged with the task of quantifying a variety of social and environmental impacts associated with rules.

Groundwater Resource Value

In 1984, EPA issued the Groundwater Protection Strategy, which established the concept of differential protection of groundwater depending on its resource value. In the preamble for the 1988 MSWLF proposal, EPA discussed its decision to "establish facility design criteria that give States the flexibility to address the value of groundwater resources in setting facility-specific design requirements." It also proposed "allowing consideration of the resource value in the corrective action and to a lesser extent, the groundwater monitoring components" of the proposed rule (53 *Federal Register* 33324). By 1990, when the final rule was under management review, the value of groundwater continued to be an issue of controversy. Several in the Office of Solid Waste believed all groundwater should be protected. Others in EPA and the Office of Management and Budget (OMB) questioned this approach, maintaining that some aquifers were of low quality and value and did not require equal protection. There was never an agreed-upon system or methodology for valuing groundwater, and the importance of protecting groundwater continued to be debated through the final decisionmaking process.

SUMMARY OF THE PROPOSED AND FINAL RULE

Several types of standards were considered in developing the MSWLF criteria, including:

- location restrictions;
- standards for the actual design (uniform or performance based) of liners (synthetic or clay), covers (vegetative, synthetic, or clay), and leachate collection systems;
- general operating practices (gas monitoring, run-on/runoff controls, access control, disease control);
- groundwater monitoring (hydrogeologic study, well construction, semiannual sampling);
- closure/postclosure care of the cover and the groundwater monitoring (ten to forty years); and
- corrective action.

Of these, location restrictions, groundwater monitoring, and corrective action were statutorily required. The variance between the regulatory options was primarily with regard to the design standards. Options included an across-the-board "uniform" design requirement, a design requirement based on a performance standard—landfills would have to ensure that the groundwater was not contaminated at a *point of compliance* (POC)—and a categorical approach that categorized landfills by particular characteristics (net infiltration, depth to groundwater, and so forth) and thereby varied design requirements accordingly. In addition, there was a "statutory minimum" option that did not include any design criteria, relying on corrective action to clean up contamination. These options are highlighted in Table 1.

While the categorical approach played a strong role during proposal (it was initially the option of choice by EPA management), it did not receive as much attention during development of the final rule primarily

Table 1. Regulatory Options Considered in Analyses.

Prevention oriented	
Final Rule	Mix of uniform design and MCL performance standard Good management practices Groundwater monitoring Corrective action Closure/postclosure care Location standards
1988 proposal	Risk-based performance standard (landfill must meet state chosen design goal of risk levels between 1 in 10,000 and 1 in 10,000,000) Good management practices Groundwater monitoring Corrective action Closure/pose-closure care Location standards
Subtitle C	Double liner, composite cover Good management practices Groundwater monitoring Corrective action Closure/pose-closure care Location standards
Categorical	Performance design criteria based upon categories of environmental settings Good management practices Groundwater monitoring Corrective action Closure/postclosure care Location standards
Reaction oriented	Statutory minimum Groundwater monitoring Corrective action Location standards
Options presented in 1988 RIA	Proposal: Same as noted above Alternative 1: Similar to Subtitle C discussed above Alternative 2: Similar to categorical approach discussed above Alternative 3: Similar to statutory minimum discussed

because of the high cost and the inability to scientifically justify category boundaries. By 1990, the two major options being considered for the final rule were a hybrid of the uniform design and the performance standard approach, which was chosen as final criteria, and the statutory minimum, an option preferred by OMB and some EPA offices.

RCRA required the Subtitle D program to be delegated to the states; it was to be a state-run program. The federal government did not expect to implement the program, to participate in permitting individual facilities, or even to mandate all requirements in each state. Instead, states were to submit program plans to EPA for approval. Authorized states would then implement the federal criteria on their own. For those states that were not approved, the landfill would be responsible for self-implementation. This state emphasis made the Subtitle D program very different from the Subtitle C program.

THE REGULATORY IMPACT ANALYSIS

The draft regulatory impact analysis was one of the most expensive RIAs conducted in the Office of Solid Waste at that time. Initial work began in 1985 with data collection and the design of the 1986 survey of Subtitle D facilities. During the course of the next five years, numerous EPA staff and contractors provided inputs, including Sabotka, ICF, Pope-Reid Associates, Inc. (later called DPRA), Temple Barker and Slone/Clayton Environmental Consultants, and American Management Systems, Inc.

A draft analysis was prepared for the proposed rule, published on August 30, 1988. The draft RIA analyzed five options (listed above). Technical background documents regarding data, risk modeling, and cost modeling were submitted to the public docket. A final analysis and an addendum were provided for the publication of the final rule. Technical background documents were also submitted to the docket at that time.

Over $1 million was spent on this project from start to finish. Major contractor efforts included data collection on waste, site characteristics, and cost and financial information; the development of the Subtitle D risk model; and the development of a Subtitle D cost model.

The RIA investigated the engineering cost of the regulatory alternatives and economic impacts on the regulated entities (municipalities and households), and examined human health risks and resource damages due to releases from landfills (the avoided risks and resource damages resulting from the regulatory options were listed as benefits of the rule).[2] Portions of each analysis are described in greater detail below.

In addition to meeting the criteria of Executive Order 12291, EPA was also required to meet the criteria of the Regulatory Flexibility Act. This act requires federal agencies to analyze the impact of regulations on small entities and consider less burdensome alternatives. For the purposes of this rule, since the majority of landfills (85%) were owned by municipalities or governmental entities, EPA included an analysis of impacts on households and municipalities for small communities.

Because the landfill universe was changing over time, the analyses were updated to reflect these changes. However, while the final draft of the RIA, completed in 1990, incorporated some of these changes, it was still not deemed representative of the current universe. For that reason, an addendum was prepared during 1991 to account for additional changes in the landfill universe. The primary changes from the RIA to the addendum were the incorporation of the increase in state and local landfill requirements, which had the effect of reducing the federal incremental cost of the regulation.

General Methodology

The RIA comprised four basic components: data collection and universe characterization, cost analyses, economic impact analyses, and risk assessment.

Data Collection and Definition of the MSWLF Universe. Most of the information on the MSWLF universe came from the 1986 survey (discussed earlier). The RIA is based on results from the first 701 survey responses. Survey data included factors such as size of landfill, expected life span, ownership, design, existence of drinking wells within a mile of the facility, and hydrogeology. To augment this information, EPA gathered data from the Geographic Information Service, weather stations, and United States Geological Service maps. The 701 survey responses were weighted according to size and location and extrapolated up to the total universe.

EPA developed leachate composition estimates by analyzing leachate data covering 212 chemical constituents at 44 operating landfills. Data came from New Jersey landfills, landfills surveyed for the presence of organic chemicals in Minnesota, and landfills submitting leachate sampling results to Wisconsin Natural Resources Department.

As mentioned earlier in this chapter, the survey indicated that over 50% of the MSWLFs in operation at the time planned to close by 1997; 80% of all MSWLFs expected to close by 2007.

Most landfills were small. Over 50% of MSWLFs received less than 17.5 TPD; over 80% received less than 125 TPD. However, large landfills (3% of all MSWLFs were greater than 1,125 TPD) handled 40% of all the waste.

The majority (over 80%) were owned by local governments (counties, cities, towns, and villages). Approximately 15% of the landfills were privately owned. The other 4% were owned by federal and state governments and universities.

Fewer new landfills were being opened each year. In the early 1970s, between 300 and 400 new landfills were being built each year. During the 1980s, however, the number dropped to between 50 and 200 new landfills.

The number of landfills was declining; in 1977, *Waste Age Magazine* (1977) estimated there were 18,500 municipal solid waste disposal units operating in the United States as of 1974. This number dropped to just over 6,000 by 1986. However, the amount of waste generated per person increased 48% between 1960 and 1988 and was expected to continue to increase (U.S. EPA 1990, ES09).

At the time of proposal, the preamble noted that according to the state census existing landfills were "distributed throughout the country, occurring in virtually every hydrogeologic setting and generally concentrated near more populated areas" (53 *Federal Register* 33318).

The preamble also noted that only 15% of landfills were designed with liners (natural or synthetic) and only 5% had leachate collection systems. Current data also indicated that only 25% to 30% of MSWLFs had some type of groundwater monitoring system.

While many landfills had residences and drinking wells (both public and private) within one mile of the facility, the majority of MSWLFs, 54%, reported having no drinking wells within this periphery.

Engineering Costs of Rule. Engineering costs were developed for each component of the requirements (liners, covers, groundwater monitoring wells, and so forth) for a variety of landfill sizes. These were then compiled into costs for each landfill category, weighted, and compiled into national cost estimates for a variety of options being considered for the proposal and final rule. Capital costs and operating and maintenance costs were developed for liners, covers, leachate collection systems, groundwater monitoring, operating practices, closure, ten to forty years of postclosure care, resource damage (cost of replacing drinking water if contaminated), and corrective action. While some factors (such as labor rates) varied significantly across regions, national averages were used.

Algorithms were developed for seven landfill sizes (10 TPD, 25 TPD, 75 TPD, 125 TPD, 375 TPD, 750 TPD, 1,500 TPD) and five landfill designs (clay liner/vegetative cover, clay liner/synthetic cover, synthetic liner/synthetic cover, composite liner/synthetic cover, Subtitle C liner/cover). These algorithms were used to develop cost curves to estimate costs based on the reported size of each landfill in the survey. In general, economies of scale played a large role in the *cost per ton* (CPT), with the CPT for large landfills being affordable, even for the highly designed Subtitle C option. The small landfills, however, had much higher CPT, even for the lower designs (see Table 2).

Some of the regulatory options were performance based; that is, a liner was required if needed to prevent contamination of groundwater. Depending on the specific site characteristics, the landfill controls needed could vary from a natural cover to a very stringent liner/cover system. To

Table 2. Estimated Total Cost per Ton for New MSWLFs. (20-year life, 40-year postclosure care period)

Landfill design	*10 TPD*	*175 TPD*	*750 TPD*
Subtitle C (off-site clay)	$136	$41	$21
Composite design (off-site clay)	$105	$31	$16
Synthetic liner Synthetic cover	$92	$25	$12
Unlined Vegetative cover	$71	$16	$7

Notes: Prepared using Subtitle D Cost Model results developed for MSWLF RIA. *Subtitle C*—Two synthetic liners, clay liner, double leachate collection systems, composite cover system, groundwater monitoring. *Composite design*—Composite liner, leachate collection system, synthetic cover, groundwater monitoring. *Synthetic liner, synthetic cover* incudes leachate collection system and groundwater monitoring. *Unlined, vegetative cover* includes groundwater monitoring.

develop national costs, EPA assigned a landfill design to each landfill that would be "protective" based on the outputs of the risk modeling runs (discussed in more detail under the Risk Assessment section).[3]

Costs for both existing landfills and new landfills were estimated. Existing landfills were not assumed to retrofit (install liners) but to conduct closure (install a cover that limited infiltration), install groundwater monitoring (if the landfill remained open past date of implementation), conduct postclosure care for ten to forty years (depending on the regulatory option), and respond to resource damages.

Because most landfills were owned by government entities, EPA used a real discount rate of 3% in determining the present value of cost streams.

New landfills were assumed to operate for twenty years. Existing landfills were assumed to close based on life expectancy reported in the survey. It was assumed that landfills that remained open past the date of implementation could amortize the costs of the existing landfill over the life span of the existing landfill and one replacement landfill. (For instance, if the existing landfill had an expected life span of five years, it was assumed the cost of groundwater monitoring, closure, and postclosure care would be amortized over the five years plus the twenty years of the replacement landfill—twenty-five years in all).

The analysis indicated that the annualized cost of the regulatory options was quite high (Table 3). Annual costs ranged from $420 million for the proposal to $3.3 billion for the categorical approach (Alternative 1). The costs were reduced significantly for the RIA for the final rule. These reductions were the result of incorporating existing state requirements into the baseline and revising the requirements. The upper-bound costs ranged from $760 million to $1 billion for the final RIA. The addendum, which only costed out two options, the final rule and the statutory mini-

Table 3. Annual Incremental Costs of Regulatory Options. (millions of dollars)

	Annual incremental cost
1988 draft RIA for proposal	
Proposal: 10-meter POC	$880
Proposal: 150-meter POC	$690
Alternative 1 (Subtitle C)	$3,340
Alternative 2 (Categorical)	$1,430
Alternative 3 (Statutory minimum)	$420
1990 RIA for final rule (upper bound)	
Final Rule	$1,040
Alternative 1 (proposal)	$970
Alternative 2 (Subtitle C)	$2,840
Alternative 3 (Categorical)	$1,230
Alternative 4 (Statutory minimum)	$760
Addendum (best estimate)	
Final rule	$330
Statutory minimum	$180

mum, again showed a reduction in costs. This was due primarily to incorporating more state requirements into the baseline. The final rule also included a small community exemption that reduced impacts significantly. The range of costs for the addendum was $180 million to $330 million.

Economic Impacts of Rule

Impact Measures. Extensive efforts were put into developing the economic impact analysis for the RIA. This was the first rule in the Office of Solid Waste (OSW), where the primary regulated entities were municipalities, not members of the private sector. Because most landfills were publicly owned, the impact analysis focused on two areas: impacts on municipalities and impacts on individual households. OSW had never conducted an impact analysis for municipalities and turned to the construction grants program for methodologies. The economic impact analysis examined three measures:

- Total annualized cost as a percentage of community expenditures (CPE)—EPA used a threshold of 1% of total community expenditures to indicate a "significant" impact.
- Total annualized cost per household (CPH)—EPA used a threshold of $210 per year (1% of the national median household income) to indicate a "significant" impact.
- Total annualized cost per household as a percentage of median household income (CPMHI)—EPA used a threshold of 1% of median household income of a community to indicate a "significant" impact.

EPA collected financial data on municipalities from a number of sources, including the 1982 Census of Governments and the *1983 City and County Data Book*. Of the more than 80,000 entities listed, EPA estimated there were about 29,000 primary[4] county, municipal, township, and other government entities and mapped these to one of the 6,034 landfills.

The impact analysis indicated that most communities would not suffer significant impacts from the regulation (see Table 4). The 1991 final RIA addendum estimated that only 392 of the 29,017 jurisdictions would have a CPE greater than 1%. It was estimated that no jurisdictions would have a CPH greater than $100 (U.S. EPA 1991a, IV2–3). The 1990 final RIA, which did not account for the most recent state requirements incorporated into the addendum, indicated a higher impact with regard to CPE: the percentage of communities with a CPE above 1% ranged from 14% for the statutory minimum to 64% for the Subtitle C option. The impacts presented in the 1988 draft RIA for the proposal showed similar CPE impacts ranging from 11% to 68% for the Subtitle C option (Alternative 1).

The average CPH was quite small. It ranged from $5 to $40 in the 1988 draft RIA to $2 to $4 in the addendum. However, the cost was not distributed evenly across households. The maximum CPH ranged from $335 per year for the categorical approach in 1988 to $52 for the statutory minimum approach in the addendum.

Table 4. Cost per Household and Cost per Community Expenditure.

	Percentage of communities with CPE greater than 1%	*Percentage of communities with CPH greater than $100*	*Average cost per household*	*Maximum cost per household*
1988 Draft RIA for proposal				
Proposal: 10-meter POC	16	0.2	$11	$119
Proposal: 150-meter POC	11	2	$8	$253
Alternative 1 (Subtitle C)	68	23	$40	$335
Alternative 2 (Categorical)	33	0.1	$17	$160
Alternative 3 (Statutory Minimum)	10	0.2	$5	$178
1990 RIA for final rule				
Final rule	27	0.1	$12	$175
Alternative 1 (proposal)	29	0.1	$11	$137
Alternative 2 (Subtitle C)	64	20.1	$3	$335
Alternative 3 (Categorical)	36	0.1	$14	$158
Alternative 4 (Statutory minimum)	14	5.9	$9	$361
Addendum				
Final rule	1.4	0	$4	$62
Statutory minimum	1.4	0	$2	$52

Financial Strength of Communities. In addition to looking at CPE, CPH, and CPMHI, EPA developed a complex system for determining and classifying the financial strength of communities. EPA used information from the *1983 City and County Data Book* to compile information on eight variables to compute the financial capability score for each government (see Table 5).

A unit score (1–5) was assigned for each indicator, based on the value of its indicator relative to all other municipalities (usually by the method of 5 for the top 20%, 4 for the next 20%, and so forth). Coverage ratio, population growth, and net overall debt per capita scores were handled differently. The composite financial capability score was developed by averaging the scores for all eight indicators. The final determination was made by rating the bottom 25% as weak, the middle 50% as average, and the top 25% as strong. This analysis was presented in the draft RIA for the proposal and received comments. While it was repeated in the final analysis, it did not receive significant attention from either EPA management or OMB and was not used during policy decisions for the final rule.

Regionalization and Waste Diversion Analyses. The universe of landfills was shifting yearly; several small landfills were closing. To incorporate this trend into the final rule RIA, EPA developed a regionalization analysis. This analysis identified which types of small landfills currently were, or would become (postregulation), more expensive than shipping to larger but more distant alternatives. The analysis compared the costs of the existing landfill to the cost of transporting the waste to a larger landfill further away that, because of economies of scale, had lower costs per ton. If the small landfill was not competitive, EPA assumed that when it closed (life spans reported in the survey), it would not be replaced by a new small landfill. Instead, the community was assumed to switch to a "regionalized landfill." EPA's analysis indicated that all landfills smaller than 175 TPD were less economically efficient and would eventually shift

Table 5. Measures Used by RIA to Indicate Financial Strength of Communities.

Financial management	Total taxes as a percentage of median household income
	Current revenues as a percentageof current expenses
	Coverage ratio (funds available for debt service payments)
	Operating expenditures as a percentageof median household income
Debt	Net overall debt per capita (direct and overlapping debt)
Demographic	Median household income
	Population growth from 1970 to 1980
	Fraction of the population below poverty level

to larger units. Over time, EPA estimated that 20% of municipal waste would be shifted to larger landfills due to regionalization.

EPA also included a waste diversion analysis for the final RIA to incorporate shifts in waste away from landfills due to source reduction, recycling, and combustion. Based on EPA projections on disposal trends, the 1990 RIA estimated an 18% decrease in the volume of landfilled waste twenty years after the effective date of the rule (U.S. EPA 1990, ES09).

Regulatory Flexibility (Impact on Small Communities). EPA looked at the impact of the criteria on small communities for the regulatory flexibility analysis (RFA). The RFA guidelines define a small government as an entity with a population of 50,000 or fewer and well over 90% of the 29,000 primary governments had populations smaller than 50,000. Thus, EPA chose to consider population cutoffs of 5,000 (more than 75% of the governments had populations smaller than 5,000) and 1,000 (45% of the governments have populations smaller than 1,000). Some of these communities already used regional landfills or, based on the regionalization analysis, were assumed to shift to them in the near future. Other communities had small landfills, which, due to economies of scale, had relatively high costs per ton and costs per household. EPA identified a "significant" impact on these small communities (that is, the cost per household being above 1% of median household income). This definition of impact, 1%, was questioned by OMB, but it was not changed.

As mentioned earlier, there had been great concern regarding small communities, within both EPA and Congress. Thus, EPA decided to incorporate design and groundwater monitoring exemptions in the final rule for small landfills in approved states if they received less than 20 TPD, there was no evidence of existing groundwater contamination, and either the landfill served a remote community with interrupted surface transportation or the landfill was located in an area that receives twenty-five inches or less of precipitation. The decision to develop the "small community exemption" came after the promulgation of the proposed rule in 1988. In developing the exemption, EPA conducted a series of cost and impact analyses for various exemption options to determine how many landfills would be affected, and impacts on risk.

For the final rule (as reported in the addendum), EPA estimated that no communities would have costs above 1% of median household income, and fewer than 400 small communities would have a CPE greater than 1%.

Risks, Benefits, and Impacts of MSWLFs

Extensive resources were invested into developing the Subtitle D risk model, which adapted EPA's liner location model.[5] The risk model, built upon the Liner Location Risk and Cost Analysis Model, was developed to address the same questions as those for land disposal units regulated under Subtitle C of RCRA. It is a dynamic model that simulates environmental fate and transport and dose response as deterministic processes, while simulating containment system failures stochastically.

OSW followed EPA's risk assessment guidance in the development of the model. The Carcinogen Assessment Group's carcinogenic potencies were used (that is, 95% upper-bound slopes based on the linearized multistage model).

EPA modeled eight *constituents of concern* (COCs), six based on human health concerns (five are carcinogens) and two for resource damage concerns. The COCs were vinyl chloride, arsenic, iron, 1,1,2,2-tetrachloroethane, dichloromethane, carbon tetrachloride, antimony, and phenol. EPA determined characteristic constituents (and their initial concentrations) of leachate by analyzing leachate composition data covering 212 chemical constituents at 44 operating landfills. Median concentrations were used across landfills. Median concentrations for 96 constituents were analyzed; 116 constituents were not detected at any facility or had been tested for at only one facility.

The risk analyses estimated the groundwater contamination caused by one set of new landfills (6,034 new landfills operating for twenty years). EPA used data from the survey supplemented with other site data to develop estimates of groundwater contamination at the nearest real drinking well (not hypothetical) and for estimated populations near facilities (existing and population increases as estimated by the Census Bureau, over time). Over 50% of facilities did not have a drinking well within one mile. Thus, immediate risk was not estimated at these landfills. Exposure at six well distances (10, 60, 200, 400, 600, and 1,500 meters) was estimated.

Leachate concentration, liner failure, and site characteristics were modeled in a Monte Carlo simulation. Results were then batched to develop scenarios of various release estimates that were then assigned to landfill categories.

Cancer Cases. Results of the risk assessment indicated that there were few cancer cases caused by MSWLFs over a twenty-year period (see Table 6). The estimated "baseline" cases—the number of cancer cases caused by twenty years of landfilling assuming no regulatory controls—were twenty-three cases in the draft and final RIAs. The addendum, which took into account the many current state requirements that already controlled MSWLFs to some degree, reported an estimated baseline of 5.7 cancer cases resulting from twenty years of landfilling. Systemic (that is, nonhealth) effects were discussed only qualitatively. The estimated

Table 6. Total and Annual Cancer Cases Caused by One Set of New MSW Landfills.

	Avoided total cancer cases	*Total cases avoided*	*Annual cases over 300 years*	*Annual cases over 300 years*
1988 draft RIA				
Baseline	23.1	NA	0.0770	
Proposal: 10-meter POC	6.3	16.8	0.0210	0.0560
Proposal:150-meter POC	6.6	16.3	0.0220	0.0543
Alternative 1 (Subtitle C)	2.6	20.5	0.0086	0.0684
Alternative 2 (Categorical)	3.2	20.0	0.0105	0.0665
Alternative 3 (Statutory minimum)	6.5	16.6	0.0216	0.0554
1990 RIA for final rule				
Baseline	23.1	NA	0.0770	NA
Final rule	7.3	15.8	0.0245	0.0525
Alternative 1 (proposal)	8.4	14.7	0.0279	0.0491
Alternative 2 (Subtitle C)	4.3	18.8	0.0142	0.0628
Alternative 3 (Categorical)	5.4	17.7	0.0179	0.0591
Alternative 4 (Statutory minimum)	7.0	16.1	0.0234	
Addendum:				
Baseline	5.7	NA	0.019	NA
Final rule	3.3	2.4	0.011	0.008
Statutory minimum	3	2.4	0.011	0.008

Note: This table provides estimates of cancer cases occurring from one set of new landfills (enough landfills to service the entire country) that operate for 20 years then close (thus representing the impacts of landfilling 20 years of waste for the nation). Cancer cases are based upon human exposure to contaminated groundwater during the 300-year period starting from the day the landfills open. The annual cases presented here are the average annual cases which would occur each year of the 300 years from the one set of new landfills (20 years worth of landfilling). This number is not comparable to the annualized costs presented in this paper.

NA: Not applicable

avoided cancer cases caused by twenty years of landfilling in the draft RIA ranged from sixteen to twenty cases, depending on the regulatory option. The final estimate of cancer cases avoided from the final rule was 2.4 cases for every twenty years of landfilling.

The baseline and avoided cancer cases were also presented in another format. The cancer cases avoided from one set of new landfills for the proposal regulatory options were presented as an average of 0.0105 to 0.220 cases per year occurring over the 300-year modeling period (baseline cases was 0.077 per year). Cancer cases were presented in this way because contamination of groundwater (and subsequent exposure due to drinking groundwater) was modeled for 300 years. However, it became clear during the consideration of the final rule, that many people were interpreting these numbers to be parallel to the cost per year (which represented costs per year during lifetime of landfill for new landfills, twenty years).[6] Because of the tendency to mix these two nonparallel presentations, total cancer cases were reported in the final addendum.

EPA chose not to discount cancer cases when presenting the total number of cancer cases. OSW has traditionally not discounted cancer cases. One reason for this is that it often takes decades for leachate from landfills to seep down into the aquifer and for the plume of contamination to travel to drinking wells. Thus, benefits (that is, cancer cases) occur several years into the future. Impacts from landfills can occur for scores of years. This analysis indicated that contamination would exist in some instances until the end of the 300-year modeling period. If discounted, these future impacts would be reduced to a negligible amount. Other offices in EPA do discount cancer cases when they are comparing total number of cancer cases to net present cost of the regulation. However, some programs, such as the Air Office, have immediate impacts and do not face the dilemma of how to consider impacts more than 100 years in the future. The issue of whether to discount benefits was a matter of discussion between OMB and EPA during the final analyses.

Resource Damage. The other benefit that EPA initially estimated was the avoided cost of resource damage due to releases from uncontrolled landfills. This, however, was a limited analysis. EPA used the same methodology to determine the number of contaminated drinking wells. The cost of sinking a drinking well outside of the plume of contamination and piping the noncontaminated water to the households within the plume was estimated. This cost was the estimate of resource damage presented. Again, because over 50% of the landfills reported having no drinking wells within one mile, and EPA only considered landfills where there were existing or expected wells in the future, there was no benefit estimated for a large portion of the landfills.

The estimated resource damage avoided was less than the annual cost of landfilling (see Table 7). This estimate did not incorporate any other value of the groundwater. It assumed that the existence value of noncontaminated groundwater was zero.

Environmental Impacts. Nonhuman health effects from uncontrolled MSWLFs were not analyzed in the draft RIA for the proposal nor in the RIA for the final rule. A surface water screening analysis was conducted between proposal and final rule (comparing surface water concentrations with national ambient water quality criteria), but limited effects were found and it was not reported in the final RIA or addendum.

Additional Benefits. Not until the very end of the regulatory development process—within a few months of final promulgation—was

Table 7. Present Value of Resource Damage and Incremental Cost of One Set of New Landfills. (millions of dollars)

	Net reduction in resource damage	*Net present value, one set new landfills*
1988 draft RIA for proposal		
Baseline RD = $2,580 million		
Proposal: 10-meter POC	$1,310	$11,899
Proposal: 150-meter POC	$980	$8,939
Alternative 1 (Subtitle C)	$2,170	$54,548
Alternative 2 (Categorical)	$2,010	$21,739
Alternative 3 (Statutory minimum)	$1,010	$6,065
1990 RIA for final rule		
(Post-closure care period of 40 years)		
Baseline RD = $2,580 million		
Final rule	$1,740	$14,839
Alternative 1 (proposal)	$1,3701	$13,021
Alternative 2 (Subtitle C)	$2,300	$46,660
Alternative 3 (Categorical)	$2,140	$18,589
Alternative 4 (Statutory minimum)	$970	$11,751
Addendum		
Baseline RD = $560 million		
Final rule	$270	$5,770
Statutory minimum	$120	$2,670

increased emphasis placed on the need to quantify nonhuman health (environmental, social) benefits. This increase in emphasis was the result of two trends.

The first was that methods for evaluating nonhuman health effects and their values were areas of interest in the 1980s. There was a growing expectation that these effects could be assessed and quantified.[7] While many scientists believe we are still far away from developing adequate tools to robustly assess ecological impacts, the perception that ecological impacts could be quantified was growing among various sectors of society. On the economic side, economists were also grappling with a variety of resource valuation challenges.[8]

The second reason for the increased interest at EPA and OMB was the increasing pressure to "justify" regulations within a quantified framework. The extensive pressure to promulgate net beneficial rules necessitated that EPA make a larger effort to quantify all benefits.

Because of these factors, EPA management decided months before final promulgation to expand the benefits analysis. At that time, there was pressure to get the final rule completed, and there was perceived to be minimal time (only a month or two) for OSW to undertake such an assessment. Thus, only a preliminary assessment was conducted. EPA investigated several areas including potential reduction in MSWLF siting costs, potential avoidance of property value reduction at leaking landfills, and the avoided reduction in the altruistic, bequest, and existence values of groundwater. Each of these is discussed below. EPA then prepared a "break-even" analysis, which identified the volume of benefits needed to make the rule net beneficial (also described below).

In addition to quantified benefits, EPA made an effort to identify nonquantified benefits for purposes of inclusion in the policy debate. These benefits, according to a draft memo from EPA to OMB (U.S. EPA 1991b), also included protecting the environment surrounding MSWLF (resulting from run-on/runoff controls and explosive gas control), promoting source reduction and recycling (by requiring the true cost of landfilling), protecting future generations from costly and unreliable groundwater cleanups due to today's landfills, and providing a protective alternative to Subtitle C for industrial wastes (allowing the agency to consider the less expensive Subtitle D management for wastes containing *de minimis* levels of hazardous constituents).

Table 8 summarizes both the quantified and nonquantified benefits and impacts associated with the MSWLF criteria.

Property Value Benefits. To estimate property value benefits, EPA investigated the potential impact on property values from contaminated water due to hazardous waste sites. A search of economic literature provided present value effects of a few Superfund and hazardous waste sites on property values, by comparing the sales prices of houses at various distances from the sites. These studies indicated that property value reductions that occurred as a result of perceptions by home buyers about risks posed by such sites ranged from a present value of $300 to $15,000. EPA contended that the effects caused by hazardous waste and Superfund sites would be similar to property value loss due to leaking municipal landfills. The RIA database indicated that there were close to six million households within a mile of a facility located at the 1,230 landfills affected by the final rule. (The other landfills were assumed to be located in states with requirements that were already protective.) It was noted that these households might be willing to invest in insurance against the lost of property values. EPA did not take the step to estimate what the total impact to property value was, it simply maintained that it was equal to or more than $120 million per year.[9]

Siting Costs. Costs to site landfills were also estimated. As mentioned earlier, concern regarding the increasing difficulty in siting new landfills was high. OSW believed that increased public confidence resulting from established federal criteria for landfills would make communities less quick to fight landfills and make siting MSWLFs easier. Reduced costs

Table 8. Benefits and Impacts Associated with MSWLF Criteria Quantified and Not Quantified in the RIA.

Quantified	
Cancer cases	Estimate of cancer cases over a 300-year period caused from drinking groundwater contaminated from one set of landfills which would handle 20 years of waste, then close.
Resource damage	Estimate of the cost to construct a new drinking well a distance away, piping clean water in, and distributing it to residences which have had their wells contaminated within a 300-year period caused by one set of landfills that would handle 20 years of waste, then close.
Not quantified	
Ecosystem impacts	The impacts to a variety of ecosystems (for instance, wetlands) from runoff and seepage from landfills were not analyzed.
Impacts on property values	While EPA conducted a preliminary review of property value reduction associated with landfills known to have contaminated the environment, property value impacts of MSWLFs were not analyzed and presented in the RIA in a rigorous way.
Economic gains associated with increased public confidence	The lack of confidence, manifested in the NIMBY syndrome, resulted in communities fighting landfills in their neighborhoods and in longer, more expensive searches for new landfill sites. Increased public confidence could reduce rejection of landfills by communities and reduce siting costs.
Value of clean groundwater	The value of noncontaminated clean groundwater (existence value) was not estimated. The analysis assumed that clean water can be obtained for only the transportation and circulation costs.
Benefits from good operating standards	These standards included gas monitoring, and disease controls which should result in fewer dangerous explosions and fewer cases of illness.

due to considering one less site (communities typically considered three sites before finding one that was acceptable) and the reduced time (it was assumed there would be a reduction of one year at 25% of the landfills) were estimated to result in a savings of $84 million per year over baseline. Savings due to one less site evaluation were assumed to range from $80,000 to $600,000 per affected landfill based on size. Time was a concern to many communities that were facing closing landfills. Estimated savings due to reduced time for siting included avoided hauling costs if an existing landfill closed and the community had to export waste to a distant regional landfill, increased tipping fees from the regional landfill, and costs of constructing a transfer station needed in some cases. Savings due to reduced time for siting were estimated for 39 of the 180 million tons estimated to be affected by the reduced siting time.

Break-Even Analysis. The break-even analysis started with the annualized incremental cost of the final rule, which was $380 million per year. The cost of the alternative approach (the required statutory minimum) was $160 million per year. Thus, the incremental cost of the final rule over the statutory minimum was $220 million per year. The RIA estimated that the reduction in resource damage was $16 million per year. The reduced siting costs as a result of the rule was estimated at $84 million per year. The remaining (unaccounted for) incremental cost of the comprehensive rule was $120 million per year. Distributed across households located within one mile of the facility, this sum works out to approximately $20 per year ($300 present value). EPA noted that $20 per year was not an unreasonable amount that these households would be willing to invest to avoid greater property value damage due to contamination from landfills.

USE OF RIA IN REGULATORY DEVELOPMENT PROCESS

The MSWLF criteria was statutorily required. Section 4010 of HSWA required EPA to revise Subtitle D criteria by March of 1988. The criteria for developing the rule did not include "net benefits." These revisions were to be those necessary to protect human health and the environment, but at a minimum, they should require groundwater monitoring and location standards and provide for corrective action. Section 4010 also stated EPA could take into account "practicable capability" of the facilities to implement the criteria. Thus, the RIA was not to be used to determine whether or not to do a rule, or even to determine net benefits; it was to be used to identify which options protected human health and the environment, to assist in the development of the options, and to differentiate between various options and suboptions regarding cost, economic impact, and risk. This point has often been lost when discussing the RIA.

1988 Proposed Rule

The proposal was developed between 1985 and 1988 by the program staff. At the same time, the program staff developed the 1988 report to Congress on solid waste disposal in the United States. The RIA was developed by the economic analysis staff (later referred to as the regulatory analysis staff), a separate staff in the Office of Solid Waste.

Proposed Rule: Regulatory Development. The RIA group worked with the program staff early on to ensure consistency of databases and to develop option modeling scenarios. As noted before, several contractors worked on the RIA.

Analyses from the draft RIA (cost, economic impact, risk) were not used much during the early stages of development of the proposal. This was due partly to the time disparities; results were not available for use early in the process. It was also due to the culture at EPA at the time, which placed a lower priority on the regulatory analyses, particularly the cost and economic impact results.

However, the data and analyses were used for some aspects of rule development, such as determining cutoffs for the categorical option (referred to in this chapter as either the categorical approach or Alternative 2 in the 1988 RIA). Cost information was also used for tweaking design/operational requirements, such as determining point of compliance distances.

The rule development process was short, with two option selection meetings being held in September and December 1986. A rulemaking package was completed for work group closure just five months later, May 1987. The option selection meetings focused on the general structure of regulatory approaches (risk-based, categorical, statutory minimum) and not on the more specific content—and thus performance—of specific options. The analysis presented for the option selection meetings was limited and largely qualitative.

Proposed Rule: Management Review

The draft RIA analyses and results were not used much during agency review up to the point of OMB review. When the proposed rule was being presented up the EPA management chain, costs were not the primary consideration. Initial RIA results showed that the proposed categorical approach was more costly than alternatives, and these results were available in early May 1987, prior to work group closure. However, there was considerable pressure to complete the rulemaking, and there had been agency participation in the decision to use the categorical approach at the option selection meeting in late 1986. Despite the high cost of the categorical approach, and the low absolute incremental risk reduction over the other options, EPA management chose it as the preferred option. However, the Office of Policy, Planning and Evaluation (EPA's policy office) did not concur at work group closure.

Proposed Rule: OMB Review. The draft proposal was submitted for concurrent "red border" review (agency review) and OMB review in July

1987. The draft RIA gained extensive attention. The cost results indicated that the categorical approach, the approach favored at that time by EPA, was very expensive. Further, the analysis was unable to provide defendable justifications for specific cutoffs between the categories. The analysis indicated that many of the landfills might be "overdesigned." The incremental human health risk reduction resulting from the categorical option over the risk-based performance standard approach (0.054 to 0.056 per year avoided cancer cases for the performance standard versus 0.065 per year for the categorical approach)[10] did not appear to warrant the extensive cost increase ($1,430 million versus $690 to $880 million).

At this time, OMB was increasing its emphasis on keeping costs of rules low and requiring "net benefits," much more so than in the early 1980s. OMB maintained this stance, despite statutory criteria that did not include cost-effectiveness. The administrator withdrew the rule from OMB, and the rule was revised. At this point, the results of the draft RIA were taken into consideration during the policy decisions for the rule revision. This revision took several months. EPA chose the risk-based performance approach, the RIA was revised, and the proposal was published in August 1988.

FINAL RULE

Final Rule: Development

After proposal, the program staff and the RIA staff worked more closely. While the initial cost and risk model runs were completed, there was an inability (due to cost considerations) to continue to conduct new major model runs. EPA used model runs to develop smaller models that could report on specific unit costs. With most of the models in place, the RIA staff was able, for the first time, to provide rough-cut estimates and quick turnaround analyses for new suboptions and design variations. Analyses were used in revising options, choosing final criteria for groundwater monitoring requirements (analyses parameters, length of postclosure care), and addressing small landfills.

The analysis was used to investigate several options for potential exemptions and cutoffs for small landfill sizes to reduce cost and economic impacts. However, as mentioned above, there was still concern about the accuracy of the analysis (for example, with regard to very small landfills). A considerable debate ensued. Potential high costs for small landfills were weighed against potential risk. In the end, EPA chose to provide an exemption for small landfills located in dry remote areas.[11] OMB encouraged the expansion of the proposed small landfill exemption to include larger landfills.

Also, due to pressure to more accurately reflect the cost of the rule, EPA revised the analysis by incorporating new information regarding the number of landfills that already had design requirements due to state criteria. The status of state and local requirements for landfills was changing rapidly. EPA's 1988 draft RIA did not address this trend and costed out impacts for a 1986 snapshot of the landfill universe. It became apparent that by 1990, this snapshot did not accurately reflect current landfill controls. EPA actually revised this analysis again in 1991, after the final draft of the RIA had been prepared. In order to avoid extensive revisions of the RIA, the results were presented in an addendum to the RIA.

Additional attention was placed on the estimated cost of corrective action. The costs of corrective action were rudimentary and based on estimates for a pump-and-treat technology. In reality, not all sites can be cleaned up with pump-and-treat technologies. While the costs and effectiveness were discussed, and current costs for Superfund sites investigated, the original analysis was not revised.

Final Rule: EPA Management

As the draft final rule underwent agency review, there was extensive discussion regarding costs, impacts, and cost-effectiveness. The RIA did not show net benefits for the preferred option, or any option, even the statutory minimum.

The Office of Policy, Planning and Evaluation and the Office of the General Council were both concerned about the high cost of the rule and small number of quantified benefits. Reinforcing their concerns was the knowledge that OMB would also object to the rule on these grounds. Numerous senior management meetings were held. Factions within EPA preferred the "statutory minimum" over the proposed "preventative approach," citing the cost of landfill controls and the minimum benefits associated with either option. OSW management maintained that it was environmentally more responsible to prevent groundwater contamination than to allow contamination and require cleanup (which was not achievable at all sites).

It became clear, after the rule had started agency review, that there were several perceived benefits outside of the traditional measure of cancer cases avoided. These benefits were not being included in the "net benefits" equation because they had not been quantified. To justify its position that the preventative approach was the more acceptable approach, OSW undertook a preliminary broader benefits analysis in the last few months of rule review (discussed above in the Benefits section). These additional analyses represented the beginning of a new approach in regulatory analyses by OSW. Up until the late 1980s, in reviewing rules, the magnitude of risk reduction, and net benefits, had not as yet been extensively questioned by OMB. OSW had not been required to quantify, in a rigorous way, ecological damages and other welfare impacts, such as reduced property values, intergenerational impacts, and the intrinsic value of groundwater. Because of the extensive pressure to present a "net-beneficial" rule, EPA prepared a short, internal "break-even analysis" (presented above). A memo discussing these additional benefits was prepared in the summer of 1991 and later submitted to OMB as part of the rationale for the rule. A discussion of some of these benefits was included in the addendum RIA.

Final Rule: OMB Review

The draft final rule underwent OMB review between September 1990 and September 1991. Despite a recent agreement that OMB would review rules for a limited number of days, OMB review extended for several months.

During that time, OMB staff provided scores of comments on the RIA and recommendations on both the cost and risk analyses. These comments included technical corrections and changes in the risk assessment to make it less conservative. Some of the comments could be responded to quickly. Others, on the other hand, would have required additional efforts beyond the original scope of the analyses. EPA responded to many of these comments.

However, there were several areas of contention where EPA was not willing to revise the analysis based on OMB comments. This was primarily because response would have required time and resources beyond what EPA was willing to invest, or recommended revisions would have resulted in an altering of the nature of the analysis. Issues discussed included the appropriateness of the 3% real discount rate (OMB supported a higher real discount rate), the definition of impact on the communities (1% of expenditures), the appropriateness of discounting cancer cases to a present value, assumptions used in the risk models, and costs and effectiveness of corrective action.

Beyond specific comments on the analysis, OMB criticized the final rule, citing the lack of quantified benefits. Looking at quantified benefits, none of the regulatory options were net beneficial, including the statutory minimum.

Final Rule: The Final Decision

In the end, the main debate was between the final rule (preventative approach) and the statutory minimum (pollute and clean up). The bigger issue, whether EPA should be promulgating a rule at all (or leave it to states to regulate), was also being debated. Because of the lack of agreement between EPA and OMB, decision meetings were held with the White House Council on Economic Advisors. At this point, several factors came into play, including statutory pressure to promulgate an MSWLF rule, court-ordered deadlines, public and congressional concern, perceived benefits of the final rule beyond cancer cases, and the need to take action on the 1988 report to Congress conclusion that MSWLFs were not safe.

Despite the fact that the RIA was not able to quantify enough benefits to show the final rule to be net beneficial, several members of upper management did believe it probably was at least as net beneficial as the statutory minimum. The following reasons were cited:

- With the continual increase in state and local regulations, the "incremental" cost of federal regulations was probably much lower than presented.
- With the cost of corrective action being much higher than estimated in the RIA, and being unachievable at some sites, the cost of the final rule was probably much more comparable to the statutory minimum than presented in the RIA.
- Other nonquantified beneficial impacts of the rule were much higher than given credit for in the discussions.

In the summer of 1991, the decision was made to go ahead with the final rule, which was finally promulgated in September 1991. In the end, staff at OMB indicate that they feel the RIA did not play a strong role in the decisionmaking process since the most cost-efficient option, according to the RIA, was not chosen. OSW staff indicate that because of concerns about costs, the analysis played a role in identifying how to keep the final rule costs down (the final rule was significantly cheaper than the proposed rule). To the extent that the analysis played a role in identifying the more inexpensive and more effective suboptions, it played a role in shaping the outcome of the final rule and in its details. EPA upper management indicated that costs were a consideration but that, in the end, other factors were also considered that weighted the decision away from the statutory minimum and to the final rule.

AREAS FOR IMPROVEMENT IN THE ANALYSIS

The MSWLF RIA was a major analytical effort for OSW. Extensive resources and time were spent on developing the risk and cost models. In retrospect, there were some areas of the RIA where more robust analysis could have impacted policy decisions. These have been discussed throughout this chapter and are repeated briefly below.

Engineering Costs

Small Landfills. Accurate cost estimates for very small landfills (1 TPD) and the small communities that used them would have provided a more informed debate on the impacts on small communities. Since the time of promulgation, EPA did go back and develop the cost for groundwater monitoring for the very small landfills during development of the revisions to the final criteria. As expected, this analysis indicated that several small landfills would have very high costs per ton. Clearly, a major improvement to both the cost and risk analyses would have been to have preliminary results available earlier in the decisionmaking process.

Siting Landfills. There was a great deal of concern about the growing costs of siting a landfill. OSW developed preliminary estimates of reduced siting costs during final consideration of the final rule. However, this was a very rough analysis. Expected impacts of the rule in improving public confidence about landfills were not rigorously investigated but appeared to be one of the pressures for getting the rule out. Indeed, the RIA did very little in capturing several of the perceived benefits of the rule: increased public confidence and increased ability to site landfills (thus avoiding the impending landfill crisis).

Corrective Action. Costs and the effectiveness of corrective action were not robustly analyzed. Better information in this area would have helped shed light on the true differences in environmental protection between the statutory minimum and the final rule. Quantifying the portion of sites where corrective action may not have been feasible due to physical constraints may have increased costs of the statutory minimum and better reflected concerns in the debate between "prevention" versus "pollute and clean up."

Value of Groundwater. The "value of clean groundwater" continued to be debated through promulgation of the rule. Cited as an issue in the 1988 draft RIA, it remained a point of difference between OMB and EPA. Up until then, OSW had not developed any estimates for the value of clean groundwater. Since then, data collection on this issue has been conducted in conjunction with the draft "corrective action" RIA. These data include information from a contingent valuation survey. Without any agreed-upon estimates of the value of groundwater, the debate remained on the level of

"It's high" versus "No, it's not." Further analyses and solid results would allow more informed and environmentally beneficial policy debates.

Model Flexibility. A major role that a regulatory analysis can play is to inform the program staff during rule development. Between the proposal and the final rule, significant work was done to develop unit cost estimates and provide quick turnaround estimates for a variety of suboptions at the request of program staff. More could have been done if the original models were designed in a more flexible way to provide unit costs. The same goes for the risk analysis as well.

Economic Impacts

As mentioned before, data on the costs to and impacts on small communities could have improved the policy debate in that area.

Risk/Benefits

Peer Review and Verification of Risk Models. Additional peer review and real-world verification of risk models and benefits analysis could have resulted in more believed results. Peer review is time consuming and expensive. Real-world verification of models is extremely expensive and time consuming. However, without having the analyses formally reviewed, EPA was constantly on the firing line from OMB for making a multitude of assumptions. Further, the debate on benefits was clouded by the contention that the risk assessment was overly conservative and did not represent useful or realistic results.

Additional Quantification of Benefits. As discussed earlier, quantitative analyses for environmental impacts and ecosystem damages were not conducted. There is still great uncertainty in this area, and a robust analysis is still beyond the scope of a typical RIA. However, the more light we can shed on this area of impacts, the more informed our policy debates will be. As it is, many environmental impacts and other benefits are often left out because no numbers are presented and thus cannot be incorporated into a quantitative benefit-cost analysis.

Other, nonhealth-related impacts, such as property value impacts, intergenerational inequities, and reduced public concern (regarding real or perceived risks) could also have been better analyzed. If there is going to be increased emphasis on quantified benefit-cost comparisons, then we need to do a better job at quantifying impacts of the rule. The benefits listed in this paragraph were discussed numerous times in policy debates both within EPA and in Congress. The more light we can shed on the actual expected impacts, the more directed the debate can be.

Expenditure of Resources: When Is Additional Analysis Justifiable?

While analyses, done properly, can shed increased light on options and concerns for decisionmakers, the cost of the analyses themselves should be seriously considered. Data collection and modeling can be extremely resource intensive and expensive. Several times during OMB review, the question was asked, "If this improvement/revision to the analyses is made, will the results be significantly changed?" When considering benefits analyses, the question was asked, "If this analysis is conducted, will the results be precise and certain enough to base a decision on?" For instance, while it was clear there were uncertainties associated with the corrective action analysis, it was also understood that precise analysis was not possible at that time. Therefore, EPA did not invest many resources into expanding the draft analysis.

Extensive resources were used in developing the Subtitle D risk model. Some question the benefits of the model, noting the uncertainties associated with the results. However, this model furthered OSW's modeling capabilities at the time and did identify general risk trends associated with the options. While extensive cancer risks were not found, lack of the modeling exercise would have resulted in even more uncertainty about risks associated with MSWLFs. While resources were spent on developing the Subtitle D cost model, the author feels these were well spent. The cost model was used for several subsequent analyses.

There were significant resources spent on data collection and conducting the MSWLF survey. The survey was conducted for use in the 1988 report to Congress as well. Much of the information from the mapping of sites has also been used in subsequent analyses. It is often difficult to determine what is the best level of resources needed to provide an adequate analysis. Because there were limited data on the national level regarding MSWLF, this information collection effort was key to developing an accurate picture of the universe. There is always the temptation to collect more data. Determining the extent of data collection that is appropriate (for instance, sample versus census) is difficult and should be decided by weighing the cost of the data collection and the potential information received from its analyses.

ACKNOWLEDGMENTS

Ken Munis of the Office of Policy, Planning and Evaluation provided editorial and technical assistance in drafting this chapter. Members of EPA's staff and management who participated in the rulemaking were interviewed and provided comments on this chapter.

ENDNOTES

[1]In 1984, EPA developed *Risk Assessment and Management: Framework for Decisionmaking*. In 1986, EPA issued the Risk Assessment Guidelines (51 *Federal Register* 33992–34054). EPA's Carcinogen Assessment Group had set out guidelines for developing carcinogenic potencies (that is, 95% upper-bound slopes based on the linearized multistage model).

[2]EPA identified some damage cases. In May 1986, 184 of the 850 sites listed or proposed for listing on the National Priority List were MSWLFs. Most damage cases were based on landfills that were old and open before the 1979 criteria, and could have accepted hazardous as well as municipal solid waste. These were discussed in the preamble for the rules but were not reported in the RIA or incorporated in any way into the quantified benefits discussion.

[3]A variety of liner types had been modeled and the extent of contamination at representative landfills was estimated. If the risk modeling indicated that the groundwater would be contaminated at the point of compliance for a MSWLF using existing designs, then a design scenario that would reduce levels of contamination below maximum contaminant levels was assigned to that landfill. This way, for the final hybrid option, where the performance standard was assumed to apply, landfills were assigned designs that would control contamination but were not overdesigned. For landfills located in states that were assumed not to have an approved program, the landfills were assigned the "uniform" design.

[4]Many of the government entities listed in the census overlapped. In these instances, EPA tried to identify the "primary" government, the one that would supply waste management services, to reduce double-counting.

[5]This liner location model was developed for EPA Office of Solid Waste by ICF, Inc. and was submitted as background material for the public docket. See *The Liner Location Risk and Cost Analysis Model: Phase II.*

[6]The annual cancer cases per year represented 1 of 300 years of cancer cases caused by one set of new landfills or the landfilling of 20 years' worth of waste. Thus, to get the total average number of cancer cases caused by 1 year of landfilling, you would multiply the annual number of cancer cases by 300 years of modeling divided by 20 years of wastes (annual cancer cases x (300 years/20 years). The other method for calculating this would be to divide the total number of cancer cases caused by one set of new landfills (representing twenty years' worth of waste) by twenty years. The annual costs reported in the RIA represent the actual costs for landfilling one year's worth of waste.

[7]Several efforts were under way to improve or expand the scientific foundations for ecosystem assessment. EPA's Office of Water, which was responsible for the Ambient Water Quality Criteria, was expanding its program to include biological criteria (that could be used to assess the biological diversity and health of a benthic ecosystem) and sediment criteria (that indicated levels of potential

adverse impact on benthic organisms). Additional information was being developed on toxicities throughout the 1980s. The Science Advisory Board had an ecological processes and effects committee to review specific agency projects.

[8]EPA sponsored an ecological valuation forum to work on developing consensus on acceptable approaches. Research was under way on contingent valuation methodologies (surveying people on their personal values for specific items, such as clean water, protected wildlife). Other methods to value resources were also being used by the Department of Interior, using recreational expenditures and other approaches to value resources.

[9]If the low end of the property value impacts, an estimate of $300 net present value impact per property (the low end of the $300 to $15,000 range cited), is annualized over twenty years at 3%, it works out to be approximately $20 per year. Multiplied by the six million households that were proximate to the 1,230 landfills affected by the rule (landfills in states that did not have stringent controls already), this works out to be approximately $120 million per year.

[10]These estimates represent the avoided cancer cases per year occurring for the next 300 years caused by one set of new landfills (20 years of landfilling). The estimated baseline total cancer cases caused by one set of new landfills was estimated to be 23.1. The total avoided cancer cases resulting from the categorical rule would be 0.065 x 300 = 19.5. The total avoided cancer cases for one set of new landfills resulting from the risk-based performance rule would be 0.054 x 300 = 16.2 or 0.056 x 300 = 16.8.

[11]EPA has subsequently revised this exemption based on a court ruling regarding the statutory minimum that required groundwater monitoring at all landfills.

REFERENCES

Federal Register. 1988. Vol. 53, No. 168. Tuesday, August 30.

U.S. EPA (Environmental Protection Agency). 1979. *Criteria for Classification of Solid Waste Disposal Facilities and Practices.* Washington. D.C.: U.S. EPA. Office of Solid Waste.

———. 1988. *A Report to Congress: Solid Waste Disposal in the United States.* Vols. 1 and 2. October. Washington. D.C.: U.S. EPA. Office of Solid Waste.

———. 1990. *Characterization of Municipal Solid Waste in the United States: 1990 Update.* June 13. Washington. D.C.: U.S. EPA. Office of Solid Waste.

———. 1991a. *Addendum to the RIA for the Final Criteria for Municipal Solid Waste Landfills.* September. EPA/530-sw-91-073B. Washington. D.C.: U.S. EPA. Office of Solid Waste.

———. 1991b. Memo from Don Clay, Assistant Administrator for Office of Solid Waste and Emergency Response, U.S. EPA, to James B. MacRae Jr., Director, Office of Information and Regulatory Affairs, Office of Management and Budget, June 14.

Waste Age. 1977. *Waste Age Magazine.* January.

10

Visibility at the Grand Canyon and the Navajo Generating Station

Leland Deck

On October 3, 1991, EPA issued a regulation requiring an Arizona coal-fired electric generating station to reduce sulfur emissions that were believed to impair visibility at the Grand Canyon. The final regulation, implementing an EPA-facilitated negotiated agreement between the plant's owners and two environmental groups, ended what one reporter described as "one of the angriest and most visible environmental struggles in the American West" (New York Times 1991). The signing agreement of the final rule, which took place on the South Rim of the Grand Canyon with both President Bush and EPA Administrator Reilly in attendance, was the conclusion of a protracted regulatory process later referred to as a "win-win" outcome by all parties (albeit for different reasons). Promulgation of the regulation marked the first time that EPA would require an industrial source to significantly change its behavior for the purpose of improving visibility. The key issues that drove the policy debate about this single facility revolved around two fundamental questions: the extent to which the Navajo Generating Station (NGS) impaired visibility at the Grand Canyon (a scientific issue) and whether the benefits of avoiding any NGS-related visibility damage at one of the "crown

---

LELAND DECK is Senior Economist with Abt Associates Inc. He was EPA's Office of Air Quality Planning and Standards project economist for the proposed and final rulemaking on the Navajo Generating Station. He was the lead technical analyst for the EPA economic analyses and was a member of EPA's negotiating team during the negotiations leading to the final rulemaking. The opinions expressed are his own and do not represent the opinions of EPA.

jewels" of the National Park system—a fully reversible aesthetic environmental attribute—outweighed the costs of doing so (a classic policy issue). In this case study, Leland Deck explores these issues by reviewing the analyses conducted by EPA and the Salt River Project (NGS's operator).

At least three aspects of the NGS case deserve special attention. First, although it was clear that NGS emissions were one source of visibility impairment at the Grand Canyon, there was considerable disagreement about the marginal impact of reducing NGS emissions. This lack of scientific understanding raises an important question about the scientific prerequisites for a quality economic analysis. Second, despite the fact that NGS's only significant contribution to visibility impairment was in the wintertime, the willingness-to-pay study that was used in the benefits analysis estimated increases in social welfare associated with summertime visibility improvement. Third, as a result of the economic analysis, more cost-effective ways of reducing emissions were discovered, leading to a 90% rather than a 70% control level and a subsequent increase in net benefits associated with the rule.

INTRODUCTION

The Navajo Generating Station is located on the Navajo Indian Reservation near Page, Arizona, and the Colorado River, between the Grand Canyon National Park and Canyonlands National Park, Utah. The plant burns a maximum of 24,000 tons of coal per day and uses naturally low-sulfur coal (0.5%) from the nearby Kayenta Mine located on the Hopi and Navajo reservations. NGS includes three generating units and has a total generating capacity of 2,250 megawatts (MW).

NGS is owned by a consortium of five electric utility companies and the U.S. Bureau of Reclamation and is operated by the Salt River Project, a public/private utility servicing portions of Phoenix and other parts of Arizona.[1] The electricity generated at NGS is distributed to the owners according to share of ownership, serving utility customers throughout the electric grid system (which includes Arizona, Nevada, and California). Much of the Bureau of Reclamation's share of electricity is used to pump water from the Central Arizona Project, a federal government-subsidized irrigation project providing water to south central Arizona farmers.

NGS was constructed between 1971 and 1976 without any sulfur emission controls. With an expected sulfur emission rate of between 0.8 and 0.9 pounds of sulfur dioxide per million British thermal units (lb SO_2/mmBtu), the plant would clearly fall well within the legal limits set to protect human health, which mandated that its emissions not exceed 1 lb SO_2/mmBtu. (NGS was constructed with controls for reducing particulate matter emissions and limiting the production of nitrogen oxides, though.) Concerns about the potential visibility damage resulting from the sulfur emissions, however, date back to the time of the original siting decision.

REGULATORY SETTING

NGS was noted as a prime example in Section 169 of the 1977 Clean Air Act Amendments (42 USC 7491), which set as a national goal "the preservation of any future, and the remedying of any existing, impairment of visibility in mandatory Class I Federal areas ... from man-made air pollution." (Federal Class I areas include 156 national parks, wilderness areas, and international parks.) The regulations implementing Section 169, published in 1980, established a four-step regulatory process leading to mandatory pollution controls on an individual source if a visibility impairment could be "reasonably attributed" to that source:

1. The federal land manager (in this case, the National Park Service) must determine whether visibility impairment exists in a Class I area.
2. The state must identify existing stationary facilities that may "reasonably be anticipated to cause or contribute" to that impairment.
3. The state must conduct a study to determine the *best available retrofit technology* (BART) for preventing/remedying visibility impairment.[2]
4. For large power plants (over 750 MW), the state must determine the emission limits representing BART following the guidelines published in the BART Guidelines (U.S. EPA 1980).

Also under Section 169, EPA was responsible for promulgating federal visibility implementation plans for any state that had failed to promulgate a visibility protection plan in its state implementation plan (SIP). As one of thirty-five states that had failed to implement visibility SIPs, Arizona became EPA's responsibility. Section 169 called for an extensive review of available remedies and encouraged the consideration of many different types of trade-offs:

> [I]n determining best available retrofit technology the State (or the Administrator in determining emission limitations which reflect such technology) shall take into consideration the costs of compliance, the energy and non-air quality environmental impacts of compliance, any existing pollution control technology

in use at the source, the remaining useful life of the source, and the degree of improvement in visibility which may reasonably be anticipated to result from the use of such technology.

That is, the decision about the appropriate remedy for an identified visibility problem was to be made by considering a wide range of cost issues, plant-specific situations, and the range of environmental impacts (both positive and negative) that would occur if abatement took place. Formal benefit-cost analysis was not required by this law, though.

In 1982, a number of environmental groups[3] sued EPA for failing to live up to its responsibility under Section 169. The suit's settlement generated a series of court orders mandating EPA action. Accordingly, by 1989, EPA had issued rulings on all Class I areas except for the Grand Canyon National Park, which had been determined by the National Park Service in 1986 to have impaired visibility. No other visibility-impaired location had to that point been determined to need any additional emission controls.

In 1989, EPA took the second step in the regulatory process for the Grand Canyon National Park when it preliminarily attributed a significant portion of wintertime visibility impairment to emissions from NGS. That "reasonably attributable" finding triggered the preparation of a BART analysis. A court-imposed schedule required proposing the BART determination by February 1991 and issuing a final ruling by October 1991. These court-imposed deadlines significantly influenced the final NGS rulemaking both by limiting the extent of research and analysis and by encouraging the involved parties to arrive at a negotiated settlement.

On February 8, 1991, EPA proposed to set BART at a 70% level of control—an emission limit of 0.3 lb SO_2/mmBtu, calculated as a rolling thirty-day average. (That limit is the same as the *new source performance standards* [NSPS] control level that would likely apply to a newly constructed similar facility.)[4] As part of its proposal, EPA also included information on two other, stricter control levels—80% and 90%—and solicited comment on them. Because the control options under consideration included options over $100 million (and thus were required to be economically analyzed under Executive Order 12291), EPA prepared a draft regulatory impact analysis (RIA) as part of the proposed rulemaking package.[5]

THE ECONOMIC ANALYSES[6]

Prior to the negotiations that led to the final rulemaking, environmental impact, benefit, and cost studies were conducted by both EPA and the Salt River Project (the operator of NGS). EPA prepared estimates of the environmental impacts, benefits, and costs of several abatement alternatives, which were presented as part of the proposed rulemaking in February 1991. Public comments were invited on all parts of the regulatory package. During the comment period, the Salt River Project formally submitted a detailed engineering-cost analysis of various control alternatives, as well as a benefit analysis. Perhaps not surprisingly, there were important differences between the EPA and Salt River Project (NGS) analyses. These differences led to discussions among the affected parties, which played an important role in reaching the negotiated settlement.

The Cost Analysis

EPA estimated the engineering costs of controlling sulfur emissions from the NGS as a part of its BART analysis. These cost estimates were included in the draft RIA prepared for the February 1991 proposed rulemaking, as well as in the *Federal Register* notice of proposed rulemaking.

In its analysis, EPA examined four technologically feasible control options (50%, 70%, 80%, and 90%), used national cost assumptions, and used standard national cost assumptions and engineering practice specifications developed by the Electric Power Research Institute.[7] The cost analysis for the proposed rule emphasized three technologies: wet flue gas desulfurization (FGD), dry sorbent injection (DSI), and lime spray drying (LSD). The DSI and LSD options were feasible for only the 50% and 70% levels of control. The wet FGD option had the lowest annual cost at all examined control levels. The other two options were included as potentially relevant options that the plant operators might consider implementing because the DSI option requires a lower capital investment than the FGD system (offset by higher annual operating and maintenance costs), and the LSD option has a lower net energy penalty. EPA cost estimates are shown in Table 1.

EPA cost estimates found that while the total annual control costs were higher for more stringent controls, the incremental (marginal) costs of controls decreased at higher levels. The average cost per ton of sulfur controlled at the 50% control level was $2,219 per ton (EPA's low estimate). The incremental cost per ton of the next 20% control (increasing to 70% control) was $1,842 per ton. The cost per ton fell to only $310 for the next 10% of control (increasing to 80%), and increased to $1,344 per ton for the following 10% of control (increasing to 90%)(U.S. EPA 1991).

The very low incremental cost of going from 70% to 80% control was caused by lumpiness in available control options. The least expensive way to build a 70% control system was to install standard-sized FGD units with higher control level capacity and simply bypass—that is, directly release—a portion of the emission stream to achieve the desired level of control. Installing and underutilizing somewhat oversized standard units

Table 1. Alternative Estimates of Annual Costs of Controlling NGS Sulfur Emissions.

	Proposal estimates			
	EPA estimates[a]		*NGS estimates*[b]	
Control level (% of the control; monthly average)	*Annual costs (millions of $1988)*	*Cost-effectiveness ($/ton)*	*Annual costs (millions of $1988)*	*Cost-effectiveness ($/ton)*
50% (0.5 lb SO_2/mmBtu)	$79 to $112	$2,219 to $3,135	$75 to $86	$2,093 to $2,400
70% (0.3 lb SO_2/mmBtu)	$92 to $128	$1,842 to $2,578	$88 to $104	$1,768 to $2,089
90% (0.1 lb SO_2/mmBtu)	$103 to $145	$1,626 to $2,280	$114 to $136	$1,792 to $2,138
Negotiated settlement				
90% control, annual average	$89.6 million ($1990)	Cost-effectiveness = $1,408/ton		

Notes: Costs in this table include both capital carrying cost plus operation and maintenance costs. Adjustment factor for $1988 to $1990 is 5.9%, based on the producer price index for construction materials and components.

[a]EPA cost estimates are for wet flue gas desulfurization (FGD) process, assuming a 30-year remaining plant life. Range of estimates derives from alternative assumptions about retrofit contingency costs and difficulty of retrofit.

[b]NGS estimates are for dry sodium FGD with baghouse (50%), wet limestone FGD and wet thiosulfate FGD (70% and 90%). Range of estimates also derives from two alternative assumptions of remaining plant life: 20 years and 35 years.

Source: SRP 1991; U.S. EPA 1991, 1996.

was less expensive than installing a custom-made system that met any desired level of control exactly. Hence, increasing the control level to fully use the installed "scrubber" capacity only increased marginal operating costs, but required no additional capital investment.

Increasing to a 90% control level required installing one additional FGD unit, with marginal costs greater than those for going from 70% to 80% control. However, the marginal cost of increasing from 80% to 90% control was lower than the average cost to control at 70%.

The apparent shape of the marginal cost curve provided an economic argument for setting BART at a more stringent level than the presumptive BART level. The EPA cost analysis, although limited to an analysis of four control levels, found a conventional "U-shaped" marginal cost curve. The 50% and 70% control options were on the downward sloping portion of the marginal cost curve, the 80% option was the minimum observed marginal cost, and the 90% option was on the upward sloping portion. If the marginal economic benefits of reducing NGS emissions exceeded the marginal costs, the efficient social outcome that maximizes net benefits would occur on the upward sloping portion of the marginal cost curve.[8]

The social benefits of reducing NGS emissions stem from improvements in visibility, and not simply from the reduction in emissions. Visibility is inversely related to ambient concentrations of sulfur; higher sulfur concentrations create worse visibility. The amount of change in visibility conditions caused by a small change in sulfur concentration varies with the amount of sulfur in the atmosphere. When sulfur levels are very low, a one microgram per cubic meter increase in sulfur concentrations can be readily apparent. At higher concentrations, the impact on visibility would be barely discernible.

The ambient concentrations of sulfur in and around the Grand Canyon originating from NGS are approximately linear with emissions from the plant. Thus, the marginal impact on visibility increases as the level of control increases. This optical property results in a demand curve for emissions reductions that is essentially flat throughout the region of the control options examined, and actually rises at the highest control levels.

As described below, EPA estimated that the marginal benefits exceeded the marginal costs at all the examined control levels. The environmental groups involved in the negotiations emphasized this point as an economic argument for requiring more stringent control than the presumptive BART.

Expected Cost Reduction from Emission Trading. The costs of the proposed rule alternatives, however, would actually be even less than the costs of the additional required controls. The sulfur reductions from controlling NGS could be used in Phase II of the acid rain allowance trading program beginning in 2000. EPA estimated that the sulfur allowances could be worth as much as $570 per ton by the year 2005. Selling the allowances could reduce the annual costs to the plant's owners by $28 million (70% control) to $36 million (90%); that is, by a quarter to a third of total costs. (Smaller cost reductions would obviously occur if the allowances traded for less.) Thus, while selling the allowances would not reduce total control costs, it would shift some of the cost burden away from the owners (and ultimately, away from their rate payers).

The Salt River Project's Cost Analysis. After the notice of proposed rulemaking was issued in February 1991, the Salt River Project (the NGS operator) submitted its own, more extensive, analysis of the costs of reducing sulfur emissions (SRP 1991). The Salt River cost estimate was prepared by an engineering design firm and was based on a more complete analysis of the plant, available control technologies, and possible regulatory control options. The Salt River analysis included a screening analysis of twenty-three technologies, including determining feasibility and preliminary cost estimates. It then proceeded to develop an in-depth cost analysis for the five best alternatives:

- Wet lime process flue gas desulfurization
- Limestone with forced oxidation FGD

- Thiosulfate enhanced limestone FGD
- Dry sodium injection without baghouse
- Dry sodium injection with baghouse

The dry sodium injection process (a specific version of EPA's dry sorbent injection process) was a cost-competitive alternative at lower control levels. The Salt River analysis argued that the sodium injection process was especially relevant for a seasonal control requirement. The relatively low capital costs became a dominant advantage if the control equipment had to be run only in the wintertime—the only time when NGS-related visibility impairment in the Grand Canyon was thought to be significant.

The overall ranges of EPA and Salt River Project cost estimates were relatively similar. The Salt River estimates had a narrower range at each control level examined in both analyses, which was consistent with a more detailed (and expensive) engineering cost study. The ranges of cost estimates for specific control technologies and emissions levels largely overlapped. The range of the Salt River estimates were generally lower than the EPA range for the 50% and 70% control stringencies, and are within, but toward the upper end of, the EPA range for 90% control. Similar to the EPA estimates, the Salt River estimates reflected a declining marginal cost of control as control stringency increases.

The Benefits Analyses

Compared with many regulatory policy questions confronting EPA, the Navajo Generating Station case offered a fairly good opportunity for preparing a quantified monetary benefits estimate. The economic benefits of visibility in general, and more specifically, of visibility at the Grand Canyon, had been the subject of academic and government studies for over a decade.[9] Further, while the statute and regulations implementing Section 169 did not *require* estimating monetary benefits, they did not close the door on doing so.

Background. Estimating the benefits that would be derived from improving visibility at the Grand Canyon required estimating the demand for improving such visibility. The total demand for visibility at the Grand Canyon included demand resulting both from visitors' direct enjoyment of visibility at the canyon and from others' desire to protect visibility at the Grand Canyon despite not visiting the region. Because the nondirect use benefits component could arise from people nationwide (and potentially worldwide), the potential size of the nondirect use component was much larger than the benefits accruing to the approximately 800,000 wintertime visitors to the canyon.

The *contingent valuation* (CV) approach, one of the only quantifiable approaches available for examining nondirect use demand, offered the best opportunity to examine economic benefits. EPA's decision to attempt to estimate benefits based on a CV approach for the NGS regulation was partially an attempt to foster a wider review of the use of an emerging and controversial technique that, if accepted, would expand coverage of benefit estimates to a wider array of effects associated with environmental policies and regulations.

CV approaches remain very controversial today and will likely remain so for some time. Many knowledgeable people who have examined the procedures used in designing, conducting, and interpreting CV analyses have expressed serious reservations about the validity of any CV results. CV is certainly not universally accepted at this time either within the economics profession or as the basis for regulatory analysis or litigation. Some analysts who believe that CV methods can potentially be used in some situations believe that CV would not be able to provide reliable results about Grand Canyon visibility because of problems arising from people's perception of the Grand Canyon as an emotion-laden symbol of America, their potentially insufficient understanding of the environmental issues, and the nature of nondirect use demand.

The disagreements over the amount of visibility degradation at the Grand Canyon caused by NGS emissions were at the core of the entire regulatory proceeding. These disagreements had many implications for the applied benefits analysis. Based on the available information, EPA had found there was a reasonable basis to link wintertime visibility impairment at the Grand Canyon with sulfur emissions from NGS. In the summer, the visibility conditions at the Grand Canyon were known to be dominated by the long-range transport of pollutants. Although there was probably some summertime contribution from local sources such as NGS, improving summertime visibility conditions (which are much worse than the typical wintertime conditions) certainly needed to be addressed by a regional visibility strategy rather than a local, plant-specific approach.

The wintertime NGS-related impairment occurred primarily when a stagnant air mass would settle over the Colorado Plateau, allowing sulfur concentrations to build within the canyon. At the worst of these conditions, a palpable haze would fill the canyon, often with a discernible haze boundary lying below the rim of the canyon. The visibility degradation under these conditions could be quite severe, making it difficult to see the far canyon walls or the bottom of the canyon from the rim. These conditions could persist for several days before the weather pattern changed, bringing cleaner air into the region.

The utilities, environmental groups, and state and federal entities involved with the problem had sharply different opinions about the frequency and severity of these wintertime haze episodes, and even more pronounced disagreements about the extent to which NGS contributed to them. Although extensive air quality modeling had been conducted, including two separate "tracer" studies where the emissions at NGS were "tagged" with chemical isotopes identifiable at monitors placed throughout the Colorado Plateau, the frequency and magnitude of the impairment remained bitterly controversial. Many of these disagreements were not resolved in the negotiated settlement. At the end of the settlement process, the owners maintained that they made the best deal possible when faced with impending regulation, never agreeing there was sufficient scientific basis to justify controlling the sulfur emissions. The environmental groups and the National Park Service argued that there was overwhelming evidence that impairment was occurring directly due to NGS emissions, with the principal uncertainty involving the ability to fully understand the extent, frequency, and duration of the episodes based on data from a limited number of years.

Although the estimates were controversial and challenged, EPA did estimate the magnitude, duration, and frequency of the visibility improvement that would occur for alternative NGS control options. Without such an estimate, it would have been impossible to estimate the quantified benefits of regulatory options. Based on a wide variety of available studies, including the extensive analyses prepared by NGS, EPA concluded that NGS-caused visibility impairment would occur approximately ten to fifteen times during the wintertime. Each episode would last between three and five days. At the peak of the worst episodes, controlling emissions by 70% would improve visibility by between 50% and 65%. The average winter day improvement (over the entire five-month wintertime period) would be 11%. The 90% control level would improve visibility on the same number of days, but the improvement would be greater: between 76% and 103% on peak days, with 14% average improvement. The central estimates of EPA benefits were based on the estimated annual environmental change derived from the EPA estimates of the changes in the average wintertime visual range conditions. The range of estimates was incorporated as sensitivity analysis.

For comparison, the Salt River Project's estimate of visibility impairment (discussed later) was a range of ten to twenty days, with a sevenfold increase (25–170 kilometers [km] visual range) on the worst days. The resulting average winter visibility impairment (calculated on a basis comparable to EPA's estimate) was an improvement of 6% to 12%.

EPA Benefits Analysis

EPA was not able to conduct a new benefits study of the wintertime visibility impairment directly related to controlling NGS emissions. There were insufficient resources available within EPA's budget and insufficient time under the court-ordered schedule to prepare a *de novo* benefits study focusing on the exact potential change in environmental quality in question. Rather, EPA prepared its monetary benefits analysis using a benefits transfer technique—using results from existing CV benefit studies in an attempt to estimate the specific benefits of interest.

After examining earlier studies, EPA selected the 1981 Southwest Parklands study to provide information on the direct visitation benefits. However, direct visitation benefits ultimately contributed only a small share to the overall benefits assessment because of the relatively small number of visitors each winter who would actually experience bad visibility conditions.

The majority of the estimated benefits came from the nondirect use value components, including option, bequest, and preservation values. EPA based most of its estimates of nondirect use value on the most recent study available at that time: in 1988 the National Park Service (NPS) and EPA had funded (through a cooperative agreement) a CV study of the value of visibility at national parks in three areas—California, the Southwest, and the Southeast. The survey involved all national parks in each region, but then especially identified three well-known national parks: the Grand Canyon, Yosemite, and Shenandoah. This CV study (hereafter referred to as the NPS Preservation Value study) was conducted by researchers with the University of Colorado and RCG/Hagler, Bailly, Inc. (NPS 1990).

Survey Methodology for the Benefits Analysis. The survey instrument development used two rounds of focus group testing, outside peer review, and a pretest. The University of Colorado conducted the study as a mail survey in 1988, with mailings to individuals living in the state with the national park, as well as to people living in two other states (New York and Missouri). The four final survey variants used the Dillman Total Design Method (Dillman 1978) for mail surveys, including a written advance notice of the survey, and up to three followup mailings. A sample of nonrespondents were selected for a telephone followup to examine potential nonresponse bias and the extent of nondetected bad addresses. The total response rate (after accounting for identified bad addresses) was 73% with a total of 710 returned surveys involving the Grand Canyon (out of a total of 1,647 responses for all parks).

The survey instrument provided the respondents with a four-color printed insert showing four visibility conditions at one national park (one set of respondents were asked about all three parks to identify possible confounding effects). The text accompanying the survey informed people that the photos represented the distribution of summertime visibility conditions at the park. Photo A, with the best conditions, represented conditions on 15% of summertime days; Photo B represented 20% of the summer days; C was 40%; and D, the worst conditions, represented 25% of the summer days. After some introductory questions on attitudes and motives for protecting national parks, the Southwest survey[10] included the page shown in Figure 1.

Later questions asked the respondents to identify the portion of their responses that represented their willingness to pay (WTP) for only the pictured park represented in the photos (such as, out of the stated value for all parks in the Southwest, what portion was the value for the Grand Canyon National Park only). Another question directly asked the respondents to consider whether their responses reflected only the value for visibility at the national parks (as the questions had asked), or if other factors ("help with other needs at the national parks") were included. The original responses were reduced to reflect only the reported percentage of the total responses directly tied to visibility alone, lowering the mean bids to 62% of the original responses.

The survey also asked respondents to identify what percentage of their responses could be explained by family use (direct use value, now and in the future), bequests ("so others may enjoy," now and in the future), and for preservation ("to have conditions as natural as possible at national parks in the Southwest, even if no one were ever to visit").

In the original survey, about 8% of the respondents gave a zero willingness-to-pay value for visibility improvement or preservation, and 3% gave unrealistically high[11] responses. These responses were examined for internal consistency with their answers to other questions and were eliminated if inconsistent. Excluding these inconsistent responses lowered the median bid of the accepted responses by 33% compared with raw responses due to exclusion of unaccepted high responses.

After excluding responses identified as invalid, the adjusted mean annual willingness to pay to improve visibility conditions was $19 for non-Arizona households for an improvement from a mean of 155 km to 200 km ($31 for Arizona households). This corresponds to an improvement in the annual mean from equal to visibility conditions on the then-current fiftieth percentile of summer days to a mean equal to approximately the then-current seventy-fifth percentile of summer days. The willingness to pay to improve visibility from 155 km to 250 km (the ninetieth percentile of days) was $24 ($41 in Arizona). The willingness to pay to prevent a degradation from 155 km to 115 km (equal to the then-current tenth percentile of days) was $21 ($35 in Arizona).

The independent variables included in the regression model used to estimate the willingness to pay for visibility changes included age (mean equals 46.6 years), household income (mean 1987 household income equals $41,400), gender (60% male), and the "before" and "after" visual

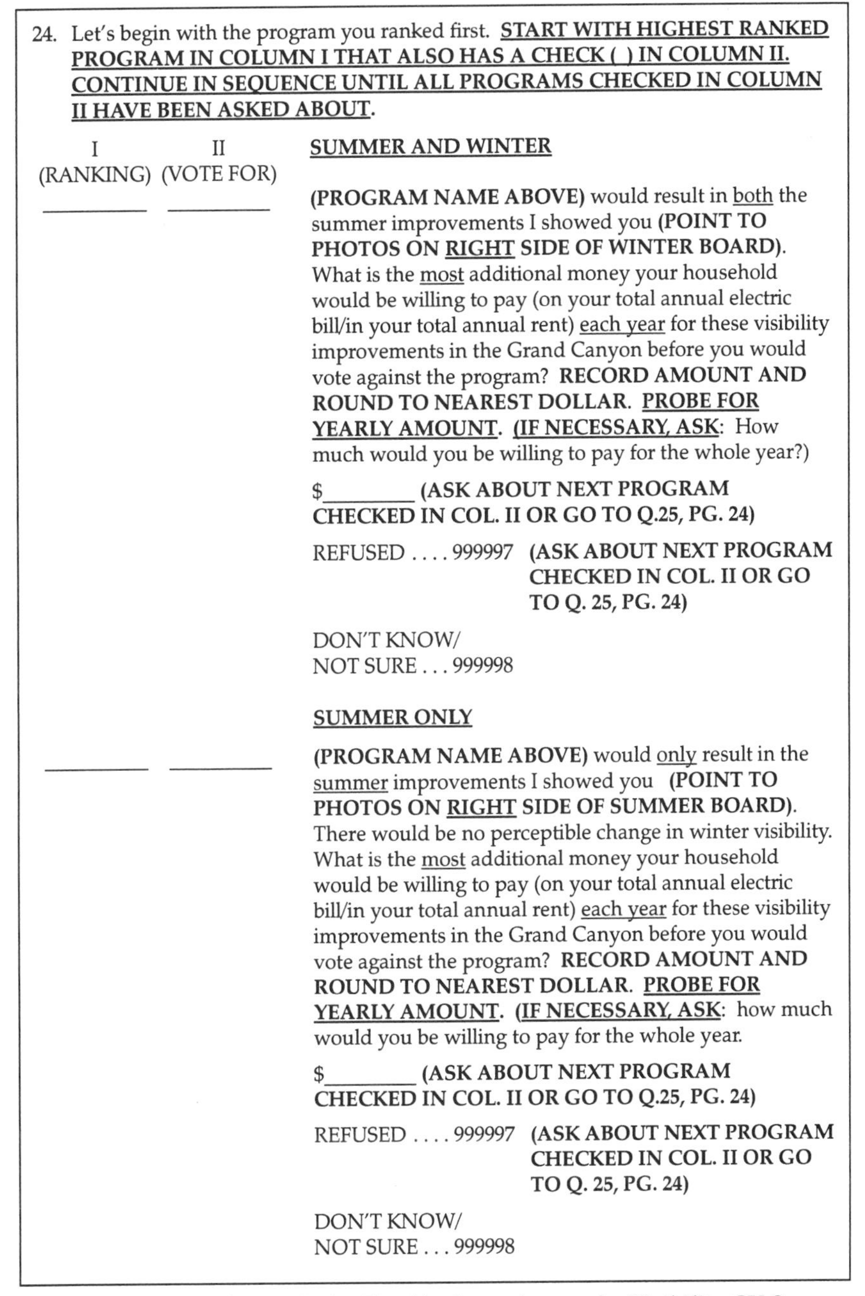
24. Let's begin with the program you ranked first. **START WITH HIGHEST RANKED PROGRAM IN COLUMN I THAT ALSO HAS A CHECK ( ) IN COLUMN II. CONTINUE IN SEQUENCE UNTIL ALL PROGRAMS CHECKED IN COLUMN II HAVE BEEN ASKED ABOUT.**

I (RANKING)	II (VOTE FOR)	
		SUMMER AND WINTER
________	________	**(PROGRAM NAME ABOVE)** would result in both the summer improvements I showed you **(POINT TO PHOTOS ON RIGHT SIDE OF WINTER BOARD).** What is the most additional money your household would be willing to pay (on your total annual electric bill/in your total annual rent) each year for these visibility improvements in the Grand Canyon before you would vote against the program? **RECORD AMOUNT AND ROUND TO NEAREST DOLLAR. PROBE FOR YEARLY AMOUNT. (IF NECESSARY, ASK:** How much would you be willing to pay for the whole year?)
		$________ **(ASK ABOUT NEXT PROGRAM CHECKED IN COL. II OR GO TO Q.25, PG. 24)**
		REFUSED . . . . 999997 **(ASK ABOUT NEXT PROGRAM CHECKED IN COL. II OR GO TO Q. 25, PG. 24)**
		DON'T KNOW/ NOT SURE . . . 999998
		SUMMER ONLY
________	________	**(PROGRAM NAME ABOVE)** would only result in the summer improvements I showed you **(POINT TO PHOTOS ON RIGHT SIDE OF SUMMER BOARD).** There would be no perceptible change in winter visibility. What is the most additional money your household would be willing to pay (on your total annual electric bill/in your total annual rent) each year for these visibility improvements in the Grand Canyon before you would vote against the program? **RECORD AMOUNT AND ROUND TO NEAREST DOLLAR. PROBE FOR YEARLY AMOUNT. (IF NECESSARY, ASK:** how much would you be willing to pay for the whole year.
		$________ **(ASK ABOUT NEXT PROGRAM CHECKED IN COL. II OR GO TO Q.25, PG. 24)**
		REFUSED . . . . 999997 **(ASK ABOUT NEXT PROGRAM CHECKED IN COL. II OR GO TO Q. 25, PG. 24)**
		DON'T KNOW/ NOT SURE . . . 999998

Figure 1. Example of WTP Script Used by Interviewers for Visibility CV Survey.

Source: DFI 1990b

range. Information on education level of the respondents was available (mean equals 14 years) but was not included in the final regression because of close correlation with household income. When the regression results were used to estimate the willingness-to-pay function for NGS, the analysis used the 1987 mean income (national mean equals $32,100; Arizona mean equals $27,900) to calculate the function. The analysis used the sample mean age and gender percentage, rather than the national values. Although the regression coefficients on these variables were negative, because both the sample mean age and gender percentage were higher than the national average, using the sample means for these variables would decrease the predicted function.

The analysis of the survey results used a willingness-to-pay functional form that was linear in the percentage increase in visual range.[12] The results of the analysis were that the total average household willingness to pay (including direct use, option, bequest, and preservation value) was $1.05 for each percentage change in visual range, or $0.68 per kilometer improvement in visual range at the baseline conditions. Residents of the Southwest had a somewhat higher willingness to pay. These estimates were about one-quarter the size of estimates from an earlier (1981) NPS contingent value study on visibility at the Grand Canyon, and somewhat higher (15% larger for the per-percent change) than a 1986 EPA study on visibility values that included a direct question about the Grand Canyon National Park.

In order to use the results from this study for analysis of the NGS BART decision, the results had to be extrapolated from responses about a substantial change in average visibility to the wintertime changes EPA believed were associated with controlling the NGS. (This is the step that makes it a "benefits transfer" analysis, as results from a study of one situation must be adjusted for use in another.) In this case, the location and context of the environmental change matched reasonably well between the original study and the policy application. However, some important specifics about the magnitude, timing, and frequency of the environmental impacts differ and had to somehow be adjusted in order to estimate the BART benefits. At the very least, this adjustment process would introduce further uncertainty into the analysis.

The smallest visibility change directly measured in the NPS Preservation Value study was an average annual visual range change of 40 km, or about 26%. EPA estimates of avoiding the NGS-related change in daily visibility conditions (for the 70% control level) on the worst days was a 50% to 65% improvement in visual range on those days. However, these worst-case conditions would occur infrequently. EPA estimated that there would be ten episodes (of varying intensity) in the winter period, with an average duration of 2.5 days and an average change in visibility during the episodes of 24%. EPA estimated the resulting improvement in average wintertime visibility conditions to be 11%. Assuming that reducing sulfur emissions from NGS would have no impact on visibility the rest of the year (when visibility conditions are generally worse due to long-range transport of pollutants into the Colorado Plateau), EPA benefit analysis estimated a change in annual mean visual range of 6%. The estimated change for the 90% control option was an 8% increase in the annual average. This estimated environmental change was used along with the values percentage increase from the NPS study to estimate the value of the change.

Other adjustments were necessary in order to complete the estimation of the benefits for the NGS control scenarios. The direct use value component from the Southwest Parklands study was directly applied to the five million wintertime visitors to the Grand Canyon. The NPS study option and bequest components of the total estimated values derived from visitation, either by the household members or by others. These values could be affected by the seasonal impact of the NGS controls. However, the exact relationship between visitation rates and these values was not clear. Two alternatives were presented: one assuming that option and bequest values were independent of the season (that is, values were based on the change in the annual mean), and the other assuming that seasonal visitation patterns were important (that is, calculated option and bequest values for the impact in each season and weighted these seasonal values by the seasonal visitation rates).

Preservation values posed a different temporal problem. The visibility impacts of NGS emissions were reversible, causing no lasting damage to the Grand Canyon. Regardless of whether the plant was controlled then, all NGS-caused impairment would cease eventually, at the time of the plant's retirement. Respondents to the survey may not have accurately considered this fact in their responses. Therefore, the estimated preservation value may overstate the preservation value associated with NGS controls. The EPA analysis presented several alternative approaches, including using the bequest component as measured, and using only half of the bequest value as a correction for possible overestimation.

EPA applied the option, bequest, and preservation values to the entire population of the United States. The value for residents of the Southwest used the NPS estimates for Southwest residents, while the rest of the population used the non-Southwest values. The resulting range of total value per household for the 70% control option was $1.00 to $1.90 ($1988, measured in 1995, the assumed first year of control). The comparable range for the 90% control level was $1.30 to $2.50. The range reflects assumptions used in the different methods for adjusting the NPS Preservation Value results to the NGS control scenario.

There would be approximately 100 million households in the United States in 1995, so the first-year total benefits ranged from $100 million to $190 million for 70% control, and $130 million to $250 million for 90% control. Population would increase over the remaining life of the plant, so the per-year total values increased even if the per-household values were to remain constant.

These estimated first-year benefits, based on the results of the original CV study and the subsequent extrapolation to the assumed NGS conditions, met or exceeded the range of estimated annual costs of control. The present value of the benefit stream, using both a 3% and a 10% discount rate, exceeded the costs in all cases. The estimated benefit-cost ratios for the present values ranged from 1.3 (for the 70% control, low benefits estimate to high costs) to 3.6 (90% control, high benefits to low costs).

Criticisms of the Benefits Analysis and the Survey. The EPA benefits analysis received extensive criticism when it was included in the proposed rulemaking package. Some commenters asserted that because EPA was not required to perform a monetary benefits analysis, it should not attempt to place a value on preserving an aesthetic aspect of a major national park. Others criticized the underlying study used in the EPA benefits analysis, as well as on the benefits transfer methods used to estimate the benefits associated with controlling the NGS.

One general type of criticism of EPA benefits estimates focused on the anticipated magnitude of the visibility impairment. The controversy about this issue was central to the rulemaking, obviously affecting the magnitude of any economic benefits associated with emissions control. However, EPA's method could be adjusted to estimate the value for any size change without making any additional assumptions. Smaller changes would clearly produce smaller benefits, and the benefits using EPA's method would not exceed costs for some estimates of NGS-related visibility changes provided by some researchers involved with the case.

EPA's analysis dealt with treating an anticipated very noticeable change in visibility on some wintertime days as a change in seasonal or annual means. The underlying NPS study was conducted for a shift of the annual distribution among four representative photos. The photos did not accurately show the pronounced shift likely to occur during peak episodes if NGS was controlled, nor did the questions specify that the improvements would be concentrated on a few dozen days. The true willingness to pay for wintertime episodic changes could be substantially different than the willingness to pay for an equivalent change in the seasonal or annual mean.

Other criticisms of EPA benefits are related more to basic methods and the application of benefits transfer than to the specifics of the case.

One important tenet of applied contingent valuation analysis is that the good in question must be adequately described and put into a relevant context. EPA needed to use a benefits transfer procedure to adjust the mismatch between the environmental change examined in the original benefits study (a shift of the distribution of the summer visibility conditions) and the NGS situation. While compared with other benefits transfer situations this may be a relatively small extrapolation, the uncertainties and controversy were quite substantial.

Both the NPS study and EPA benefits transfer procedure implicitly assumed that the preferences for visibility at the Grand Canyon among nonrespondents (27% of the original questionnaires) were no different than the preferences among the respondents. A limited telephone follow-up survey with seventy-two telephone respondents found that nonrespondents were somewhat less likely to visit national parks and had a somewhat lower income (nearer to the national average). This suggests that the willingness-to-pay responses among the nonrespondents may have been smaller than the bids by the respondents. On the other hand, 70% of the phone respondents said they would pay something to prevent degradation, and 64% said they would pay something for an improvement, suggesting that the nonrespondents should not necessarily be assigned a zero value.

Another criticism is the NPS study's use of a mail survey instrument. Mail survey techniques have been under considerable attack recently in the contingent valuation community. For example, the 1993 National Oceanic and Atmospheric Administration (NOAA) "Blue-Ribbon Panel" reviewing contingent valuation methods concluded that "the Panel believes it unlikely that reliable estimates of values could be elicited with mail surveys" (NOAA 1993). However, professional opinions are divided on this issue, with some practitioners offering a strong defense of mail surveys. The debate involves a trade-off among costs, accuracy, and defensibility. Mail surveys, while not inexpensive to prepare and conduct, are much less expensive than in-person interviews.[13] The cost differences are greater for national, large-scale studies, where the economies of scale of the mail-based approach are substantial. Given chronically scarce EPA resources, mail-based techniques are often the only CV instrument method that can be used in agency-sponsored studies. The NPS study was conducted using conventional mail-based techniques, including pretesting, open-ended "why did you say that" questions, and procedures that maximize the response rate. However, the use of a mail instrument remains an open question.

The NPS study was also criticized for the selection of photos provided to the respondents. The photos were selected on the basis of visual range, a common measure of visibility used in most valuation studies.

Visibility is a very complex subject though, and there are many different aspects of quantifying visibility. While visual range is a measure of how *far* a person can see, it doesn't attempt to measure how *well* a person can see. Visual range measurements can also be affected by sun angle, cloud cover, color and reflectance of distant objects, and so forth. The scientific community dealing with visibility issues stopped emphasizing visual range some time ago and now relies on measures that are more closely related to the effects of pollutants (or other factors) on the physical transfer of light through the atmosphere. The most direct measure of this is extinction (essentially the amount of light lost while traveling from the object to the viewer). Contrast, a measure of how well an object can be seen, is closely tied to extinction. Extinction and contrast can be more accurately measured than visual range, and the measurement is unaffected by lighting conditions.

While the photos in the NPS study were correctly selected using visual range as the metric, they were incorrect when measured using contrast or extinction. The point contrast measured in the photo presented as the second-best conditions was actually slightly better than the photo presented as the best conditions. This shows the difficulty of any single measure of visibility. The two photos show noticeable (to a lay observer) differences in visibility conditions. The differences arise from the photos being taken at different times of the year, with different sun angles present (all photos were taken at 3:00 P.M.) and the coloration of the clear sky. This is an example of the challenge that CV practitioners face in accurately depicting the environmental good in question to the respondents. At a minimum, problems such as these increase the uncertainty of the results and provide another avenue for criticism. At worst, these problems could yield biased results, with an unknown direction of bias.

Salt River Project/NGS Benefits Analysis

Prior to the 1991 proposed rulemaking, EPA informed the interested parties in the NGS case of the benefits analysis it was conducting.[14] In response to this, and realizing that benefits could legally play a part in the decisionmaking process, the NGS operators prepared a new study. That study (see SRP 1991 and DFI 1990a) was designed to directly assess the visibility improvements that the operators of NGS maintained were likely to occur if the plant was controlled. The study was submitted as a part of the NGS formal comments on the rulemaking. Many of the salient features of the EPA benefits analysis and the NGS study are shown in Table 2.

The NGS benefits study was a contingent value study conducted using an in-person interview method. The research team developed the

Table 2. Comparision of the Contingent Valuation Studies.

Survey component	*EPA/ National Park Service study*	*Navajo Generating Station pilot study*
Date survey taken	August–November, 1988	July–August, 1990
Type of survey instrument	Mail survey using Dillman Total Design Method, with advance letter, three follow-ups, and telephone follow-up.	In-person interviews of randomly selected residences
Number of valid responses	710	183
Sampling frame	Five states: Arizona, California, Missouri, New York, Virginia. Respondent addresses selected from national database drawn from drivers licence, car and voter registrations, and four other sources.	Two urban locations: St. Louis County, Missouri (excludes city of St. Louis) and San Diego County, California. Respondents selected from a random assignment of blocks and a quota scheme per block.
Response rate	58% unadjusted; 73% adjusting for nondelivered mail; 76% including responses derived from phone follow-up.	Not applicable. Only completed surveys recorded. No return visit or other follow-up procedure was used.
Respondent demographics	Mean income: $41,400 Mean education level: 14 years 59% male Median age: 44 Years	Median income bracket: $40–$75,000 (St. Louis), $25–$40,000 (San Diego) Median education: Some college 44% male (St.Louis), 47% (San Diego) Median age bracket: 36–55 years
Depiction of visibility impairment at the Grand Canyon	Grand Canyon, Yosemite, and Shenandoah NPs included. Four photos of visibility condition distribution, with text specifying "Visibility on about 15% of days" [good days], next 20%, next 40%, and worst 25% of days.	Only Grand Canyon National Park included. Three pairs of photos in each of summer and winter seasons, with text specifying number of days represented by each photo.
Visibility change	Three questions: two to improve mean annual visibility conditions (for visual range change from 200 to 250 km, and for 155 to 250 km), and one to prevent worsening of the same conditions (155 to 115 km).	Series of five changes: Year round (mean annual visual range change from 154 to 259 km), winter only, summer only, twenty winter days, ten winter days (winter day change from 22 to 170 km).

continued on next page

Table 2. Comparision of the Contingent Valuation Studies—*Continued*

Survey component	*EPA/ National Park Service study*	*Navajo Generating Station pilot study*
Payment vehicle	Payment card in response to: "Most your household would be willing to pay every year in increased prices and taxes"	Open ended question in response to: "What is the most additional money your household would be willing to pay (on your total annual electric bill/in your total rent)"
Percent $0 or refuse to answer (raw results)	Largest visibility change: 16% Smallest visibility change: 19%	Largest visibility change: 40% Smallest visibility change: 90%
Analytical results: Study conclusions of per household WTP for relevant visibility change.	Mean: $1.00 to $1.90 per household for 70% NGS control; $1.30 to $2.50 per household for 90% NGS control	Median bid (recommended): $0 ($0) for 20 (10) day change 10% trimmed mean: $0 ($0) for 20 (10) day change 5% trimmed mean: $0.20 ($0.12) for 20 (10) day change Mean bid: $1.59 ($1.41) for 20 (10) day change

Sources: NPS 1990, DFI 1990b, SRP 1991.

survey instrument using focus groups and pretests. Because of the regulatory schedule and resource constraints, the study was implemented on a pilot-test basis only. Surveys were completed in two locations (St. Louis and San Diego), with a total of 183 valid responses.

The visibility conditions were presented using large display boards with photographs depicting separate conditions in the summer and winter. For each season, the distribution of conditions using three different photographs representing typical wintertime conditions on the 50 worst days (20 days in summer), the 50 middle days (70 in summer), and 50 best days (120 in summer). A "before-and-after" set of three photos was used to show the change in the distribution in the summer and in the winter. Thus, the number of days in the distribution was kept constant, while the photos of the typical conditions changed. Two sets of display boards depicting the winter conditions were created using alternative photographs. One set of wintertime photos was of layered haze conditions under clear skies, and the other was layered haze with a thick overcast. The study used a split sample, with half the respondents seeing each winter board. There were no statistical differences between the results. The NGS results discussed below are for the combined set of responses.

The NGS study asked respondents about five different changes in visibility conditions. The five changes included:

- A year-round change (a 73% improvement in average visual range)
- A summer-only change (a 73% improvement)

- A winter-only change (a 69% improvement in winter average)
- An improvement on twenty winter days (visual range improving from 25 to 170 km, with the resulting annual average increasing by 5%, and the winter average increasing by 12%)
- An improvement of ten winter days, resulting annual average increasing by 2.5%, and the winter average increasing by 6%)

The estimated percentage improvements reported here were calculated using visual range data and daily distributions for the photos presented to the respondents. This method is different from how the EPA benefit analysis calculated the changes in the seasonal change from NGS controls, but similar to the way EPA used the information presented to the respondents in the NPS Preservation Value study in estimating the willingness-to-pay function.

The NGS benefits analysis did not explicitly match any one visibility control scenario with the controls at the NGS. Rather, it selected the changes to cover a range of possible improvements in order to provide an adequate basis for extrapolating the results to the visibility improvements that might occur. Note that the smallest visual range change estimate (ten winter days) was less than half the size (measured in terms of annual or winter average) of the EPA estimates for the 70% control option.

The portion of the script used by the interviewers that elicited the willingness-to-pay information is shown in Figure 2. Table 3 shows the basic study results presented by the NGS researchers.

The results for the scenarios most relevant to the NGS control scenarios, the ten and twenty winter days, were dominated by the number of zero bids. Respondents first ranked the improvement alternatives according to their own preferences (most had similar preference orderings) they were then asked to give valuation estimates beginning with the most preferred program. If a respondent gave a zero answer for one program, a

Table 3. Reported Visibility Benefit Values, NGS Pilot Study. ($1988)

Study results	*Summer and winter*	*Summer only*	*Winter only*	*20 winter days*	*10 winter days*
Number of valid bids	183	182	182	181	181
Mean bid	$27.78	$15.71	$6.34	$2.38	$2.28
5% trimmed mean[a]	$20.20	$10.51	$2.92	$0.50	$0.46
10% trimmed mean[a]	$16.15	$8.11	$1.25	$0.02	$0.00
Median	$10.00	$0.00	$0.00	$0.00	$0.00
Percent zero bids	40%	58%	79%	89%	90%

Notes: The two "trimmed means" deleted the highest and lowest nine bids (5% trimming) or eighteen bids (10% trimming). "True zero bids" and "Don't know" responses are not included in responses.

Source: SRP 1991, Table 6-2.

WHAT IS THE VALUE OF PROTECTING VISIBILITY
AT NATIONAL PARKS IN THE SOUTHWEST?

New air pollution controls being considered for the protection of visibility at national parks in the Southwest could mean higher prices and higher taxes throughout the country. The next questions concern how much obtaining improvements and preventing worsening in visibility at national parks in the Southwest would be worth to your household.

These questions concern only visibility at national parks in the Southwest and assume there will be no change in visibility at national parks in other regions. Other households are being asked about visibility, human health and vegetation protection in urban areas and at national parks in other regions. For these questions, assume you could be sure that any change would occur next year and continue forever, and all households now and in the future would also pay the most it is worth to them to protect visibility.

Q-12 With additional air pollution controls, average visibility conditions in and around all national parks in the Southwest could improve. What is the most your household would be willing to pay every year in increased prices and taxes to have average visibility improve from Grand Canyon Photograph C to Photograph B at all national parks in the Southwest? (Circle best answer)

$0.00	$2	$8	$25	$60	$150	$400
$0.50	$3	$10	$30	$75	$200	$500
$1.00	$4	$15	$40	$100	$250	$750
$1.50	$5	$20	$40	$125	$300	MORE THAN $750

Q-13 What is the most your household would be willing to pay every year in increased prices and taxes to have average visibility improve from Grand Canyon Photograph C to Photograph A at all national parks in the Southwest? (Circle best answer)

$0.00	$2	$8	$25	$60	$150	$400
$0.50	$3	$10	$30	$75	$200	$500
$1.00	$4	$15	$40	$100	$250	$750
$1.50	$5	$20	$40	$125	$300	MORE THAN $750

Q-14 It is also possible that some additional air pollution controls may be needed just to keep visibility at national parks in the Southwest from getting worse. What is the most your

Figure 2. Example of WTP Questions for Visibility CV Survey—*Continues on next page*

Source: NPS 1990.

household would be willing to pay every year in increased prices and taxes to prevent average visibility at all national parks in the Southwest from becoming like Photograph D for Grand Canyon rather than like Photograph C. (Circle best answer)

$0.00	$2	$8	$25	$60	$150	$400
$0.50	$3	$10	$30	$75	$200	$500
$1.00	$4	$15	$40	$100	$250	$750
$1.50	$5	$20	$40	$125	$300	MORE THAN $750

Q-15 Please provide any information that helps explain your answers to Questions 12, 13, and 14 above. You may also use the back page of the questionnaire.

____

____

____

Q-16 We understand it may be difficult to determine the most you are willing to pay for changes in visibility at national parks. Would you say your answers to Questions 12, 13, and 14 are: (Circle number of best answer)

1 VERY ACCURATE?
2 WITHIN THE BALLPARK?
3 SOMEWHAT INACCURATE?
4 PROBABLY VERY INACCURATE?

Figure 2. ***Continued***

zero was recorded for all his or her less preferred programs. Hence, the number of zero bids increased steadily for less preferred programs.

The high number of zero bids made the results very sensitive to procedures for handling outliers. When all the valid data were included, the per-household results from the NGS study were slightly larger than the EPA estimates for a larger visibility change. It was difficult to directly compare the results of EPA benefits transfer analysis with the NGS study because of differences in the way the visibility conditions were presented. One way of making a comparison was to convert the NGS ten- and twenty-day wintertime changes into their impact on annual mean visibility. As reported earlier, this conversion gave a range of 2.5% to 5%. Thus, the NGS mean bids were for $2.28 to $2.38 for an annual mean change of 2.5% to 5%. The EPA results ranged from $1.00 to $2.50 for an annual mean change of 6% to 8%.

However, when potential outliers were adjusted for by using a trimming procedure[15] (including the median bid, which is 50% trimming), the NGS results fell dramatically. The authors of the NGS study suggested that the 5% trimming figures were a crude upper bound, and the 10% trimming figures were a crude lower bound, of the range of a point estimate for the true unobserved willingness to pay. Thus, the authors of the NGS study concluded that the range of best estimates of the value were significantly less than EPA estimates.

Using the NGS authors' recommendations on both trimming and the likely visibility improvements, the NGS benefits analysis estimated that first-year (1995) benefits would be $1.4 million for the 70% control option and $2.3 million for the 90% control option.

The method of handling the relatively few positive bids in this study obviously had a pronounced impact on the results. After eliminating only the "protest" bids, the benefits estimates using the resulting mean bid for the winter visibility improvements in the Grand Canyon project were similar to EPA's estimates, and the benefits exceeded the costs. Following the authors' recommended trimming procedures lowered the benefits by an order of magnitude (5% trim), or reduced the benefits to zero (10% trim). The extreme sensitivity to the tails of the bid distributions seen in the NGS study (literally becoming a make-or-break issue) was unfortunate, however, because there were only heuristic arguments supporting any specific procedure for dealing with outliers. Some form of trimming is often recommended as an appropriate procedure in CV analysis. At a minimum, outlier analysis is a useful test for stability of results to outlying data. Trimming is also used as one approach to guard against undue influence of "protest responses" or responses given to deliberately "game" the outcome.

The underlying cause of the sensitivity was the large number of zero bids. An important question was whether these were "true zero" bids (that is, they reflected the respondents underlying true willingness to pay for the ten- or twenty-day scenarios), or if they were an artifact of the survey instrument. There were a number of possible reasons to suspect that the true values of some of these respondents might differ from zero. The basic valuation question being asked was difficult, and there was considerable uncertainty in people's responses. There may have been a type of starting-point bias problem, where the initial large changes were anchoring peoples responses. After answering the questions beginning with a relatively large annual change, with a mean response of $27.78 and a range from $0 to $360, the respondents may not have taken the time to separate the differences between a $1.00, $0.50 or $0.25 bid. In the NGS case, however, that was the policy-relevant range. Scenario rejection may also have occurred, with people wondering if a ten-day change (and no effect the rest of the year) was really possible.

The outlier sensitivity may also have been influenced by the small number of respondents in the pilot study. Out of the 181 accepted bids in the ten-day change, only 18 (10%) gave a nonzero answer. Trimming 5%

off each tail, eliminated nine of the positive bids (and nine zeroes), and trimming 10% eliminated all 18 positive bids. This problem may have diminished in a full implementation of the study.

IMPACT OF THE ECONOMIC ANALYSES ON THE POLICY PROCESS AND THE FINAL RULE

During the negotiations that led to the eventual outcome, EPA staff and representatives from the Salt River Project (including Salt River employees and the contractors from the engineering firm that prepared the cost estimates) spent considerable time examining the source of differences in the cost and benefits analyses. Although the ranges of cost estimates were relatively similar, the differences in cost estimates for nearly identical systems were large enough to warrant further attention. For example, EPA's "best" estimate for the 90% control level was $103.5 million per year for the wet FGD system, while the Salt River estimate was for $127.4 million.

Impacts of Cost Analyses

While numerous specific differences in the engineering assumptions were identified, one difference was identified that became very important in the final negotiations. In order to meet the proposed monthly emission standard, the Salt River engineering contractor argued that good engineering practice required surplus control capacity, in the form of redundant scrubber units. The surplus scrubbing units were necessary to provide sufficient control capacity whenever each generating unit was operating. The additional scrubbing units would allow for routine and periodic scheduled maintenance, along with unanticipated scrubber malfunctions and upsets, to occur without requiring a shutdown of a generating unit. This was less an issue at the 70% control level, where the unit size of commercially available scrubber units created some excess capacity. But at the 90% control level, providing the backup capacity required an additional scrubber unit, creating an important part of the difference between the cost estimates. The low end of EPA cost estimates for 90% control did not include the extra scrubber unit.

Isolating this difference in the cost assumptions helped identify a possible compromise between the involved parties. The design engineers' objective of "good engineering practice to avoid unit shutdown" required that emission rate limits with shorter averaging times have larger amounts of standby scrubber capacity, while longer averaging times would require less excess capacity. The final agreement negotiated between the plant owners and the environmental groups provided for a 90% reduction, but calculated using an annual average instead of a rolling monthly average. The NGS operators indicated that faced with this limit they would likely install sufficient control capacity to provide 93% to 95% control when all scrubbers were operating at full capacity. The extra margin of control above 90% provides the plant a sufficient cushion for meeting the annual average that will allow individual scrubber units to go off-line for relatively short periods (a few days) for maintenance or malfunction without curtailing electricity generation.

The negotiating parties agreed to the 90% annual average, which was not an option considered at the time of the proposed rulemaking by EPA, NGS, or the environmental groups. This compromise was accepted by EPA and became the emission limit adopted into the Arizona Federal Implementation Plan. Thus, during the course of the direct negotiations, the control limit changed from the initially proposed 70% control, monthly average standard to a 90% annual average. The substantial cost reduction that resulted primarily from switching from a 90% monthly average (examined as an option at the time of proposal) to a 90% annual average reduced the total cost to less than the originally estimated 70% monthly cost. The 90% annual average would likely cost more than a 70% annual average, although that particular control option was neither examined nor advocated by anyone during the negotiation process.

Other important provisions in the agreement included scheduling routine maintenance on the generating units in the winter (providing the advantage of periodic shutdown—a zero emission rate—in the season of most concern), and a 29-month delay in the implementation date for two of the three generating units (the compliance date for the final unit to be controlled was delayed only 1 month). The NGS engineering contractor prepared a revised cost estimate that included the agreed-upon emission limits, averaging time, and schedule. The revised estimated annualized costs were $89.6 million ($1990), lower than the EPA's cost estimates at the time of proposal for the recommended 70% reduction. This is the basis of the "win-win" characterization of this negotiated rulemaking cited by the plant operators, EPA, and the environmental groups: more environmental protection (although delayed) at a cost lower than the original EPA proposal.

The cost savings identified in the negotiation process were primarily the result of shifting from a monthly average emission limit to an annual average. In order to attain a monthly standard, the plant would likely have to achieve a somewhat higher average daily control level (approximately 2% more) than necessary to achieve the annual average. The marginal cost of this small increase in control levels was quite high. The marginal cost of going from a monthly to an annual average standard was also substantial for 70% control (although smaller than the marginal cost at 90%). Ideally, a benefit-cost analysis would want to compare these marginal costs with the marginal benefits gained from the annual-to-monthly averaging time improvement, but obviously the available benefit analyses could not support such a level of resolution. The willingness of the parties involved in the negotiation to agree to the 90% annual average suggests that they were generally satisfied that the marginal social gain of going from 70% monthly to 90% annual was worth the small increase in cost, but they were reluctant to incur the additional cost of a stringent monthly standard.

Impacts of Benefits Analyses

While the cost analyses ultimately became a central factor in the final outcome, the benefits analyses played a much smaller role. After the NGS benefit analysis was formally submitted, the results were reviewed by both EPA and the Office of Management and Budget (OMB). During the ensuing negotiations, the benefits analyses were discussed several times. Partly because the various parties to the negotiations could not agree on the magnitude of the likely visibility improvements, issues about the differences between the benefits analyses were never resolved. There were also serious issues about fundamental methods of contingent valuation, proper treatment of the data, and interpretation of the range of uncertainty around the results. Ultimately, the parties agreed to disagree over the benefits estimates, as they did about the magnitude of the visibility changes, and the negotiations proceeded. As described above, OMB waived the requirements for an RIA, and EPA never formally responded to comments on the benefit analysis.

Neither of the two benefits analyses prepared for this rulemaking fully reflect state-of-the-art contingent valuation techniques. Measuring economic benefits is difficult under the best of circumstances, and certainly any estimates of the benefits of nondirect use values for an aesthetic improvement at a well-known national park will have a broad uncertainty range. Although the results of these two benefits analyses are sufficiently different to reverse the benefit-cost conclusion, it is not clear whether the differences are anticipatable variations in results for a difficult-to-measure environmental good, the result of structural differences in study design or implementation, or even a reflection of the influence of the sponsor. The Grand Canyon visibility problem presented a prime example for applied benefits analysis. First, there is a long history of available visibility valuation studies. Second, a recently conducted study of approximately the right issue, and the specific location, was available for benefits transfer and a new pilot study was designed specifically for the policy question at hand. Third, there were no legal barriers to considering benefits analysis. However, in the end the competing estimates effectively became a standoff.

Future Policy Implications

There are now other visibility issues confronting EPA that may call for benefit-cost analysis of visibility impairment issues. There are other power plants on the Colorado Plateau that may be causing attributable visibility impairment to the Grand Canyon or other Class I areas in the "Golden Circle" of national parks in the region. Future BART determinations and regulatory actions similar to the NGS decision may be necessary to resolve those issues.

The Clean Air Act Amendments of 1990 created the Grand Canyon Visibility Transport Commission (GCVTC) to address broader visibility issues on the Colorado Plateau. Similar transport commissions for other regions are also authorized as needed. The GCVTC is charged with recommending an integrated approach for protecting visibility from both local emissions and long-range transport of pollutants. Such an approach is necessary to improve summertime visibility in the region, which is currently significantly worse than in the winter. The summertime visibility is likely degraded by emissions from throughout the western United States, northern Mexico, and western Canada, requiring a comprehensive approach that may affect controls on existing emission sources, siting of new facilities, and provision of a "clear air" corridor.

The GCVTC released a draft integrated assessment report produced by a contractor and requested public comments on the report. The revised final report was submitted to the GCVTC in May 1996.

IMPLICATIONS AND INSIGHTS FOR ECONOMIC ANALYSIS

Perhaps the most important lesson from the Navajo Generating Station rule about the role of economic analyses in EPA's regulation promulgation process is the potential for markedly improving the outcome through a negotiated settlement. The negotiation process led to a better understanding of the sources of the cost estimates, and eventually led to a less costly approach that provided more environmental protection. The negotiations failed, however, to bring agreement on the environmental impact and economic benefits issues.

Cost Issues

As is often the case, the final phases of this rulemaking process came down to cost issues. The plant's representatives ultimately concluded that regulation was inevitable and were interested in finding the least painful solution. In this case, the default assumption was that this meant installing the presumptive 70% level of control at the plant on as long a schedule as possible. The environmental groups involved in the case, though, were reluctant to settle for "half a loaf." The estimated marginal cost of going beyond 70% control was less than the average cost of the 70% option, and the estimated marginal cost curve was still downward sloping at the 70% level. As long as the marginal cost curve was declining, the environmental groups argued it made little sense to stop at 70%.

Given the range of uncertainty inherent in estimating the costs of most large-scale retrofit operations, it was tempting to attribute the remaining differences in the competing cost estimates to specific assumptions and parameters used in the two cost analyses. However, by working through the cost analysis on a parameter-by-parameter basis, the plant's representatives (and their consultants) and EPA staff from both the Air Office and the Office of Research and Development (and their consultants) were able to identify one important difference in the analyses that had a marked impact on the costs. The willingness of all parties to work through the costs analyses at such a detailed level—including both staff and senior management level personnel—was a necessary step toward the final outcome.

The different assumptions about backup capacity became an important element in the ongoing negotiations, which were making very slow progress at that time. It introduced a new issue into the possible settlement: a longer-term averaging time. The new wrinkle helped get the negotiations moving again. All sides were interested in finding a compromise by this time, and this opportunity provided a break in the negotiations.

Benefits Issues

Both the physical and the economic benefits analyses had been tabled by the time the negotiations concluded, with strongly held unresolved differences among all the parties. The EPA's benefit analysis was significantly stronger than many EPA RIAs. The analysis was based on a recent contingent valuation study of the right issue in the right location. It is no accident that the underlying contingent valuation study was available. Although not explicitly linked to the Navajo Generating Station regulation, the research (funded through a cooperative agreement with the University of Colorado) was clearly funded in part because of the potential relevance to several visibility-related policy issues. Although the regulatory analysis had to use a "benefits transfer" procedure to match the specific range of changes in environmental conditions examined in the study with the policy-relevant change, this is a far smaller gap than in many other situations.

Design of the NPS CV Study. One of the methodological objections raised about the EPA analysis was the use of a mail survey instrument. Subsequent to the NGS analysis, the NOAA panel recommended[16] against using mail surveys because "the Panel believes it unlikely that reliable estimates of values could be elicited with mail surveys" (NOAA 1993). Mail surveys have been criticized for a lack of an adequate national sampling frame, difficulty in ensuring the appropriate person completes the survey, difficulty in obtaining a sufficient response rate with a potential for selection bias, and a belief that respondents may pay less attention to the details in a mail survey.

Mail surveys do have proponents though, who argue that while some of these are substantive issues, appropriate methods do exist for minimizing the impact of the problems. Each potential problem must be examined in detail to determine if the likely result will be increased uncertainty in the results or a bias in the responses. A certain amount of increased uncertainty can be tolerated and partially offset by larger sample sizes, more variants, and so forth. Bias in the responses leading to biased conclusions is a much more serious potential problem. The biggest argument in favor of mail surveys is one of cost. Anecdotal evidence suggests that modern national mail surveys may be conducted at as little as a tenth the cost per response as other methods for national surveys. Well-done mail surveys using state-of-the-art mail survey techniques can be substantially less expensive to prepare, administer, and collect than other methods. This potentially permits larger sample sizes and more survey variants to be used, or may allow a CV study to be conducted on a lower overall budget. Given EPA's chronic lack of resources for economic research and analysis, there is a powerful incentive to carefully reexamine the identified problems of mail-based surveys to determine whether the problems are so severe as to preclude the use of mail, or if the reduced cost for reduced certainty is an acceptable trade-off. If mail-based surveys are effectively eliminated, CV analysis is likely to become a "sport of the rich," available to well-funded parties (such as industry or wealthy citizens) but not widely used by either public agencies or poorly funded parties (such as environmental advocacy groups or low-income citizen groups).

Other aspects of the NOAA panel guidelines were not met by the NPS study, even when evaluated in terms of the purpose of the study (estimating the value of a change in mean annual visibility at three major national parks). Specific potentially relevant NOAA panel guidelines that were not met include use of a referendum instrument (the NPS survey asked a direct willingness-to-pay question using a bid card), an adequate description of the specific program was not offered, a "no answer" option was not provided, transaction values ("warm glow," or the direct satisfaction from the act of giving to a worthy cause) effects (that is, may not have been adequately deflected), and there was neither advance approval by the potentially affected parties nor full release of the data.[17] Other guidelines were not met in relation to the specific environmental change associated with controls on NGS (such as a clear statement that the visibility changes would be an interim improvement rather than a permanent improvement), but this is a limitation on the application of the NPS study and not on the original study per se. Guidelines from the NOAA panel that were addressed (whether adequately or not is a judgment call) include efforts to minimize nonresponses, pretesting for variant effects and of the CV questionnaire, reporting summaries of the data, using open-ended questions ("Why did you say that?"), using willingness-to-pay surveys in general, pretesting of photographs, and cross-tabulations of results.

It must be emphasized that the NPS study was conducted in 1988, four years before the NOAA panel published their recommended guidelines for CV use in natural resource damage assessment cases. The art and science of conducting and interpreting CV studies have been rapidly evolving throughout the 1980s and 1990s, and few older CV studies meet the current "standards." The NPS/EPA study was designed in part in response to shortcomings noted in earlier studies of the demand for visibility at national parks and was in many ways an improvement on those earlier efforts incorporating advances in the practice of CV.

The benefit analysis prepared on behalf of NGS was a pilot study, conducted on a limited basis with a total of 200 residents of two cities. The information elicitation method used in the NGS study (in-person interviews conducted by trained survey personnel with randomly selected respondents) is an example of the type of elicitation method recommended by the NOAA panel. The small sample size was likely influenced by the cost of conducting such a study. While the researchers who prepared the study always characterized their study as the first phase of a larger, national study necessary to prepare an adequate analysis, the full national study was never conducted. Thus, it was difficult to assess how much reliance should be given the pilot study results. In addition, the high number of zero responses given for the policy relevant changes, and the recommended "trimming" procedures used on the responses, increased the uncertainty on the study results.

Design of the CV Research Issues. One missed opportunity for furthering the state of knowledge about applied CV studies was especially evident. The research team for NGS kept EPA and NPS staff informed of the study developments throughout the study period and invited active participation by the government agencies. During the development process, and prior to conducting the pilot study, the idea of conducting a simultaneous test of the both the mail-based method and the in-person survey instrument was proposed. This would have provided a good opportunity to conduct a side-by-side experiment examining the impact of the two methods. The simultaneous studies could have provided the respondents with similar background information, photographs, and questions. A properly designed experiment could have provided valuable insight into the question of what are the practical differences in the results of mail-based versus in-person surveys.

In order to accomplish such an experiment, certain inevitable differences in the survey protocols would have created some degree of uncertainty in the comparability of the results. While the differences could have been minimized through careful design, certain limits between the two protocols would likely remain. For example, it would have been impractical to mail the large photo-board displays used in the in-person interviews, and the complicated pattern of skipping questions used in the in-person survey would be difficult to use in a mail instrument.

Although the idea of such an experiment was discussed at the time, it did not occur for a number of reasons. In order to conduct such an experiment, EPA and the NGS research team would have had to agree on the proper depiction of the policy-relevant environmental changes. As discussed above, such agreement never occurred throughout the entire regulatory proceeding. Second, EPA was unable to provide any additional funding for economic analysis for this project. Finally, the operators of NGS, who were funding the NGS research team and facing their own budget concerns about conducting a national analysis, were not particularly interested in funding this research. While the project would potentially have had considerable value in furthering research in benefits estimation methods, such motivations were largely external to the parties directly involved with the regulations. For a complicated set of reasons, the research was not conducted, and an opportunity was missed.

ENDNOTES

[1]The Bureau of Reclamation owns the largest share of the plant: 24%. The other owners are the Salt River Project (Arizona, with 22%), the Los Angeles Department of Water and Power (21%), the Arizona Public Service Co. (14%), Nevada Power and Light (11%), and the Tucson Electric Power Co. (7%).

[2]A BART analysis effectively consists of two parts: first, examining the impairment situation and identifying a realm of possible actions; and second, determining the appropriate course of action that becomes identified as BART for that facility.

[3]The environmental groups participating in the lawsuit included the Environmental Defense Fund (EDF)—the principal negotiator for the litigants through much of the judicial and regulatory process. However, during the final phase of the process, the principal lawyer for the EDF was unable to fully participate due to health problems. The Grand Canyon Trust, an Arizona-based environmental organization not included in the original suit, took over as lead negotiator for the environmental litigants.

[4]The BART regulations identify the NSPS level as the "presumptive BART," but give flexibility in setting BART at different levels if the situation so dictates.

[5]Because the BART decision in the final rulemaking had an estimated cost of less than $100 million, the Office of Management and Budget exempted this rule from the requirements of an RIA, and EPA did not finalize its draft analysis.

[6]Details of the analyses and the process behind the NGS regulatory decision are reported elsewhere—in the *Federal Register* notices (see especially the proposal notice at 44 *Federal Register* 5173 and the final notice, 56 *Federal Register* 50172), an EPA report and its draft RIA (U.S. EPA 1980, 1991), and NGS's benefit and cost analyses (SRP 1991)—and are not fully reported here. Additionally, there is an interesting discussion emphasizing the negotiation process written by two of the key participants on behalf of NGS in Rappoport and Cooney (1992).

[7]As a part of the policy debate leading up to the Clean Air Act Amendments of 1990, EPA had examined a wide variety of control technologies for limiting sulfur emissions from coal-fired power plants. That research led to the development of a desktop computer model that estimated plant-specific costs and effectiveness of a range of proven control technologies. The Integrated Air Pollution Control System Design and Cost Estimating Model (IAPCS—Version 3.0, copyright: PEI Associates, Inc.; U.S. EPA 1986a) requires input data on the existing equipment at the plant, including estimates about the remaining life of the plant, construction costs in the area. The model incorporates methods and assumptions obtained from utility industry sources (such as the Electric Power Research Institute) concerning capital financing, conventional engineering practices, design efficiencies, and so forth. Additional engineering data are required for specific control options, especially for emerging control options.

[8]The mathematical conditions for maximizing net benefits include the partial derivative of net benefits with respect to control level being zero, and the second derivative of net benefits being negative.

[9]Some of these valuation studies had been in policy settings and had undergone peer review by EPA's Science Advisory Board (in setting national ambient air quality standards) and by the National Acid Precipitation Assessment Program.

[10]The other surveys used identical language, substituting "California" or "the Southeast" and individual park names as appropriate. The variant that included all three regions and parks was worded appropriately.

[11]The highest value shown on the bid card used was ">$750," which was treated as a response of $1,000. Any bids that exceeded 1% of reported income,

and respondents who chose ">$750" for all three visibility changes, were examined individually for consistency with other answers.

[12]WTP = f(ln(VR2/VR1)), where VR2 is the improved visual range condition, and VR1 is the original visual range.

[13]Because of the need to show the respondents photographs, telephone-based surveys could be used only in visibility studies if combined with a mail component. A phone-mail–phone-mixedmedia survey instrument could be used to provide respondents with the photos prior to a telephone interview.

[14]A preliminary version of the benefits transfer approach and results were presented at an Air and Waste Management Association conference on visibility (Rowe, Chestnut, and Deck 1989).

[15]"Trimming" refers to any procedure that identifies specific observations for deletion from a data set. A common trimming procedure deletes the highest and lowest 5% or 10% of all the observations or deletes observations identified as atypical in some other way. Because there is no single approach to identifying candidates for trimming that is generally agreed upon, subjective decision rules must be used.

[16]With the exception of the use of a mail instrument rather than a personal interview, the underlying contingent valuation study generally meets the NOAA panel's "General Guidelines," although the guidelines were not published at the time of the survey.

[17]Because the original NPS study was done under a cooperative agreement with the University of Colorado, EPA never had direct control, nor possession, of the original data.

REFERENCES

Dillman, D.A. 1978. *Mail and Telephone Surveys: The Total Design Method.* New York: John Wiley and Sons.

DFI (Decision Focus Incorporated). 1990a. *A Review and Critique of the Applicability of Visibility Valuation Studies to a Navajo Power Plant BART Decision.* January. Los Altos, California: DFI.

———1990b. *Development and Design of a Contingent Value Survey for Measuring the Public's Value for Visibility Improvements at the Grand Canyon National Park.* September 1990. Los Altos, California: DFI.

Levy, D.S., J.K. Hammitt, N. Duan, T. Downes-LeGuin, and D. Friedman. 1991. *Comments on Contingent Valuation of Altered Visibility in the Grand Canyon Due to Emissions from the Navajo Generating Station.* EPA Docket A-89-02A, IV-D-177. Washington, D.C.: U.S. EPA.

Mathai, C.V. 1995. The Grand Canyon Visibility Transport Commission and Visibility Protection in Class I Areas. *Environmental Manager.* December. 1: 20–31.

National Academy of Science. 1990. *Haze in the Grand Canyon: An Evaluation of the Winter Haze Intensive Tracer Experiment.* October. Washington, D.C.: National Research Council.

NPS (National Park Service). 1990. *Preservation Values for Visiblity Protection at the National Parks.* Prepared with U.S. EPA and the University of Colorado and RCG/Hagler, Bailly, Inc. Washington, D.C.: NPS.

New York Times. 1991. Utilities to Take Steps to Cut Haze at Grand Canyon. August 9, 1991. p. A1.

NOAA (National Oceanic and Atmospheric Administration). 1993. *Report of the NOAA Panel on Contingent Valuation.* By Arrow, K., R.Solow, P.R.Portney, E.E.Leamer, R.Radner, and H.Schuman. *Federal Register* 58-4601 (January 15 .

Rowe, R.D., L.C. Chestnut, and L.B. Deck. 1989. Controlling Wintertime Visibility Impacts at the Grand Canyon National Park: Preliminary Benefit Cost Analysis. In *Visibility and Fine Particles: Transactions of the Air and Waste Management Association International Specialty Conference.* Pittsburgh: Air and Waste Management Association.

Rappoport, D. Michael and John F. Cooney. 1992. Visibility at the Grand Canyon: Regulatory Negotiations Under the Clean Air Act. Arizona State Law Journal. 24(2): 627–42.

SRP (Salt River Project). 1991. *Navajo Generating Station (NGS) BART Analysis.* Prepared by Decision Focus Incorporated. for the Salt River Project. Los Altos, California: DFI.

U.S. EPA (Environmental Protection Agency). 1980. *Guidelines for Determining Best Available Retrofit Technology Analysis for Coal-Fired Power Plants and Other Stationary Facilities* EPA-450/3-80-009b. Research Triangle Park, North Carolina: U.S. EPA.

———. 1981. *The Benefits of Preserving Visiblity in the National Parklands of the Southwest.* Volume III of *Methods Development for Environmental Control Benefits Assessment.* Report prepared for EPA by W.D. Schulze, D.S. Brookshire, E.G.Walther, and K. Kelley. Washington, D.C.: U.S. EPA.

———. 1986a. *Users Manual for the Integrated Air Pollution Control System Design and Cost Estimating Model.* Version II, Vol. 1. Contractor report by PEI Associates, Inc. Research Triangle Park, North Carolina: U.S. EPA.

———. 1986b. *Establishing and Valuing the Effects of Improved Visibility in the Eastern United States.* Report prepared for EPA by G.A.Tolley and others. Washington, D.C.: U.S. EPA.

———. 1991. *Draft Regulatory Impact Analysis of a Revision of the Federal Implemenation Plan for the State of Arizona to Include SO_2 Controls for the Navajo Generating Station.* Washington, D.C.: Office of Air Quality Planning and Standards, U.S. EPA.

———. 1996. *Approval and Promulgation of Implementation Plans: Revision of the Visibility FIP for Arizona.* In 56 *Federal Register* 50172 (October 3).

11

Agricultural Pesticides and Worker Protection

Louis P. True Jr.

In August 1992, EPA issued a major revision to its worker protection standard for agricultural pesticides. The new rule, designed to protect workers from pesticide exposure, established a comprehensive set of requirements, applicable to all pesticides. Perhaps the most significant provision from both health and cost perspectives was the requirement to largely exclude workers, for specific time intervals, from recently treated areas. Many controversies surrounded this rule. Despite the widespread view within EPA that the existing regulations were inadequate to protect workers, there was significant uncertainty about the health effects of occupational pesticide exposure, and therefore about the benefits of the rule. Many of the costs would be borne by fruit and vegetable growers, which included a large number of small farmers. Opponents of the rule were well positioned in Congress and, through the Department of Agriculture, within the executive branch. As Louis True, notes in this case study, although the regulatory impact analysis was not particularly influential in the rule design and, in fact, did not involve a terribly complex economic analysis, it did help shape certain aspects of the final rule, including the addition of an administrative exception process for permitting worker entry to recently treated areas when economically justified. The fact that the economic analysis did not embody detailed information on the practical effectiveness of certain provisions—

LOUIS TRUE, an independent consultant, was a senior manager in EPA's Office of Pesticide Programs. He led the EPA rulemaking effort for the Worker Protection Standard, including protracted negotiations with OMB and USDA that led to promulgation of the final rule in 1992.

such as how workers would respond to requirements to wear personal protective equipment—at least partially explains its limited role in policymaking. The case study also points to areas where future research could be useful in designing a more performance-oriented rule.

INTRODUCTION

The worker protection standard (WPS) for agricultural pesticides (40 CFR 170 and 40 CFR 156, Subpart K) establishes a comprehensive set of duties, most of which fall upon farmers and other agricultural employers, designed to protect employees from pesticides to which they may be exposed in their workplace. This system of duties includes protecting applicators and others during the application of pesticides; keeping workers out of recently treated areas for specific time intervals; providing safety training, decontamination supplies, gloves, coveralls, protective eyewear, and other personal protective equipment; providing warnings about applications; and responding to medical emergencies. In order to make its requirements enforceable, the WPS directs pesticide manufacturers to revise several thousand pesticide product labels to include the requirement that users comply with the new regulation. The WPS also requires the labels to be systematically revised to include required safety measures specific to each pesticide, based on its toxicity.

Before the new rule was issued, farmers and other employers had only the typically vague, confusing, and unenforceable instructions on older pesticide labels to guide their efforts to protect their workers and themselves. These labels often varied substantially in their requirements due to the absence of a standard approach. Even when appropriate protections were indicated, the existing labels commonly failed to specify those requirements clearly. For example, labels calling for use of a respirator rarely fully specified the type of respirator, when and how it was to be cleaned, how often any filter elements were to be replaced, that it should fit the employee, or the person responsible for supplying it. In contrast, the new WPS created a single source of general instructions on these and related practices that EPA considered critical for the success of any program to provide a safe workplace in the presence of toxic substances.

The WPS represented a novel approach to EPA's regulation of the behavior of pesticide users. For the first time, growers and other pesticide users seeking to comply with the Federal Insecticide, Fungicide and Rodenticide Act (FIFRA; 7 USC. 135) needed to obtain information that was not present on the labels of the pesticides they used. As a result, the new rule created a significant challenge to EPA and the agricultural community in general in disseminating information on the new requirements. Outside of pesticide regulation, such general rules had been applied to many agricultural operations. Examples include the Occupational Safety and Health Administration's (OSHA's) Field Sanitation and Hazard Communication Standards, which have some similarities to the WPS. Such general regulations, however, were unfamiliar to growers with respect to pesticide use. The magnitude of the change made the development of the WPS controversial throughout the affected agricultural industry, among the state governments responsible for most enforcement of federal pesticide law, and among organizations representing farmworker interests. That controversy has continued as the rule has been implemented and included the enactment of special legislation delaying implementation of some provisions.

At least eight years before the new WPS was issued in final form, EPA had concluded that the existing requirements for the protection of agricultural workers from pesticides were seriously deficient. EPA was concerned both that the existing method of regulation was not designed to provide adequate protection and that it was often ineffective in achieving what it was designed to do. For example, the old regulation had been intended to prevent workers from being exposed by entering treated areas soon after application of a pesticide. However, most of the resulting "reentry intervals" under the old rule were vague, and nearly all permitted workers to enter fields immediately after spraying, provided only that they wore ordinary work clothing. In light of more recent scientific understanding of potential exposure under these circumstances, such a provision seemed patently defective. Other provisions of the old rule had been found to be too vague to be enforced. In addition, while the information on farmworker poisonings was limited, available data showed that they were continuing to occur. This was the general setting for the beginning of the rulemaking process that led to the 1992 final rule.

The limitations of available baseline data on the likely human health effects of occupational exposure to the several hundred active ingredients in agricultural pesticides, along with limited data on certain compliance costs, would prove troublesome for EPA. In order to convincingly assess in a formal regulatory impact analysis (RIA) whether the benefits of the proposed new rule exceed compliance costs, EPA would need to estimate the extent to which the WPS would avert injuries and illnesses. Ideally, the agency would also be able to develop some measure of the value of such avoided cases to society in a way that could be compared to compliance costs. Unfortunately, available poisoning statistics and epidemiological studies provide only a limited basis for such estimates, as discussed in some detail below. While EPA possessed toxicity tests on laboratory animals for many agricultural pesticides, the data were incomplete for most.

Finally, there is no accepted methodology for modeling risks due to the widely varying exposure of millions of agricultural workers, on hundreds of thousands of farms and other operations, to a varied mix of hundreds of pesticide active ingredients. The difficulty in quantifying many of the expected health benefits of the rule would prove to be the single biggest controversy within the federal government in issuing the final revisions to the WPS.

Legislative, Regulatory, and Policy Context

The principal legislative authority for EPA's regulation of pesticides is FIFRA. Under this statute, EPA exercises premarket licensing, or registration, authority over all pesticides sold in the United States. Among the terms of any pesticide product registration is the content of pesticide product labeling, including any precautions or requirements for the use of the pesticide. These labeling requirements frequently included provisions to protect handlers (principally those who mix, load, or apply pesticides); some also included protections for persons who work in areas treated with pesticides. Section 12(a)(2)(G) of FIFRA makes it "unlawful for any person ... to use any registered pesticide in a manner inconsistent with its labeling," and other sections of FIFRA provide authority for civil and criminal penalties for violations. This is the principal authority under which EPA regulates the behavior of pesticide users to prevent "unreasonable adverse effects" from the use of any pesticide—FIFRA's basic standard for registration. EPA has relied upon placing a statement referring to the WPS on pesticide product labels to make the rule enforceable. (This may seem an oddly indirect basis for enforcement, but FIFRA does not make noncompliance with a regulation such as the WPS a prohibited act subject to sanctions.)

The 1992 revisions to the WPS grew out of the agency's concern about the safety of the agricultural workplace with respect to workers' exposure to pesticides directly and to pesticide residues in treated areas. The WPS was originally promulgated in 1974 (39 *Federal Register* 16888; May 10, 1974). In a 1983 review of regulations, the provisions of the WPS were judged inadequate to protect agricultural workers due to concerns about both the enforceability and the adequacy of the coverage of the 1974 regulation. The review also cited continued reports of worker poisonings by pesticides. As a result, EPA published an advanced notice of proposed rulemaking (49 *Federal Register* 32605; August 15, 1984) announcing EPA's intention to revise the regulation and soliciting comments to help identify weaknesses in the existing regulation and ideas for addressing such weaknesses. Wide differences of opinion were evident in the comments EPA received, although most commenters agreed that some revisions were warranted.

In 1985 and 1986, EPA undertook a regulatory negotiation for the revision of the WPS. A temporary advisory committee was established, consisting of members chosen to represent the principal parties interested in WPS revisions. These included representatives of worker advocacy groups, grower groups, the agricultural chemicals industry, and state regulatory agencies. Like all negotiated rulemakings, the goal was to develop a proposed rule judged adequate and workable by all of the diverse interests involved. After several meetings at which a number of options were discussed and a variety of positions and concerns explained, the worker representatives stopped participating, without offering a specific explanation. While that ended the attempt to reach a consensus, EPA held several additional meetings with the remaining members of the committee to further discuss options and issues associated with potential revisions.

EPA then returned to the more typical process of rulemaking by developing, internally, a proposed regulation (53 *Federal Register* 25970; 40 CFR 170, July 8, 1988). The 1988 proposal included expansion of existing provisions to cover more agricultural employees on farms, forests, nurseries, and greenhouses, including those employees handling pesticides for use on such sites. It required specific notifications to workers concerning pesticide applications, use of personal protective equipment, and restrictions on entry to treated areas. The proposal also included new requirements concerning decontamination facilities for workers, emergency medical assistance to poisoned workers, monitoring of pesticide handlers, and training. In addition, EPA proposed the creation of special labeling regulations (53 *Federal Register* 25970, 40 CFR156, Subpart K) to guide revisions to individual pesticide product labels to reflect new requirements related to the WPS, including required personal protective equipment and restricted-entry intervals (periods after application of a pesticide during which entry to the treated area is restricted).

EPA followed publication of the proposed regulation with public meetings in at least fifteen agricultural areas of the country to explain the proposed changes and elicit comments. Three hundred and eighty written comments were filed, totaling more than 2,000 pages. Commenters included agricultural producers, pesticide manufacturers, worker representatives, state government agencies, the U.S. Congress, the U.S. Department of Agriculture (USDA), and others. The comments addressed all major proposed provisions and ranged from requests for complete exemption from some provisions to criticism of EPA for not going far enough to protect workers. The final regulations were issued in 1992 (57 *Federal Register* 38102; August 21, 1992).

The final rule covered most of the points in the proposal but differed in several particulars. The regulations staggered implementation of their provisions over several years in order to provide time to effect required pesticide label changes, develop implementation materials, and establish training and enforcement mechanisms. Most of the provisions applicable to agricultural employers were to be complied with by April 15, 1994. A number of grower groups and many state departments of agriculture (the state agencies usually responsible for enforcement of FIFRA) expressed concerns about the implementation schedule and other aspects of the rules. In response, Congress enacted legislation in April 1994 that further delayed implementation of many employer responsibilities until January 1, 1995. The parts of the regulation applying to employers have been in full effect since that date.

Scope of Rulemaking

The WPS creates duties principally for two categories of persons: manufacturers of certain pesticides, who must alter pesticide labeling, and employers (including labor contractors) of agricultural workers and agricultural pesticide handlers, who must take steps to protect certain of their employees and other persons.

Pesticide Products Affected. The regulation applies to registered pesticides labeled to permit use in the production of agricultural plants on farms, forests, nurseries, and greenhouses. This covers most of what is normally considered agricultural use. Not covered are uses on livestock, rangeland, and pastures; most uses on crops after they have been harvested; noncommercial and nonresearch uses, such as in home gardens; uses in parks and public gardens; uses to control vertebrate pests and building pests; and uses not directly related to growing agricultural plants, such as right-of-way herbicide applications. Total agricultural pesticide usage is approximately 800 million pounds of pesticide active ingredient per year (U.S. EPA 1992a, III-1), the bulk of which is within the scope of the WPS.

Persons Protected. The regulation is intended principally to protect approximately 3.9 million persons (U.S. EPA 1992a, III-5,7), including:

- 2,216,000 agricultural workers (such as harvesters and weeders) on about 560,000 agricultural establishments (farms, forests, nurseries, and greenhouses using pesticides in ways covered by the WPS), and
- 1,652,000 pesticide handlers (including mixers, loaders, applicators, and flaggers) on about 8,500 commercial pesticide handling establishments and many agricultural establishments.

THE REGULATORY IMPACT ANALYSIS

The RIA covered the full range of WPS provisions and attempted to address all major expected costs of implementation and the benefits associated with the rule. Given the limitations of the available information on health-related benefits, however, the opportunities for rigor were greater for the costs than for the benefits.

Starting around 1987, the analysis went through several versions, including a thorough 1992 revision responding to comments from USDA. (That agency often comments on EPA pesticide decisions. In the case of regulations, FIFRA's Section 25 provides USDA a special opportunity for review and comment.) The revised RIA was the subject of extensive deliberation among EPA, USDA, and the Office of Management and Budget (OMB) until the issuance of the final rule and RIA in August 1992. Among the more hotly disputed issues were:

- EPA's attempt to estimate the baseline of health effects among agricultural workers as a result of occupational exposure to pesticides,
- the degree to which the expected benefits of the rule (principally averted health effects) could be quantified,
- the need for flexibility for agricultural employers in meeting the requirements for training workers in pesticide safety, and
- the potential impact of restricted-entry intervals on agricultural production.

The RIA for the final rule was prepared under the direction of an EPA staff economist (approximately one-half person-year) with the assistance of DPRA Incorporated under contract. Contract funding totaled $71,500. Other EPA staff carried out the bulk of the risk analysis for the benefits assessment and assisted in the analysis of compliance costs (approximately one-half person-year).

Problem Definition

The final form of the RIA was shaped in large part by the key policy goals of the regulation, some major uncertainties in the economic analysis and data, and the key players with economic interests affected by the WPS.

Key Policy Goals. Like any RIA, the WPS analysis was intended to assist in determining whether the results of the regulation would be worth what society would have to pay for them. In the case of the WPS, the RIA sought to characterize the nature and size of the problem being addressed (principally the current burden of health effects resulting from exposures to pesticides in the agricultural setting, among workers, pesticide handlers, and some others), the degree to which the contemplated regulatory provisions would avert such incidents, and at what cost.

Major Analytic and Data Problems. Uncertainties exist for both the benefit and cost assessments for the WPS, although the uncertainties are more extensive for the benefits.

WPS benefits consist principally of health effects avoided or reduced in severity through behavior changed as a result of implementation of the rule. Any assessment of such benefits must begin by estimating baseline rates of health effects. Relevant data consist of animal studies of pesticide toxicity, field studies of occupational exposure, epidemiological studies, and poisoning incident reports. These sources are typically incomplete and limited for agricultural pesticides. For example, most agricultural pesticides do not yet possess a laboratory-based toxicological or field exposure database that EPA has reviewed and found consistent with its current standards as defined through EPA's "reregistration" program. In the case of data on the effects of actual human exposure, sources include data on pesticide-related hospital admissions, emergency room treatment surveys, studies of migrant worker health, and, especially, reports on the incidence of physician-treated pesticide poisonings in California. These data on human experience are problematical with respect to both acute effects—those occurring soon after exposure—and delayed effects. Acute effects of pesticide exposure often mimic other illnesses and are thus difficult to attribute to occupational exposures. Current understanding of the nature of many of the potential delayed effects of pesticide exposure, such as cancer, indicates that they can occur long after exposure, can often be caused by other than exposure to pesticides, and can otherwise frustrate attempts to establish a causal link to specific occupational exposures. Nonetheless, existing data were strong enough to allow EPA to conclude that the health effects among those exposed occupationally to agricultural pesticides were significant.

In the case of WPS implementation costs, the main uncertainties stem from such sources as varied estimates of the size, seasonal employment pattern, and rate of turnover of the agricultural workforce; wide variations in workplace circumstances that may affect implementation cost or practicality; and the potential effects on crop yield or quality of some provisions of the rule.

Major Economic Actors. The principal groups with economic interests affected by the WPS include growers (agricultural employers), commercial pesticide-handling establishments serving agriculture, pesticide manufacturers/distributors, and the families of agricultural workers.

Institutional and Political History. Growers and their representatives have been sensitive to the regulation of occupational matters by the U.S. Immigration and Naturalization Service, the Department of Labor (wages and hours, housing, and so forth), and OSHA (workplace safety). Of particular relevance to the WPS are OSHA's regulations, including the Hazard Communication Standard and the Field Sanitation Standard, affecting occupational safety on some agricultural establishments. These regulations were, and are, highly controversial among agricultural interests.

More generally, Congress has shown its concern about regulations affecting small agricultural employers through an annual rider to OSHA's appropriation forbidding expenditures to enforce most OSHA standards on farms with ten or fewer employees (and without a seasonal labor camp). EPA has made estimates indicating that about 94% of WPS-covered establishments may meet this criterion (U.S. EPA 1992b). Roughly one-half of the agricultural workers may be employed on such establishments. Since the WPS contains no similar exemption, it can be viewed as a significant departure for many agricultural employers. On the other hand, the change is incremental with respect to pesticides, since pesticide labels already contain a variety of requirements with no exception for smaller operations.

Regulatory Approaches Considered

In developing the WPS, EPA considered a wide variety of options, including those concerning:

- scope (for instance, whether forestry operations should be subject to the rule and which categories of pesticide handlers should be covered),
- whether systematic blood monitoring should be required for certain pesticide handlers,
- how responsibility for compliance should be assigned,
- whether to rely upon the states to develop their own requirements or to create a national minimum standard,
- treatment of small versus large agricultural establishments, and
- more detailed options for many individual provisions.

But, for cost analysis in the formal RIA, this complex range of options was summarized into three major options. These are listed below with the estimated total (not incremental) first-year cost of pesticide user compliance (U.S. EPA 1992a, IV-2):

- *Low option*—Less costly and less protective. Compared to the final rule, this option would have required shorter restricted-entry intervals, permitted unlimited entry during those intervals (if personal protective equipment was worn), not required employers to supply or maintain personal protective equipment, reduced required warnings to workers, eliminated decontamination and training for workers, and eliminated the requirement for employers to provide emergency assistance ($55.3 million).

- *Revised rule*—As promulgated ($190.1 million).
- *High option*—More costly and more protective. Compared to the final rule, this option would have established longer restricted-entry intervals, prohibited any entry during those intervals, required additional warnings to workers, required the provision of additional personal protective equipment/decontamination supplies and training, and required establishment of a blood cholinesterase monitoring program for all commercial pesticide handlers ($364.9 million).

Regulatory Options Not Considered

EPA policies generally encourage consideration of market mechanisms, or incentive-based alternatives to command-and-control regulations, to achieve regulatory objectives. In the instance of the WPS, the agency had no explicit legislative authority to create market mechanisms. Neither, however, were there specific prohibitions in FIFRA. In any case, options such as mandatory insurance to cover occupational illness, or other market-based mechanisms, were not given significant attention during the rulemaking.

It should be noted, however, that many market-oriented schemes would seem to face intractable measurement problems in the case of the WPS. Such systems would require, at least in the ideal, the capacity to recognize health effects stemming from specific occupational pesticide exposures. That is, ideal market mechanisms for worker protection would create incentives for actions by an individual employer that achieve the goal of reduced injuries or illnesses due to pesticides. To create these incentives, society's costs associated with occupational exposure to pesticides, including the attendant health effects and other consequences, should be borne by the employer permitting the exposure. For such mechanisms to work, the health consequences of individual employer actions must therefore be measured routinely. In practice, however, such measurement is extremely difficult or infeasible. This is especially true of delayed health effects, some of which become evident only many years after exposure and which often have significant incidence rates due to other causes. The measurement problem is exacerbated by the fact that workers may move among employers and may be exposed to pesticides and other potential causative agents other than through their occupations.

In short, many of the potential real-world health effects from occupational pesticide exposures are difficult to identify as such, much less link to specific exposures that may have caused them. Therefore, the social costs associated with those illnesses are difficult to internalize to the production decisions on individual agricultural establishments.

On the other hand, it is conceivable that less idealized market mechanisms could be more workable. Possible approaches might be tied to surrogate, intermediate outcomes, such as measures of exposure, or even to overall pesticide use. While these approaches also present measurement problems, these may be more manageable. Still, targeting such intermediate phenomena in market mechanisms would necessarily weaken the link to the ultimate objective and reduce the likelihood of achieving desired outcomes. For example, basing an incentive system on worker exposure is in theory attractive and could become more feasible if future technology permitted routine exposure monitoring. This would likely require an elaborate and continuous monitoring system, due to the commonly episodic nature of agricultural pesticide exposure. Even perfect exposure monitoring, however, must still be coupled with methods for estimating health effects due to the measured exposure. These methods remain uncertain and controversial.

An even less attractive approach has been suggested that would be designed to reduce the use of pesticides overall through taxes, fees, or other mechanisms. Such an approach may or may not predictably reduce pesticide-related health effects among agricultural employees, depending on how the reduction is distributed among pesticides, crops, application timing and techniques, and perhaps other factors. Even if these factors could be incorporated into the incentives, such a system would not itself distinguish between two farms with identical pesticide use patterns but that differ greatly in the measures (such as those required under the WPS) taken by the employers to minimize worker exposure. Total pesticide use is thus a further step removed from the ultimate objective of reducing health effects. In addition, use reduction could cause society to forgo benefits of pesticide use with little assurance that the sacrifice brought sufficient progress, if any, in reducing health effects.

In short, market-based alternatives to command-and-control regulation in this area inevitably collide with the difficulty of reliably estimating the full range of health effects stemming from occupational exposure to agricultural pesticides. On the other hand, approaches based on exposure, while imperfect, might nonetheless encourage cost-effective exposure reduction and could perhaps be relied upon, particularly if combined with a safety net of basic requirements for safe work practices like those in the WPS.

While traditional market mechanisms were not considered for the WPS, EPA did attempt to design the rule to employ performance (as opposed to specification) standards in a number of areas to give employers the flexibility to reduce worker risks cost-effectively.

Finally, there is a sense in which the WPS may have indirectly created market-like incentives by increasing the likelihood of successful civil actions claiming that employer negligence led to injury of employees through exposure to pesticides. In fact, during the regulatory negotiation on the WPS in 1985 and 1986, representatives of agricultural employers voiced special concern about "private right of action" emerging from the WPS. While the WPS holds employers responsible for compliance, it contains no provisions concerning civil suits and none appeared in EPA drafts. Moreover, FIFRA contains no authority for such provisions. Still, the agency's legal staff acknowledged during the negotiations that courts could take the WPS to create a standard for defining employer negligence in civil cases. Concern about vulnerability to civil suit is still mentioned by some as an important motive for employer compliance with the WPS.

Cost Analysis

The cost estimates in the RIA were developed for each of the three major options—high option, revised rule, and low option—listed above using the following approach:

1. Eight major pesticide use-site categories were identified as potentially affected by the rule. These included five crop categories—feed and grain; cotton; tobacco; other field crops; and vegetables, fruits, and nuts—and three types of operations—nurseries and greenhouses, forestry operations, and commercial pesticide-handling establishments.
2. Eight cost factors were assessed under each use-site category, as applicable. These factors (groupings of provisions of the rule) were:
   - *restricted-entry intervals* for each pesticide (typically ranging from twelve to seventy-two hours) that limit worker entry to recently treated areas;
   - *personal protective equipment*—which may include coveralls, chemical-resistant gloves, protective eyewear, or respirators—to be worn by pesticide handlers and certain workers;
   - *notification;* that is, warnings to employees of the timing and location of pesticide applications and restrictions on entry;
   - *training* in pesticide safety, covering hazards of pesticides, personal and family safety practices, WPS provisions, and, as necessary, correct use of personal protective equipment;

- *decontamination,* including provision of soap, water, towels, and, in some cases, additional supplies;
- *emergency assistance,* including transportation to medical care;
- *rule familiarization* by employers to learn how to comply; and
- *monitoring:* the costs of blood cholinesterase monitoring for selected pesticide handlers, intended to detect accumulated exposures to certain classes of pesticides; estimated for the high option only (proposed but not included in the final rule).

3. Unit cost estimates were developed, often in considerable detail, for units designed to be appropriate for each use site and factor. For example, unit costs were estimated for restricted-entry intervals on the basis of an available study of income losses per acre per year for the use of pesticides with the longer WPS restricted-entry intervals, by crop. Similarly, costs for certain personal protective equipment and decontamination were estimated on a per-person basis. Rule familiarization costs were estimated for each establishment.
4. Unit costs were then multiplied by estimates of affected units (agricultural establishments, acres of certain crops treated with specific classes of pesticides, workers, and so forth) and summed to yield total costs for an option. As appropriate, fixed or infrequent costs were isolated from continuing costs in these estimates. Results were displayed by use site and cost factor, and also in other ways, including costs on a per-establishment basis.
5. Incremental costs were then derived by subtracting, from the total costs, estimates of the costs of those actions that were already required or that may otherwise be already incurred due to existing state and federal regulations (such as California's existing pesticide regulations and OSHA's Field Sanitation Standard), existing pesticide label requirements (such as personal protective equipment for use of some products), and voluntary compliance.
6. A separate estimate was developed of the one-time costs to pesticide registrants (manufacturers and distributors) to modify pesticide product labels as required by the rule. This was based on EPA's estimate of the number of registered pesticide products requiring label revisions and a cost to registrants to make such changes of $1,000 to $2,000 per product.
7. The estimated incremental cost stream (first year and out year, expressed in constant $1991) was then also converted into annualized values.

Key Areas of Uncertainty. The cost analysis relies heavily on estimates of numbers of hired and nonhired (typically family member) workers and pesticide handlers who may be subject to the rule. Estimates of these populations vary widely. In addition, no data were available to guide estimates of how many farmworkers performed duties as pesticide handlers; EPA assumed that each agricultural establishment that used pesticides had two handlers. Only limited data were available, as well, on the distribution of hired employees among agricultural establishments.

Limited, proprietary data were available to provide estimates of pesticide usage by agricultural crop sector. Among the most important limitations of the available data was that on the potential effects of restricted-entry intervals on the yield or quality of individual crops; the single study that was found was used to make these estimates.

In addition, the analysis was unable to determine with confidence the cost of compliance with many provisions due to the range of options often available to employers in adjusting to the rule. For example, employers may choose to reschedule applications of pesticides, select pesticides with shorter restricted-entry intervals, or employ nonchemical pest control measures to avoid some costs attributed to restricted-entry intervals. Such choices may, in turn, create some losses through increased cost of pest control, reduced efficacy, or both. Similar management choices are available to agricultural employers in complying with other provisions of the WPS, and the results of these decisions are similarly difficult to predict. The RIA does not attempt to anticipate such complex, interacting, but expected adaptations to the regulation. EPA therefore concluded that its estimates were likely to exaggerate some compliance costs. Indeed, the RIA states EPA's belief that some growers will switch to less hazardous pest control methods or chemicals because of WPS requirements, and that the change is likely to reduce both compliance costs and poisonings below levels estimated in the RIA (U.S. EPA 1992a, VII-4). In general, EPA reported that it had tended to employ estimates that seemed likely to exaggerate costs, although the agency provided no estimates of the likely degree of such exaggeration.

Sensitivity Analyses. Limited sensitivity analysis of costs is included in the RIA. EPA performed an analysis of the effect of various discount rates in calculating annualized costs of compliance. The range of rates employed varied from 0% to 10% and yielded annualized costs (in constant 1991 dollars) from $53.9 million to $56 million per year, respectively, showing that results are insensitive to the choice of a discount rate (U.S. EPA 1992a, IV-12–18).

The RIA also displays the costs of various provisions of the rule individually to show which are the most significant contributors to total costs. The varied costs imposed on several agricultural sectors are similarly displayed. For example, the per-establishment, incremental, annualized compliance cost (using a 10% discount rate) ranged from $36.20 for feed and grain to $368.57 for vegetable, fruit, and nut operations.

In addition, EPA carried out a form of sensitivity analysis on the effect, on its cost estimates, of its assessment that most agricultural-establishment compliance costs were variable (proportional to the number of WPS-covered employees). EPA did this by assuming for part of its analysis that two of the cost factors, training and notification, were entirely fixed (per-establishment) costs. EPA regarded this as a worst-case assumption in its assessment of the WPS impact on small entities, required under the Regulatory Flexibility Act. As expected, this analysis showed some economies of scale expressed as per-worker costs. These differences varied among use sites. They were greater, for example, for feed and grain operations than for fruits, vegetables, and nuts (U.S. EPA 1992a, Tables VI-2, VI-8). EPA concluded that, even in this exaggerated analysis, neither of these economies of scale was large enough to justify further reductions in protection for employees on small agricultural establishments. Table 1 summarizes the compliance costs by cost factor (that is, by rule provision). (Other tables in the RIA summarize costs by crop sector and display per-establishment costs in each sector.)

Table 1. Compliance Costs of the Worker Protection Standard (WPS). (in millions of $1991)

Cost factor	*Incremental first year*	*Incremental out year*
Establishments		
Restricted-entry interval	21.1	21.1
Personal protective equipment	17.9	9.5
Notification	15.7	5.0
Training	6.9	2.3
Decontamination	12.4	8.9
Emergency assistance	0.01	0.01
Rule familiarization	6.0	1.0
Subtotal:	80.0	47.8
Commercial handler firms		
Training	0.06	0.05
Decontamination	0.4	0.4
Personal protective equipment	1.6	1.1
Emergency assistance	.0002	.0002
Rule familiarization	0.2	0.01
Subtotal:	2.3	1.6
Registrants		
Labeling changes	12.0	0
Grand total:	94.3	49.4

Source: U.S. EPA 1992a, Table IV-2.

Benefits Analysis

The RIA's benefits analysis depends almost entirely on estimating the adverse health effects that compliance with the rule could avert. This required EPA to:

- characterize existing, baseline health effects due to occupational exposure to agricultural pesticides, and
- estimate the degree to which compliance with the rule would reduce the existing rate of health effects in the future.

This entailed use of EPA's experience in performing risk assessments of the uses of individual pesticides. It also involved evaluation of the literature on relevant epidemiological and related research. Greatest attention, however, was given to systematic reports of pesticide poisonings, especially those collected by the state of California.

There is greater uncertainty associated with the benefits of the WPS than for the costs of the rule. As a result, the benefits analysis focuses far less on distributing benefits to categories of pesticide use sites or individual provisions of the rule than was possible on the cost side. This is also due to the intrinsic difficulty of determining the relative contribution of each provision of the rule in avoiding adverse health effects. For example, when a pesticide applicator avoids poisoning himself or herself, it is unclear to what degree that is a result of WPS training, as opposed to the maintenance of personal protective equipment, the availability of decontamination supplies, or other features of the WPS. The RIA also briefly discusses potential indirect benefits, such as the advantages that may flow from improved worker knowledge and health. These include possible increases in productivity and lower employer liability, insurance, and legal costs.

Key Areas of Uncertainty. The principal source of uncertainty in assessing the potential benefits of the WPS concerns the incidence of poisonings attributable to occupational exposure to pesticides applied to agricultural plants. The WPS addresses occupational exposure to hundreds of often dissimilar, usually highly bioactive substances, many of which may cause widely varying health effects in workers. The data available on these individual substances vary substantially in quality and quantity.

This inconsistency in the laboratory and field data available for older pesticides is being addressed through EPA's "reregistration" program. While this effort was accelerated by the 1988 amendments to FIFRA, it remains far from complete. Moreover, even as this program advances, science and regulatory policy will undoubtedly continue to change, rendering the databases on growing numbers of pesticides incomplete once again. In short, some uncertainty about the effects of individual pesticides is inevitable.

The use of, and thus the potential occupational exposure to, agricultural pesticides are also varied and complex. Much of it takes place outdoors, is principally dermal, is known to be episodic, and is difficult to predict generally. In contrast, some other exposure scenarios for contaminants may be more easily analyzed or modeled on the basis of averaged exposure. For example, contaminants of water, ambient air, the air of some enclosed work spaces, or even residues of pesticides in the food supply, often involve some dispersion or blending of contaminants with environmental media and thus some averaging of exposure.

In addition, agricultural work commonly involves exposure to multiple pesticides. Dozens of pesticides may be used on a single crop and individual workers may be exposed to many crops during their employment. While detailed risk assessments are often performed for individual uses of specific pesticides, there is no accepted methodology for estimating the occupational health effects of exposure to most combinations of pesticides, much less for assessing the effects of all agricultural pesticides, as a group. It seems conceivable that this problem could have been addressed by testing combinations of pesticides on laboratory animals. In practice, however, the numerous potential combinations and interactions of the pesticides would have imposed enormous additional testing costs at a time when complete testing and evaluation for each pesticide individually were proving (and remain) very difficult, costly, and time consuming. Even if pesticide combinations were tested, there would remain the unexamined potential for important interactions between pesticides and other substances to which workers may be exposed.

Despite these problems, it is clear that EPA possessed a large collection of laboratory animal studies on pesticide toxicity and more limited data on human exposure in the field. While this information was significantly incomplete by EPA's current scientific and regulatory standards, it provided a substantive basis for expecting a wide range of health effects in sufficiently exposed humans. In addition, data available on actual health effects in agricultural workers, despite their limitations, served to corroborate some of the concerns raised through animal testing. EPA also enumerated several effects that may be caused by occupational exposure to pesticides, but for which data did not permit detailed analysis or numerical estimates. Finally, EPA noted the likelihood of additional, currently unrecognized effects.

Data Problems. Most relevant data on actual human health effects are based on either epidemiological studies or poisoning reporting systems. Like all epidemiological studies, those available for agricultural pesticide exposure face problems, such as confounding exposures to other risks and imperfect records or recall about exposures and circumstances. Epidemiological studies are also limited in their ability to discern infrequent effects or increases in effects with significant background rates. On the other hand, with current scientific understanding, only long-term epidemiological studies can detect real-world health effects that exhibit delayed onset, such as cancer. That is, reports of individual, delayed-onset health problems cannot usually establish the essential link to exposures that may have occurred years before.

Acute effects, however, appear quickly and often arise from higher and more obvious exposure situations. As a result, acute effects are more readily investigated through systems relying on individual reports. Of such reporting systems reviewed by EPA, the collection of reports from treating physicians in California is by far the most systematic, reliable, and comprehensive. Even these valuable data, however, present many problems in interpretation. For example, extrapolation of California's experience to national estimates is made difficult because of differences between California agriculture and that in much of the rest of the country. These differences include the prevalence of arid conditions in California that can affect the persistence of some pesticide residues on crops. In addition, California has a relatively elaborate system of pesticide use regulation that is thought to affect exposures and that therefore further complicates extrapolation. Also, like any reporting system of its type, there are problems of completeness of reporting.

A more basic uncertainty, however, is the fact that many symptoms of acute pesticide poisoning mimic those with other causes. As a result, many incidents may be treated symptomatically by physicians, without establishing possible causation by exposure to pesticides. Moreover, many workers exposed to pesticides while working in California may not, for many reasons, seek medical care at all, or may not seek it from a physician within California. The California data were nonetheless the best available and were relied upon extensively by EPA in assessing the incidence of acute poisonings.

In the case of delayed health effects, EPA possessed inadequate data to prepare unassailable estimates of the frequency of occurrence. The RIA did include, however, "illustrative," and, in EPA's view, probably conservative, estimates of the possible annual frequency of several such effects attributable to occupational exposure to agricultural pesticides:

- *Cancer*—Cases were estimated based on an extrapolation from an EPA assessment of the risks of the pesticide Captan to field workers and the fact that about one-third of the pesticides tested and evaluated so far had been found by EPA to be oncogenic in test animals.

- *Serious developmental defects*—About one-third of tested pesticides were found by EPA to cause developmental defects in the offspring of treated test animals. Case studies tended to corroborate these effects in individual humans exposed to pesticides. In addition, epidemiological evidence suggested that congenital birth defects were more common among agricultural workers than others. With this background, EPA calculated the range of values for the annual incidence of such effects among agricultural employees if only 1% of the serious developmental defects estimated to be of unknown cause were attributable to pesticide exposure.
- *Stillbirths*—This class of effects was estimated based on limited epidemiological data indicating a higher rate among persons exposed to pesticides.
- *Persistent neurotoxicity*—EPA used published estimates of the fraction of poisonings with organophosphate pesticides that lead to longer-term neurological or psychiatric symptoms. This was combined with an extrapolation of physician-diagnosed systemic poisonings with organophosphate pesticides in California agriculture.

Other delayed health effects of pesticides have been observed in test animals, and EPA concluded that they too may well be experienced as a result of occupational exposure. Data were found to be too limited, however, to permit numerical estimates.

Sensitivity Analyses. EPA examined a variety of methods for extrapolating the California data on acute poisonings to the nation. Resulting national estimates proved to be quite sensitive to the approach taken. EPA ultimately settled on the ratio that California's hospital admissions, due to occupational pesticide exposure in agriculture, bear on the estimated national rate of such admissions. More difficult was the task of estimating what fraction of the total acute poisonings of interest are physician-diagnosed in order to estimate how many are undiagnosed and therefore unreported. Several approaches were considered, but there was great controversy on this point within the government, based on the uncertainty of the available information. EPA was ultimately unable to gain agreement from the other agencies on a method and was forced to forgo any numerical estimate in the final RIA for what seemed certain to be the most numerous health impact of the rule. The RIA nonetheless concluded that undiagnosed poisonings were an important category of outcomes. Table 2 summarizes the RIA's findings on health effects.

EPA estimated that full compliance with the revised WPS would avert 80% of these health effects. This estimate was based upon EPA's assess-

Table 2. Summary of Health Effects Attributable to Occupational Exposure to Agricultural Plant Pesticides.

Health effect category	*Estimate (per year)*
Acute	
Hospitalized poisonings	300 to 450
Physician-diagnosed (nonhospitalized) poisonings	10,000 to 20,000
Undiagnosed poisonings	Unquantified[1]
Delayed-onset[2]	
Cancer cases	6 or more
Serious developmental defects	20 to 52 or more
Stillbirths	56 to 222
Persistent neurotoxicity cases	150 to 300
Others	Unquantified[3]

Notes: The following summarize some of the nonquantitative assessments of the RIA:

[1]This class of acute effects was characterized as "a significant number...very likely to be large."

[2]This category of health effects was characterized as having "potentially important numbers" overall, with qualifications on precision of largely illustrative estimates.

[3]These other instances of delayed onset health effects included spontaneous abortions, infant mortality, sterility, infertility, impotency, and systemic effects to the heart, circulatory system, skin, lungs, respiratory system, liver and kidneys. For these effects, it was noted that "some or all of these effects may be occurring."

Source: U.S. EPA 1992a, V-20–33.

ment of the overall effectiveness of major WPS provisions and the available studies on the efficacy of personal protective equipment, engineering controls such as enclosed tractor cabs, training, and decontamination practices in reducing exposure. These studies suggest efficacy ranging from near 100% for some provisions under ideal conditions to lower estimates for other provisions. EPA arrived at its overall estimate by considering both the practical realities of the agricultural workplace that might limit the efficacy of some provisions and that many of the provisions would reinforce each other when implemented in combination through the WPS.

The RIA makes no attempt to undertake a quantitative benefit-cost analysis by assigning monetary value to the expected reduction in health effects. It does, however, qualitatively discuss the value of some benefits, such as potential reductions in insurance costs, lost time from the workforce, reduced medical expenses, and increased well-being and productivity. EPA concluded that such benefits "cannot be adequately quantified with available data" (U.S. EPA 1992a, V-29).

Similarly, the final RIA reveals no attempt to calculate compliance costs per poisoning avoided. This would require that regulatory compliance costs be allocated to each of the categories of expected health effects. In earlier versions of the RIA, however, EPA attempted such analyses for all categories of potential health effects for which the agency felt useful rough estimates could be made. In EPA's view, display of an allocation of implementation costs to even these incomplete and conservative estimates of effects demonstrated that the regulatory costs for avoiding each such incident were unlikely to be unreasonably high. While EPA held that these admittedly crude results shed important light on the deliberations, the agency was unable to convince USDA and OMB of the value of this approach. One of USDA's major concerns was that any allocation of costs to categories of effects would be arbitrary. In addition, USDA and OMB resisted many of the underlying numerical estimates of incidents. Despite this dispute within the government on what analyses would be included in the document, the RIA ultimately concluded that "the benefits to society from avoided incidents of acute, allergic, and delayed adverse effects from occupational exposures to agricultural-plant pesticides exceed the costs attributable to this final rule" (U.S. EPA 1992a, V-29) .

Other Analytical Considerations

The RIA also considered possible economic effects of the WPS on the production and prices of agricultural commodities as well as on small economic entities. While any explicit analysis of distributional issues was absent from the RIA, EPA did attend to some related issues.

Economic Impacts. Section 25 of FIFRA requires EPA, in promulgating a regulation, to consider such factors as possible effects on the production and prices of agricultural commodities, effects on retail food prices, and other effects on the economy. In the case of the WPS, EPA compared the costs of compliance with the WPS with other agricultural production costs and with revenues. WPS incremental out-year costs were found to represent less than 1% of user expenditures for agricultural pesticides and less than 0.1% of the total value of the crops affected by the regulation (U.S. EPA 1992a, II-3–4). As a result, EPA regarded the WPS as unlikely to measurably affect commodity supplies or prices, or have any other discernible effects on the economy as a whole.

Effects on Small Economic Entities. EPA conducted an analysis, as required by the Regulatory Flexibility Act, of the potential effects of the WPS on small entities, many of which are subject to the WPS (U.S. EPA 1992a). It found that the compliance costs of most WPS provisions were largely variable; that is, proportional to the number of workers covered on each establishment. It also noted that the WPS exempts establishment owners and their immediate families from many provisions, thereby significantly reducing costs in the establishments with the fewest workers.

Other WPS options and exceptions may also favor small establishments. EPA therefore concluded that "the rule has avoided or mitigated, to the extent feasible, potential disproportionate burdens on small entities and is structured to provide a nearly equitable burden on both small and large entities" (U.S. EPA 1992a, VI-1).

Distributional Issues. No explicit analysis of distributional issues is apparent in the RIA. It could be argued, however, that such questions are appropriate, since occupational risks are not borne by consumers of agricultural products at large, who are thought to enjoy lower prices, improved quality and variety, and perhaps other advantages as a result of pesticide use in agriculture. While consumers may be exposed to residues of pesticides in their diets, this exposure is considered to be far lower than that experienced by workers. In short, the risks and benefits of agricultural pesticide use are distributed differently.

In addition, the agency was aware of the well-known disadvantages faced by many farmworkers, including those in health care, education, language, a variety of employment issues, and economic security generally. As a result, it is fair to say that considerations of what has been called environmental equity or environmental justice did play some role in EPA decisionmaking. Examples of provisions of the final rule that were influenced by such factors include notification (which requires that workers be given access to information about pesticide use without having to ask their employers for it), multilingual pesticide labels and warning signs, and the assignment of principal compliance responsibility to employers. The agency did not, however, include any specific analysis from this perspective in the RIA.

CRITIQUE AND OBSERVATIONS

A critique of the RIA is best separated between two perspectives: reactions to it upon its issuance and those from a contemporary perspective.

Reactions at the Time of the RIA

USDA provided extensive comments on the rule and, in particular, on the RIA. That agency's concerns about the substance of the rule led to several changes. One of these was EPA's review of all existing long reentry intervals (greater than seventy-two hours) on pesticide product labels. (*Reentry interval* is the older term for restricted-entry intervals.) This was done to ensure that the intervals would remain appropriate under the new WPS regulatory scheme. EPA also added an administrative exception process to the rule to permit EPA to grant entry to recently treated areas, with special protections for workers, to perform specific, time-sensitive, and economically vital activities. A training grace period was added to relieve burden on employers by giving them more time to comply after a new employee begins work. The rule was also modified to address USDA concerns about the need for crop advisers to have ready access to treated areas. The quality criteria for decontamination water were revised to ease employer compliance. Other USDA comments did not result in changes to the rule.

USDA's comments on the draft RIA resulted in a thorough revision of the document, including the development of far more detailed cost analyses and, ultimately, a less quantitative approach to benefits assessment. In the revision, EPA employed additional data obtained from USDA and other sources. Important revisions to the cost analysis included changes to the estimates of training costs to better reflect costs beyond those already required by OSHA's Hazard Communication Standard. USDA comments also caused EPA to perform a more detailed analysis of the degree to which some costs of the decontamination provisions of the WPS may be fixed (on an establishment basis). USDA suggested that the RIA show costs of lost wages, productivity losses, and reduced competitiveness due to restricted-entry intervals. EPA responded that it found no evidence of these phenomena in California, where restrictions on entry to treated areas similar to those in the revised WPS had been in effect for years. EPA therefore concluded that such effects were unlikely in response to the new national standard.

Much of USDA's criticism, however, was directed at the draft RIA's benefits estimates. USDA questioned whether the rule met the benefit-cost test, indicating that (based on earlier cost estimates) it would take 239,000 avoided hospitalizations, at $580 per incident, for the rule to show positive benefit. EPA responded that it had not relied solely upon reductions in hospital admissions for acute poisonings to justify the rule. Rather, EPA regarded a broader measure of benefits as essential, including reductions in nonhospitalized cases, reductions in lost work time, reduced medical treatment and insurance expenses, and increased well-being and productivity. Prominent among USDA's criticisms of the draft benefits assessment was its concern about the supportability of a published estimate, cited by EPA, that 300,000 acute occupational pesticide poisonings occur annually in U.S. agriculture. In its response, EPA rejected the specific bases for USDA's critique of the estimate, but acknowledged the difficulty of making a reliable extrapolation, as discussed above. Largely as a result of USDA's concern, no specific estimate of total acute effects (both diagnosed and undiagnosed) appears in the benefits discussion of the revised RIA.

Many additional comments were received from other parties after the 1988 proposal, but most concerned the substance of the rule rather than the RIA. Many prompted departures from the proposed rule. For example, information from the cut roses industry led EPA to publish, along with the WPS final rule, a proposed administrative exception to permit, under special limitations, harvesting of cut flowers and cut ferns during restricted-entry intervals. EPA issued a final, temporary version of this exception, limited to cut roses, in June 1994.

Critique from Today's Perspective. Given EPA's statutory and informational constraints, the most obvious alternatives were considered during development of the rule. As discussed earlier, it is difficult to see how market mechanisms could have readily been applied given the infeasibility of adequately measuring associated exposures or health effects.

On the other hand, the rule would probably be improved, and compliance costs reduced, by allowing more discretion in compliance through greater use of performance standards (as opposed to specifications of behavior). This could be useful, even if the measures of performance were intermediate (for instance, using surrogates based on reduced exposure rather than avoided illness or injury). The WPS is not, however, devoid of such approaches.

For example, the WPS provides exceptions to the requirement that certain personal protective equipment be worn when enclosed cabs are used on pesticide application equipment. The agency's intention was to create incentives for the development and adoption of engineering controls, such as effectively protective enclosed cabs. The rule sets a level of performance for the air filtration systems on such cabs, not engineering design specifications.

In many other areas, however, such performance-based approaches proved elusive. For example, restricted-entry intervals are probably the most important provision of the rule in their ability to improve the protection of workers, simply by keeping most workers out of recently treated areas. On the other hand, they also appear to be most troublesome and costly to growers. The RIA found that the restricted-entry provisions made the largest contribution to out-year WPS costs, representing about 40%. This, combined with EPA's acknowledgment that the science underlying its process for setting interim intervals will require refinement, suggests that replacement for the restricted-entry interval approach should receive a high priority. Field residues of pesticides are known to dissipate or break down after application at a variety of rates, often depending on weather, soil type, sunlight, and other factors. As a result, any fixed intervals, such as the restricted-entry intervals set by the

WPS or individually by EPA, are likely to be sometimes longer than necessary and at other times inadequate for the protection of workers.

In the preamble to the WPS final rule, EPA discusses these concerns in detail and expresses the hope that better methods will be developed. These might involve the use of inexpensive and reliable field test kits and protocols to determine actual residues that might threaten workers. There have been promising advances in such methods, based on immunoassay techniques. This technology could allow, particularly for the most hazardous pesticides, a reentry scheme based on site-specific exposure risks, rather than a fixed interval such as the restricted-entry interval. Similarly, development of more practical worker exposure monitoring technology (such as chemical-specific exposure badges or the equivalent) might permit EPA to give employers wider discretion in where to send workers, when, and for how long, while holding them accountable for exposures that do occur.

With respect to overall balance in the WPS RIA, EPA's investment in the detailed modeling on the cost side seems excessive given the profound uncertainties on the benefits side. To a degree, the cost analysis was responsive to specific criticisms, and the anticipation of more, from USDA and OMB. On the other hand, most of the additional detail did not actually seem to contribute significantly to decisionmaking on the substance of the rule, nor is it clear how it could have.

Finally, the paucity of comprehensive, reliable data on baseline health effects proved to be the largest single obstacle in governmentwide deliberations on this rule. Additional investment at the national level in developing such information could assist future efforts to select cost-effective regulatory approaches. However, part of the problem that EPA faced is unlikely to yield soon, if ever, to additional research.

If history is a guide, agencies responsible for public health regulation will often have concerns based on emerging but not yet fully understood risks. This is particularly true with respect to exposures to bioactive synthetic chemicals like most pesticides. An example is provided by recent controversy (and concern expressed by EPA officials) over "estrogenic pollutants," including many pesticides. Some scientists have adopted the view that low levels of exposure to such compounds may disrupt endocrine functions. Even if this phenomenon does not prove to be the problem some believe, experience suggests that others will take its place. Some of these may be substantiated and incorporated into standard risk assessments, as appears to have happened in recent years with delayed neurotoxicity concerns. But until that happens, such uncertain but potentially important issues will remain difficult to address analytically. Perhaps, however, they should not be ignored in regulatory analysis because of this uncertainty. History may argue instead for prudence. This might take the form of granting a measure of the benefit of the doubt in regulatory analysis for uncertain but reasonably based health effect concerns—more than that displayed in the final WPS RIA benefit assessment. This could be particularly appropriate when the interventions under consideration appear to be fairly inexpensive for the affected industry, as seems to have been the case with the WPS.

Relation of the RIA to Policy Processes and Decisions

As described at the beginning of this chapter, EPA began the WPS rulemaking after finding that the existing regulation was badly deficient as a mechanism to provide across-the-board protection from exposure to agricultural pesticides. This general concern was reinforced by EPA's slow pace in evaluating older pesticides individually through its reregistration program, which affords an opportunity for adjustments to the specific worker protection requirements for each pesticide product. Little more was needed to convince EPA staff that the rule needed strengthening if EPA was to continue to maintain that agricultural pesticides were not causing "unreasonable adverse effects," as FIFRA requires to justify any pesticide registration. While some provisions of the revised rule remain controversial, few critics have disputed this basic conclusion by EPA. Thus, the RIA was not critical during the earliest stages of the rulemaking.

During deliberations within EPA, before the draft final rule was sent to USDA or OMB, the issues typically involved basic workability and reasonableness, driven by the enormous range of situations found in American agriculture. As a result, EPA staff tended to focus its impact assessment less on formal, quantitative cost analysis for the RIA and more on the informal assessment of the practical effects of individual provisions of the rule, on both exposure and agricultural practices. EPA invested considerable effort in the analysis of draft provisions in an attempt to find and fix those that might lead to unreasonably burdensome, ineffective, or nonsensical efforts to comply. Besides, EPA had already satisfied itself, as expressed in the RIA for the 1988 proposed rule, that the total incremental cost of the rule would likely be so small a fraction of total costs for agriculture that significant economic disruptions were unlikely, and that the potential risk reductions for workers were likely to greatly outweigh cost. The detailed analysis and refinement leading to the final RIA largely confirmed both this overall judgment and the operating assumptions of the staff working on the rule: that fruit and vegetable operations would bear the greatest burden of compliance due to their frequent use of both pesticides and hand labor, while also offering the greatest opportunity for risk reduction.

Even the most noteworthy reduction in compliance burden between the proposed and final rules—elimination of the requirement for cholinesterase monitoring of certain pesticide handlers—was a decision based largely on nonquantitative, pragmatic concerns about feasibility and effectiveness, not on specific cost estimates developed for the RIA. Another key example concerns deliberations over risks to workers entering recently treated fields. In theory, personal protective equipment could perform almost as well, in reducing exposure, as the obvious alternative of excluding workers from those fields. As a result, this might seem to be an opportunity for a straightforward cost-effectiveness analysis. That is, the cost of the use of personal protective equipment could be compared to any losses in yield or quality stemming from keeping workers out during restricted-entry intervals. EPA was persuaded through comments and its own knowledge, however, that the routine use of personal protective equipment for often heavy hand labor in fields, such as during harvesting, was not a realistic or acceptable option. The agency was concerned not only that the required personal protective equipment would not be correctly used or effective in such circumstances, but that its use in typical field conditions could cause additional risk though heat-induced illness. This concern and similar practical assessments proved more influential in shaping the final rule than the formal, quantitative cost estimates in the RIA. In short, the more formal aspects of the RIA cost analysis evolved within EPA as the rule did but did not always play a key role in decision-making.

The RIA played an especially central role, however, during the end game of promulgating the final rule, when the debate involved EPA, USDA, and OMB. As indicated earlier, the benefits assessment of the RIA received more attention at this stage than any other facet of the effort. While there was also deliberation on specific provisions of the rule at that time, those discussions, again, tended to involve informal, nonquantitative analysis of issues of practicality, technical feasibility, or relative convenience.

On the other hand, concern about the potential cost to growers of the WPS restricted-entry intervals was critical to the final form of the rule. The concern was that the government may not adequately understand the costs of compliance with these requirements in all facets of the production of all crops, in all regions and circumstances. This led directly to the addition to the final rule of the administrative exception process for "early entry" at Section 170.112(e). This provision is designed to be a safety valve in the event that the agency's understanding of the effects of the rule were significantly flawed in narrow cases. It allows EPA to make timely adjustments to prevent excessive and unforeseen economic consequences of WPS implementation.

Finally, the draft RIA's use of extrapolated poisoning and epidemiological data in the benefits assessment proved to be highly sensitive to the agricultural community in general and to USDA in particular. USDA argued that the real uncertainties involved rendered EPA's quantitative estimates unjustified. EPA, however, typically must make pesticide registration decisions based on uncertain data on risks. For potential health effects, the information usually consists solely of laboratory animal data. Thus, EPA's basic registration decisions are usually made before any significant human exposure has occurred to provide even the imperfect poisoning data that exist for historical pesticide use. To an agency so accustomed to extrapolating from rodents to humans in such decisions, it seemed natural to use available human data to make rough quantitative estimates of adverse effects in evaluating regulatory actions to manage those risks. This conflict in basic approach consumed more time than any other in final deliberations and significantly delayed promulgation. The controversy nonetheless had scant effect on the provisions of the final rule.

On balance, the effort to systematically assess costs and benefits in the RIA was useful in deciding the general direction of the rulemaking, but also served as a lightning rod for concerns that delayed final promulgation.

CONCLUSIONS

The RIA did develop information and analyses that supported decisionmaking. The early results, during development of the 1988 proposed rule, corroborated EPA's starting assumption by providing rough and probably exaggerated cost estimates that did not appear unreasonable to EPA given the large expected benefits of a strengthened WPS regulation. In the late stages of developing the final rule, the formal estimates in the RIA played a narrower but still useful role—for example, by focusing attention on the basis for EPA's understanding of the economic impact of restricted-entry intervals.

If this rulemaking were to be relived, it might profit most from more study, earlier, of the impact of a range of entry restrictions after pesticide application (restricted-entry intervals). This would no doubt have proved difficult since most growers have many options for responding to potential restrictions of that type. For example, they can reschedule the pesticide application; reschedule or, sometimes, mechanize the labor tasks; or change pest control method or pesticide. Given this range of available choices, whose feasibility will often depend on crop, regional, or other variables, it is difficult to determine how decisionmakers would respond to a hypothetical change, how often, and with what effect. With the WPS in effect since January 1, 1995, EPA is receiving requests for exceptions to the entry restrictions. These petitions should be a source of such data based on experience. The requests to date, however, seem to contain little convincing information of this kind. Nonetheless, future requests may prove the best source of facts to compare to the RIA's estimates as field experience with WPS compliance accumulates.

As discussed above, many of EPA's decisions on the details of the provisions of the final rule turned on basic questions of their likely workability, relative convenience, or practical effectiveness. Such questions continue to be the focus of controversy since the final rule was issued. While the informal investigation of such issues may be regarded as a form of economic analysis, the more formal and quantitative approaches typically used in RIAs are not necessarily well suited to address such detailed concerns during rulemaking. At least, that seemed to be so in the case of the WPS. Field tests of proposed provisions, on the other hand, may have offered an attractive option for helping EPA to better understand these important aspects of implementation costs. Volunteers could have been sought to represent the range of affected establishment types and attempt to implement all or parts of the proposal, or alternatives to it, with EPA help and observation. Workers would have needed to play a key role in such experiments to help assess effectiveness of the proposed rule. It could be argued that the regulatory negotiation process undertaken for the WPS, had it been successful, could have provided just such insight to EPA's decisions. While that is no doubt true to a degree, representatives of national organizations (typical participants in regulatory negotiations) are not always able to know and convey the full range of circumstances encountered on individual establishments, much less how those circumstances may affect implementation of the rule in detail. In any case, their views are unlikely to be as informative as real-world attempts to carry out proposed provisions. Without such information, EPA was left with the written comments, supplemented by discussions with interested parties before and during the comment period (*ex parte* communications are not permitted after the close of the comment period), and thought experiments based on all that information and EPA's own knowledge of agricultural practices.

It is sobering, however, to suggest any elaboration of a rulemaking that took eight years to complete. In order to avoid additional delay, it would presumably have been necessary to carry out any field tests before or during the comment period. In any case, with such information, the government could have been more confident in the quality of its decisions, and in the cost-effectiveness of the regulation. On the other hand, the cost of such an information-gathering process could have been significant. This is especially true if the tests were numerous enough to assess a wide variety of the circumstances to which the rule must apply, including various crops, practices, seasons, and regions. It is possible, however, that information gathered in this way, and perhaps the resulting regulations, would have been more widely accepted because interested parties would have felt more involved in creating them. This, in turn, could have accelerated and smoothed some of the implementation difficulties EPA encountered with the WPS.

Finally, an improved and widely accepted understanding of the likely health effects due to occupational exposure to agricultural pesticides would improve the quality of the policy debate surrounding this area of regulation. In addition to basic laboratory studies of the effects of individual pesticides and field studies of exposure, this need extends to establishment of accepted techniques for addressing exposure to multiple toxins, epidemiology of delayed effects of pesticides, more comprehensive reporting of pesticide poisoning incidents, and methods for extrapolating from reports of physician-diagnosed poisonings to all that occur nationally, including those that are not physician-diagnosed. On the other hand, while more information on health effects would have enriched deliberations during the assessment of potential WPS provisions, the option of getting more was not free with respect to either resources or time. Even if resources had been available, unless the investment were made far in advance, the cost in time could have taken the form of substantial delay in issuance of the WPS and a resulting loss of benefits. Moreover, in the case of the WPS and similar regulations, the effects on health, even if known precisely, do not necessarily permit better decisions in the absence of consensus on how to value those effects of the rule. In short, further delay in the issuance of the WPS to gather additional health effects information was probably unjustified.

REFERENCES

U.S. EPA (Environmental Protection Agency). 1992a. *Regulatory Impact Analysis of Worker Protection Standard for Agricultural Pesticides.* August 11. EPA Docket OPP-300164. Prepared by EPA's Biological and Economic Analysis Division with support from DPRA, Inc. Washington, D.C.: U.S. EPA. Office of Pesticide Programs.

———. 1992b. *Worker Protection Standards: A Summary of the Public's Comments and the Agency's Response.* August. EPA Docket OPP-300164. Washington, D.C.: U.S. EPA.

12

Vehicle Inspection/Maintenance

Todd Ramsden

On November 5, 1992, after two intense years of policymaking activities, EPA published its final rule on vehicle inspection/maintenance (I/M) programs. In compliance with the Clean Air Act Amendments of 1990, the I/M rule was designed to address the in-use air pollution of motor vehicles and to ensure that vehicles are operated in accordance with standards. According to the regulatory impact analysis that was performed, the rule represented a well-balanced and rational approach, significantly reducing pollution at a low cost. At the same time, and despite the fact that the RIA did contribute to lowering the costs of the final rule, the I/M rulemaking process has been criticized on a number of fronts. First, it has been argued that the RIA did not really serve as an analytical tool for guiding the development of policy. That is, EPA has been criticized for narrowly focusing on a predesignated approach rather than comparing a wide and inclusive set of policy options. Second, it has been suggested that a significant amount of the rule's costs were not accurately represented in the RIA, such as the cost of inconveniencing car owners. Additionally, other technologies that may have achieved virtually the same results at lower cost were not fully investigated. Ideally, the RIA would have at least reflected the flavor of the negative political reactions to the rule, which, ultimately was revised by the Clinton adminis-

TODD RAMSDEN was an analyst with EPA's Office of Policy, Planning, and Evaluation during the development of the proposed and final rulemaking for enhanced inspection/maintenance. He is currently pursuing a doctorate at the University of Michigan's School of Natural Resources and the Environment.. The opinions expressed in this paper are the author's and do not necessarily reflect the views of any organization.

tration in 1995. In this case study, Todd Ramsden reviews the history and content of the I/M rule, and explores the issues involved with and the controversies surrounding the regulatory impact analysis and rulemaking process in general.

INTRODUCTION

In the United States, motor vehicles are a significant source of urban air pollution, contributing one-third of emissions of nitrogen oxides (NO_x), one-quarter of the emissions of volatile organic compounds (VOCs) , and well over half of carbon monoxide (CO) emissions (U.S. EPA 1994). Ozone air pollution (a major component of urban smog) has proven itself to be one of the most intractable urban air pollution problems. Located in the troposphere (ground-level), ozone is a health and ecological concern, leading to respiratory tract inflammation and ailment and reducing tree and crop growth. (In the stratosphere, ozone is a vitally important gas: it helps form a protective shield that filters out harmful ultraviolet radiation.) Tropospheric ozone is formed when ozone precursors, mainly VOCs and NO_x, react photochemically in the presence of sunlight, typically occurring on hot summer days. Carbon monoxide, which interferes with the oxygen-carrying capacity of the blood and causes cardiovascular problems, results from incomplete fuel combustion typically occurring during the colder winter months. Automobile use, which results in significant emissions of VOCs, NO_x, and CO, thus contributes to air pollution problems throughout the year.

Beginning in the late 1970s, automobile inspection and maintenance (I/M) programs have been used to reduce the emissions of these pollutants by ensuring that vehicle emission control systems are in proper working order. Actual emissions of vehicles on the road have been discovered to be, on average, one and a half to two times higher than the design value to which the vehicles were certified, with some vehicles having emissions fifty times higher (McConnell and Harrington 1992). Recognizing that these so-called "in-use emissions" from vehicles were an ongoing problem, Congress, as part of the 1990 Clean Air Act Amendments (CAA, or the Act), required that vehicle inspection/maintenance programs be updated. These revisions are the central focus of this case study. It should be emphasized that CAA requirements narrowed EPA's policymaking focus from addressing in-use vehicle emissions to the more narrow task of developing more effective I/M programs.

To comply with Section 182 of the Act, EPA needed to promulgate regulations regarding enhanced inspection/maintenance programs.[1] To accomplish this, EPA first had to develop a model I/M program and determine the emission reductions resulting from this program. Since state I/M programs had to achieve at least this level of emissions reductions, EPA also needed to decide how to assign emission reduction credits to state programs that were not the same as the model program. Finally, EPA had to decide which program elements would be required of all I/M programs.

Clean Air Act Requirements

The Clean Air Act requires *national ambient air quality standards* (NAAQS) to be established for air pollutants for which air quality criteria have been issued (these air pollutants are referred to as "criteria pollutants"). For a given criteria pollutant, Section 109 of the Act requires that primary NAAQS be set at a level "allowing an adequate margin of safety... requisite to protect the public health."[2] EPA has set a NAAQS for ground-level ozone at 0.12 parts per million (ppm) averaged over one hour and not to be exceeded more than three times in a three-year period. For carbon monoxide, the ambient air quality standard is 9.1 ppm.

The Clean Air Act, as amended by the Clean Air Act Amendments of 1990, classifies areas that are not in attainment of the national ambient air quality standard for ozone or carbon monoxide. This classification system designates ozone "nonattainment" areas as "marginal" to "extreme" based on their design values (defined as the fourth highest monitor reading over a three-year period), requiring different deadlines for achieving attainment depending on the severity of their ozone problem (see Table 1). Section 182 of the CAA requires certain, specified control programs be put into place for sources within each ozone nonattainment category. Similarly, Section 187 requires various control programs for CO nonattainment areas. Ozone and CO control programs include application of *reasonably available control technologies* (RACT) to major stationary sources, implementation of automobile inspection and maintenance programs, requiring 1.5-to-1 pollution "offsets" for new stationary sources, and adoption of transportation control measures designed to reduce vehicle miles traveled.

In addition to the control measures specifically required for each nonattainment category, Section 172(c)(2) requires ozone nonattainment areas to meet *reasonable further progress* (RFP) targets. Section 182(b) specifies that to achieve RFP goals, "moderate" and worse nonattainment areas must reduce emissions of VOCs by 15% by 1996 compared to 1990 levels. "Serious" and worse areas must also reduce VOCs by an average of 3% per year (averaged over a rolling three-year interval) starting in 1996 and continuing until the attainment date. NO_x reductions may be substituted for VOC reductions to meet this 3% per year goal.

To demonstrate that attainment will be achieved by the target dates, Section 110 of the CAA requires that each state submit a *state implementa-*

Table 1. Ozone and CO Nonattainment Area Classifications and Attainment Dates.

Pollutant	*Area classification*	*Design value (parts per million)*	*Attainment date*
Ozone	Marginal	0.121 up to 0.138	Nov. 15, 1993
	Moderate	0.138 up to 0.160	Nov. 15, 1996
	Serious	0.160 up to 0.180	Nov. 15, 1999
	Severe	0.180 up to 0.280	Nov. 15, 2005
	Extreme	0.280 and above	Nov. 15, 2010
Carbon monoxide	Moderate	9.1 up to 16.5	Dec. 31, 1995
	Serious	16.5 and above	Dec. 31, 2000

Source: Clean Air Act (42 U.S.C. 7401-7626), Sections 181(a) and 186(a).

tion plan (SIP) "which provides for implementation, maintenance, and enforcement of such primary [national ambient air quality] standards in each air quality control region (or portion thereof) within such State." In addition to describing the state programs developed in response to the specific requirements of the Act, the SIP must also show what measures the state will take to meet RFP requirements for each nonattainment area. To enforce this requirement, Section 179 allows the EPA administrator to apply sanctions to any state that fails to submit a plan or submits a plan lacking any of the elements required by the Act. If a SIP has not been corrected within eighteen months of a finding of inadequacy, a state may be denied federal highway grants or may be required to obtain two-to-one emission offsets before gaining permit approval for a new pollution source in a nonattainment area.

Mobile Source Pollution Control

Based on the requirements of the 1990 Clean Air Act Amendments, EPA has adopted a three-part strategy to reduce air pollution emissions of motor vehicles: cleaner cars, cleaner fuels, and in-use compliance. Continuing the regulatory trend that has spanned three decades, EPA issued technology-based standards for automobile manufacturers, designed to reduce the tailpipe and evaporative emissions of new cars and trucks. Examples of such regulation are Tier I tailpipe standards, on-board diagnostic equipment, and on-board vapor recovery systems. Since automobile emissions are partly a function of the fuel used, EPA also promulgated rules requiring the use of oxygenated fuels and reformulated gasoline to reduce the emissions of CO and ozone precursors, respectively. Inspection and maintenance programs, designed to ensure that vehicle emission control systems are working properly, are the primary tool used to achieve in-use compliance.

EPA's first I/M policy was established in 1978 in response to the requirements of the 1977 Clean Air Act Amendments. Essentially, the inspection program consisted of a test of a vehicle's tailpipe emissions with the engine running at idle or at a combination of idle and 2500 rpm. Recognizing that existing I/M programs were not adequately addressing in-use compliance, Congress called for an overhaul of I/M policy as part of the 1990 Clean Air Act Amendments. The 1990 amendments required changes to the existing I/M programs (referred to as *basic* inspection/maintenance programs) that would be required in marginal and moderate ozone nonattainment areas and in moderate CO nonattainment areas. For serious and worse ozone nonattainment areas and serious CO nonattainment areas, the Act required adoption of enhanced inspection/maintenance programs in urban areas.[3]

CONTEXT OF THE I/M RULEMAKING PROCESS

According to Section 182(c)(3) of the amended Act, state-enhanced I/M programs shall apply to light duty cars and trucks and, with limited exceptions, must include at a minimum: use of computerized emission analyzers, on-road vehicle testing, expenditure of at least $450 on repairs before a waiver can be issued to a failing vehicle, and enforcement through registration denial. Additionally, state programs require annual testing of vehicles and operation of a centralized testing facility unless the state demonstrates to EPA that the state's proposed decentralized and/or biennial program, together with other features, will achieve equivalent emission reductions. Section 182 requires that each state develop an enhanced I/M program that complies with guidance to be issued by EPA and submit to EPA a revised SIP that incorporates this program within two years of the enactment of the amendments (namely, by November 15, 1992).

Section 182(c)(3) of the Act requires state enhanced I/M programs to "comply in all respects" with EPA's guidance. It further directs EPA to base its guidance on a performance standard.[4] To meet the requirements of the Administrative Procedures Act regarding such a binding performance standard, EPA decided to promulgate a formal rulemaking for inspection/maintenance programs including public notice and comment. The rulemaking encompassed both basic and enhanced I/M programs. Since the enhanced I/M program has been significantly more controversial than the basic program, the remainder of this paper will address issues related to the development of the enhanced I/M performance standard.

Focus of the Rulemaking

To set a performance standard, EPA decided it needed to develop a "model" enhanced I/M program and then determine what emission reductions would result from operation of such a program. The development and use of a model program to determine compliance led to three significant policy decisions: what specific test procedure would define the model program, what testing network type would be assumed for the model program and which network types would be allowed, and how emission reduction credits would be assigned to programs that did not exactly match the model program. Another important policy decision concerned which specific program elements would be required of all enhanced I/M programs, although to a great extent, this was predetermined by the Clean Air Act. Beyond these issues, EPA needed to decide to what extent auto manufacturers needed to provide repair information (especially regarding computer codes for a vehicle's emission control unit), the degree to which remote sensing (on-road testing) would be used, and when state programs needed to be implemented.

A great deal of EPA's efforts revolved around developing a new testing procedure. The idle-exhaust tests in use at the time worked reasonably well for older, carbureted cars, but they did not work well with modern vehicles with fuel injection and computer controlled engine operations. Furthermore, these inspection tests were easily defeated, allowing dirty cars to falsely pass the test. To address these issues, EPA sought to develop a model test procedure that would fully and accurately test newer cars with advanced emissions control systems while limiting the potential for cheating by inspectors and motorists.

The other central question EPA faced regarding the form of the model program was the type of testing network in which the test procedure would be performed. In general, two testing network types exist. In a centralized inspection network, vehicle testing is performed in larger, high-volume, test-only stations run by either a government agency or by a single contractor. On the other end of the spectrum are decentralized systems where emission tests are performed by numerous, governmentally licensed private businesses, such as gas stations or automotive service shops. Typically, these decentralized inspection sites combine testing with other services, often automobile repair. Within these pure test network types fall hybrid systems that mix elements of both. One type of hybrid system is a multiple party, decentralized network in which participants conduct inspections only, relying on higher volume to remain profitable. Another type of hybrid system uses multiple, test-only contractors, each with a defined territory.

From EPA's perspective, the most relevant component of network type was whether or not the network provided separation of testing and repair. Based on historical performance data and the results of covert audits, EPA found that decentralized, test-and-repair networks were not as effective as centralized systems where the test function was performed separately from repair. EPA believed that there was an inherent conflict of interest where businesses that service vehicles also perform emission inspections. Specifically, to maintain good relations with repair customers, EPA believed that decentralized inspection stations that also repaired vehicles might falsely pass vehicles that should fail, reducing the emission benefits of the I/M program. As such, EPA virtually defined "centralized" to mean "test only." In the draft regulatory impact analysis (RIA) that accompanied its proposed I/M rule, EPA noted that "where programs are referred to as either 'centralized' or 'decentralized,' this indicates only whether or not the functions of testing and repairing are separated and performed by different entities (U.S. EPA 1992a)."

Aside from potential conflicts of interest, choice of test network affects the performance of an I/M program more directly. On the cost side, there is a possibility that centralized test networks may be more inconvenient to vehicle owners, resulting in greater time costs. This might arise in two ways. Operation of large, centralized facilities is likely to mean that fewer stations are available, thereby increasing driving distance for many vehicle owners to have the test conducted. Perhaps more problematic is the need for owners of failing vehicles to bring their cars to another location for repair and then back for followup testing. This might result in a "ping-pong" effect as drivers shuffle their cars back and forth between test center and repair shops. In contrast, decentralized testing networks may have higher inspection costs due to the generally higher cost of labor in decentralized systems (since many of the inspectors are highly trained and paid auto mechanics) and due to the lower vehicle throughput in decentralized systems.

Affected Stakeholders

As with many of EPA's rulemakings, there were numerous parties interested in the I/M policy process. The rule affected states, motorists, automobile manufacturers, automobile dealers, inspection test equipment manufacturers, inspection stations, automotive repair and service shops, and environmental groups. These stakeholders were represented by numerous lobbying groups. Groups like the Motor Vehicle Manufacturers of America, representing auto manufacturers, had somewhat limited interest in this rule. (MVMA was most concerned with promulgation of a test which provided significant emission reductions and correlated well with the Federal Test Procedure used to certify new vehicles for sale.) Other parties were most interested in one or both of two aspects of the I/M rulemaking: design of the model test program and choice of inspection network.

Choice of the model test procedure had the greatest impact on test equipment manufacturers. This question affected makers of BAR90 computer emissions analyzers, dynamometer manufacturers, and companies involved in the production of remote sensing devices. The choice of test networks was another element of the rulemaking. Whether or not test and repair functions would be allowed in one facility affected principally automobile dealers (represented by the National Automobile Dealers Association) and auto repair and service shops (represented by the National Automotive Service Association).

DEVELOPMENT OF THE I/M RULE

As discussed above, EPA's major focus in the development of the I/M rule was the creation of a model I/M program. The model program was designed to serve as the basis of a performance standard that state I/M programs would need to meet. To gain approval of its state implementation plan (SIP), a state needs to demonstrate, using EPA's mobile source emission factor model (that is, MOBILE4.1), that the state's proposed program would reduce emissions at least as much as EPA's model program. EPA would make this equivalency determination for the particular pollutants of concern (carbon monoxide, VOCs, or NO_x) compared to reductions required by specific milestone dates which depended on the nonattainment classification.

Perhaps due to its congressional mandate or perhaps due to its institutional preferences, EPA took a very narrow view of what options were available to it. Although the decision to regulate I/M programs through the use of a performance standard presumably increased flexibility, EPA did not consider other methods of regulating I/M programs. This was largely a function of the CAA requirements, but it also reflects EPA's inclinations on how to regulate in-use emissions of vehicles. The prescriptive nature of the Act did not readily allow more innovative, perhaps market based, approaches.

Furthermore, it must be stressed that even the use of a performance standard provided less flexibility than might be assumed. EPA sought to address the shortcomings it saw in currently operating I/M programs. As seen below, to do this EPA proposed (and later made final) a very complex, high technology inspection process performed in a centralized testing network where the inspection and repair functions were separated. While this test methodology was used to set a performance standard, the high hurdle this model program set virtually ensured that the specific technologies and testing network that EPA proposed would need to be used. Thus, although the Act calls for EPA to develop a performance standard to measure individual state I/M programs, and the Agency ostensibly complied, the model program EPA used to set a performance standard became a *de facto* design standard.

EPA's Proposed I/M Rule

To accurately test newer vehicles with modern emissions control systems, EPA felt that testing must include cycles of acceleration and deceleration (that is, transient testing) to better simulate real world driving conditions. Therefore, EPA developed a test procedure that involved the use of a chassis dynamometer (a device that combines rollers and weights allowing a stationary vehicle to be "driven" under load). In its Notice of Proposed Rulemaking for inspection/maintenance (U.S. EPA 1992b), EPA proposed this transient testing procedure, labeled the IM240 test, as the basis of the model program for enhanced I/M. The IM240 test is based on the Federal Test Procedure, a driving cycle used to certify new vehicles. By using this transient test, enhanced I/M programs can test NO_x emissions, which can only be accurately checked under loaded conditions, as well as the traditionally tested emissions of VOCs and carbon monoxide.

Coupled with this state-of-the-art exhaust testing, EPA's proposed model program included two tests to inspect the integrity of a vehicle's evaporative emission control system.[5] A pressure test, officially known as the "Evaporative System Integrity Test," was designed to ensure that the evaporative system had no leaks. The "Evaporative Performance Test," commonly referred to as the *purge test*, was designed to test the ability of the vehicle to deliver gasoline vapors captured in the canister to the engine. Although these tests do not lead to lower emissions of carbon monoxide, they are thought to lead to significant improvements in control of evaporative VOC emissions, which EPA believes are a major component of overall vehicle hydrocarbon emissions.

EPA began to develop this set of inspection procedures even before final passage of the 1990 Clean Air Act Amendments. EPA started a vehicle test program in Maryland in 1989 using a procedure that included IM240 and evaporative emissions testing. In February of 1990, EPA began more extensive testing of the IM240 procedure in Indiana. The data from the vehicles tested in these two prototype enhanced I/M programs provided the basis of the emissions reductions benefits found in the regulatory impact analysis that accompanied the proposed rule.

In the proposed rule, EPA established IM240 testing with evaporative system pressure and purge testing as the basis of its model program. Specifically, the proposed model program required IM240 testing of 1986 model year and later vehicles, two-speed idle testing of vehicles from the 1981 to 1985 model years, and idle testing of pre-1981 vehicles. Pressure testing would be required for 1983 and later model year vehicles and purge testing required for 1986 and later model years. Inspections under the model program needed to be performed annually in a centralized inspection network that separated the test and repair functions. Additionally, the model program required on-road testing (such as remote sensing) of at least 0.5% of the subject vehicle population as well as checks of on-board diagnostic equipment for vehicles so equipped. EPA estimated that by adopting the model program, a typical urban area would "experience a 28% reduction in emissions of VOCs, a 31% reduction in CO emissions, and a 9% reduction in NO_x emissions from highway mobile sources by 2000 when compared to what the area would experience without an I/M program (U.S. EPA 1992b, 31062)."

In its proposed rulemaking for inspection/maintenance, EPA took comments on the use of different test options as the basis of the model program. The IM240 test, with its associated evaporative emission tests, was considered to be the "high" test option. EPA also proposed a "low" and a "medium" option. The low option was essentially a program yielding the same emission benefits as the best I/M programs currently in use under the requirements of the 1977 Clean Air Act Amendments. The medium option improved on the low option by adding pressure testing of the evaporative system.

As discussed further below and summarized later in Table 3, a nonattainment area employing a low-option test procedure, on average, would reduce vehicular VOC emissions by 10% compared to a base case in which no emissions testing was conducted. Use of the medium option test procedure would reduce VOC emissions by 20%. In contrast, the high-option I/M test procedure would reduce VOC emissions by 28%. Moreover, the low and medium options, by not incorporating loaded-mode dynamometer testing, could not test for, or reduce, NO_x emissions.

In addition to the high, medium, and low options, EPA took comment on the use of Acceleration Simulation Mode (ASM) tests such as those advanced by the state of California, SAVER, Inc., and the Atlantic Richfield Company (ARCO). ASM testing is based on a loaded mode, steady state procedure which would allow evaluation of NO_x emissions. ASM tests could also be conducted in conjunction with the pressure and purge evaporative emissions tests. EPA also took comment on allowing the use of other in-use vehicle controls to make up the shortfall states might have if they chose to implement an I/M program that was different from the model program and did not provide as much emission benefits. In proposing these options, EPA sought to give states additional flexibility to choose a more cost-effective route.

As it did with the form of the model test procedure, EPA took comment on a variety of proposals regarding the choice of I/M network type. Although the proposed model program assumed operation of the inspection test in a traditional centralized network, EPA proposed conditions that would allow the use of decentralized and hybrid network types. For pure decentralized I/M networks, EPA proposed that test-and-repair programs be granted "provisional" equivalency to centralized programs for the purposes of the initial SIP demonstration. Once in operation, both decentralized and centralized programs would be required to demonstrate that they were meeting the performance standard. If test-and-repair programs were to be implemented under this procedure, the state needed to include in its SIP revision the authority to switch to a test-only system should the test-and-repair program fail to meet the performance standard. In regard to hybrid systems, EPA proposed that decentralized systems that employed test-only inspection stations be granted "presumptive" equivalence to centralized, test-only networks.[6]

In summary, EPA's proposed regulatory approach regarding the enhanced I/M program was to set a performance standard that allows states flexibility in tailoring I/M programs to their individual needs. For instance, even though the proposed model I/M program relied on annual testing, EPA strongly recommended that states institute a biennial program. This could be accomplished while still meeting the model program's performance level by changing other requirements of the proposed I/M program to increase its effectiveness (such as extending the requirements for pressure or purge testing to earlier model year vehicles). EPA considered different options as the basis of the model program used to set the performance standard. Additionally, it considered different amounts of credit to assign to nonmodel I/M programs. In this way, actual I/M programs would be given credit based on the effectiveness of the particular procedures and elements as determined by EPA and built into the MOBILE emission factor model. In short, EPA considered alternatives to try to define the best model program that effectively reduced emissions, was technologically feasible, could be reasonably implemented (including cost), and could be enforced. As far as EPA was concerned, its model program did not need to be the "best" in all respects; theoretically, the use of a performance standard provided the flexibility for other programs to be instituted.

EPA's Final I/M Rule

On November 5, 1992, EPA published its final rule on I/M programs (U.S. EPA 1992d, 52950-52989). As was discussed in its proposed I/M rule, EPA promulgated a final rule that established transient IM240 testing, combined with evaporative pressure and purge testing, to be performed in a centralized inspection network as the basis for its model program. Thus, EPA selected the high technology option it had proposed as the basis of its model program.

EPA received abundant comments from numerous stakeholders following publication of the proposed rule. Although many parties were critical of the proposal and the regulatory impact analysis that accompanied it, EPA defended its choice of high-tech IM240 testing in its final rule. The model program EPA forwarded included all the elements required by the Clean Air Act. Furthermore, EPA believed its RIA supported its decision. As discussed in greater detail below, the RIA showed that high-option, IM240 testing was very effective at reducing vehicle emissions (for instance, the model program would reduce vehicle VOC emissions nearly 28%) and was more cost-effective that any other option, including a no-change option (excepting the use of IM240 in a biennial testing system, as noted below).

Included in the final I/M rule were provisions related to the type of inspection network. Citing the concerns of such groups as the Natural Resources Defense Council and the American Lung Association, EPA removed the provisional equivalency of decentralized, test-and-repair systems in its final rule. EPA contended that this decision was supported by state agencies charged with implementing enhanced I/M programs, "virtually all [of which] urged EPA to eliminate provisional equivalency from the final rule (U.S. EPA 1992d, 52973)." Instead, based on data regarding the effectiveness of decentralized programs, EPA decided to grant only 50% of the emissions reductions credits to programs that were operated in a decentralized, test-and-repair network. However, EPA decided to keep its provisions regarding presumptive equivalency in the final rule. As such, I/M testing performed in decentralized, test-only networks was automatically assumed to be as effective as similar testing completed in centralized networks.

EPA's Regulatory Impact Analysis of the I/M Rule

EPA performed an RIA as part of its decisionmaking process to determine the costs, benefits, and impacts associated with the various inspection/maintenance policy options. As part of this investigation, EPA estimated the costs of the various I/M strategies to motorists in terms of test costs, repair costs, inconvenience costs, and foregone fuel expenditures due to fuel economy improvements. Contrasted to these costs, EPA estimated the emission reduction benefits resulting from fixing poorly maintained vehicles. Based on these estimations, EPA determined the cost-effectiveness of the chosen I/M alternatives which could be compared to the cost-effectiveness of other air pollution control programs. EPA also performed a rudimentary benefit-cost analysis of the model I/M program using rules of thumb for the benefit of reducing air pollution emissions. Finally, EPA calculated the impact the preferred I/M procedure—transient IM240 exhaust emissions testing with pressure and purge evaporative emissions testing—would have on small entities, notably emissions inspection and automotive repair shops.

EPA had completed a draft version of its regulatory impact analysis by late 1991. A revised version of this RIA, incorporating much of the same data, was available when EPA published its proposed rule in July 1992. When EPA published its final rule, the core information in the final RIA remain unchanged from that contained in the earlier RIA available when the I/M rule was proposed. In fact, all of the data presented below and in Tables 2 through 5 are the same as those found in both the RIA that accompanied the proposed rule and the RIA that accompanied the final rule. Since for the purpose of evaluating the costs, emission reductions, and cost-effectiveness ratios of the different I/M options, the RIA that accompanied the proposed rule is virtually the same as the final RIA, discussion and analysis of EPA's regulatory impact analysis is consolidated into a single section, presented below.

Methodology

In its analysis of the costs of an I/M program, EPA focused on the costs borne by motorists. EPA assumed that ideally the inspection fee that motorists would be charged would cover the cost of the inspection (including labor and equipment costs), cover the cost to the state to provide oversight and management, and provide a reasonable profit to private inspection operators. Thus, the overall costs borne by motorists were assumed to encompass the total costs of operating an I/M program. (As discussed below, EPA also included an analysis of the impact of the I/M rulemaking on small businesses such as inspection and repair facilities.)

To estimate the costs and benefits of operating an enhanced I/M program, EPA considered how a theoretical model inspection facility would operate and perform. EPA focused on a model test facility, and specifically an individual inspection lane, as the unit of analysis. As part of the modeling of costs, EPA considered the cost of the inspection equipment, labor involved in the test, and test lane throughput to determine the cost of conducting an individual test for each of the high, medium, and low test options. Based on the ability of each test option to identify failing (that is, high polluting) vehicles, EPA estimated the cost of repairing the failing vehicles and the emission reduction benefits of these repairs. As part of its analysis, EPA modeled the costs and emission reduction benefits of a baseline scenario (with no changes to the existing I/M programs) and each of EPA's proposed enhanced I/M options. EPA, however, did little in the way of sensitivity analysis to see how changes to its underlying assumptions would affect the outcome of the analysis.

In conducting its cost analysis, EPA was hampered by a lack of data regarding the three proposed model test procedures. The cost of the equipment and the time required to administer the IM240 test were not known since the test was both new and quite a bit more extensive than any currently operating test. Likewise, EPA had data on the effectiveness of existing test programs and network types but did not have sufficient data on the effectiveness of well run programs using such features as computerized analyzers, registration denial, high minimum repair expenditures before providing waivers, on-road testing, and rigorous program quality assurance. The analysis of I/M program costs thus required EPA to make a large number of assumptions. The information used to make these assumptions came mainly from three sources: data from existing I/M programs, EPA's best judgment based in large measure on its experience conducting on-going emissions testing at its Motor Vehicle Emission Laboratory, and data from special I/M testing programs.

As part of its development of a model I/M program, EPA conducted two vehicle testing programs. The first of these was a somewhat limited IM240 pilot program conducted in cooperation with Maryland in 1989, which ultimately tested about 600 vehicles. A larger test program in Indiana began in February 1990 that performed emissions tests on about 8,300 vehicles. While together these efforts constitute a fairly large testing program, they still represent a fairly small sample size, especially considering that much of the analysis depends on the number of failing vehicles, not the number of tested vehicles. The small sizes of the samples for different test failure and repair classifications introduces some amount of uncertainty into the analyses EPA conducted.

To conduct its benefits analysis, EPA considered the reductions of three air pollutants—VOCs, NO_x, and CO—resulting from the operation of various I/M programs. The emission reductions associated with operation of each of the three proposed model I/M programs as well as a baseline scenario were modeled using EPA's mobile source emission factor model, MOBILE4.1. EPA used the MOBILE4.1 model to determine the gram per mile emissions of the vehicle fleet based on a number of inputs, including type of I/M program in place. EPA developed the I/M portion of the MOBILE model using information taken from its Motor Vehicle Emissions Lab's emission testing program coupled with data from the Maryland and Indiana I/M test programs.

As is often the case with EPA's regulatory impact analyses, the health benefits that would accrue from reduction of the three targeted air pollutants were not monetized. Instead, EPA used general rules of thumb regarding the benefits of reducing air pollutants. Since, as noted above, nonattainment areas were required by the Clean Air Act to reduce VOC (or NO_x) and CO emissions by a certain percentage, EPA assumed that any emission reductions not obtained from implementing an I/M program would need to come from some other air pollution control program. Specifically, EPA assumed that for any ton of pollutant not reduced through an I/M program, a stationary source control program would need to reduce an additional ton of this pollutant. The marginal cost of stationary source controls were considered to be $5,000 per ton of VOCs or NO_x reduced. Thus, the cost-effectiveness of an I/M program (in dollars per ton of pollutant removed) could be compared to this $5,000 per ton target figure.

Cost Analysis[7]

As indicated above, the cost of implementing an I/M program can be calculated by considering three main components: the cost of performing the emissions test, the cost of performing repairs to fix vehicles that failed the test, and the savings associated with fuel economy improvements of repaired vehicles. In addition to these costs, however, such a program imposes an inconvenience cost on vehicle owners. I/M programs require motorists to bring their vehicles to inspection stations for testing and, if the vehicles fail, the owners must have them repaired and re-tested. This can require a fair amount of time, especially if a car fails the test and must be taken to a separate repair facility.

Inconvenience costs aside, the direct cost of performing the test is dependent on several items. The first is the cost of the new equipment needed to perform the test. For IM240 this cost is significant—EPA estimated it to be about $140,000. This is quite a bit more expensive than the $10,000 or less needed to buy the equipment required by many of the basic I/M programs in place at the time. Another relevant cost is the cost of the labor performing the test. EPA assumed that this was quite a bit higher for decentralized programs compared to centralized programs ($21 per hour versus $6 per hour) since decentralized stations often employed highly paid auto mechanics as test inspectors. Another major factor affecting the cost of the emissions test is the overall vehicle throughput as this determines how much of the cost of capital expenditures (such as for test equipment) is amortized over each test. For centralized test lanes EPA assumed throughput to be about 23,000 vehicles per year. Based on data from existing test programs, EPA estimated throughput at decentralized stations to be about 1,000 vehicles per year.

The costs of repair consisted of two elements: the cost of repairing problems discovered using exhaust tests (2500 rpm/idle or IM240) and the cost of repairing problems associated with the vehicle's evaporative emission system. In either case, total repair costs depended on the cost of the average repair resulting from failing a particular test and the failure rate associated with that test. EPA used data from existing programs to estimate the failure rate of 2500 rpm/idle tests and the average cost of fixing a vehicle flagged by that test (estimated to be $75). Using data from its two test programs and its best judgment, EPA estimated the failure rate and the cost of fixing a vehicle that failed the IM240 yet passed the two-speed test (presumed to be twice the cost of a two-speed detected repair, $150). EPA also used the test program data to estimate the failure rate and repair costs resulting from pressure and purge testing ($38 and $70, respectively).

Paralleling the cost of repair, fuel economy benefits are associated with repairs discovered due to evaporative emissions testing and exhaust emission testing. Evaporative emissions test-related repairs result from both purge testing and pressure testing. Fuel economy improvements result both from increased engine efficiency and performance due to the repairs as well as due to combustion of fuel vapors that would have otherwise been emitted as running losses had it not been for the repairs. Using data from its two testing programs, EPA estimated fuel economy improvements for pressure and purge testing to be 6.1% and 5.7%, respectively, leading to an average of 5.9% for evaporative emission-related repairs. Again based on data from its test programs, EPA estimated fuel economy improvements associated with IM240 discovered repairs to be 12.6%. These figures compare to the 8.0% fuel economy improvement resulting from repairs initiated by 2500 rpm/idle testing.

Using this information, EPA estimated the test and repair costs and the fuel economy benefits for the country as a whole for the year 2000. To do this, EPA considered all of the nonattainment areas that would be required to implement some form of inspection/maintenance. (In 1992, nearly 64 million vehicles were registered in areas that would need to implement I/M programs. Due to growth in the vehicle fleet, the number of vehicles subject to I/M in the year 2000 will be higher.) As part of its cost evaluation, EPA first estimated the costs of operating all of the I/M programs under the then-current inspection regime (that is, no regulatory changes regarding I/M). These results were compared to an I/M system of enhanced areas using IM240, pressure, and purge testing and basic areas with a similar mix of centralized and decentralized stations as was currently the case. (For this evaluation, EPA ignored inconvenience costs.)

As seen in Table 2, EPA estimated that the total cost to motorists nationwide of operating I/M programs would be $894 million if EPA did not make any changes to the existing I/M system. With changes made to basic I/M programs and with biennial IM240 testing in enhanced I/M areas, EPA estimated that the total cost of I/M programs nationwide would be $541 million. It is interesting to note that EPA predicted the net cost of operating a mixed system of enhanced and updated basic programs would be about $350 million less than operating a national I/M program with no changes made to the current system. While this may seem counter-intuitive, it is important to recognize that EPA based its estimates of I/M program costs on the costs borne by motorists. Since revised inspection procedures—especially those recommended for enhanced I/M areas—lead to repairs that in turn lead to improvements in fuel economy, reduced expenditures for fuel can offset higher test and repair costs, allowing the total cost of a more stringent I/M program to be less than the cost of the current program.

Benefits Analysis

As noted above, EPA did not perform an assessment of the environmental and health benefits associated with its proposed I/M program and instead focused on the emission reductions resulting from I/M.[8] EPA compared the emission reductions on two different levels: changes to vehicle fleet emission factors associated with different I/M program options, and the

Table 2. Annual I/M Program Costs for the Year 2000. (millions of dollars)

Program type	*Test cost*	*Exhaust test repair cost*	*Evap. test repair cost*	*Exhaust test fuel economy savings*	*Evap. test fuel economy savings*	*Net cost*
Unchanged I/M system						
Centralized	182	140	n/a	(92)	n/a	230
Decentralized	565	252	n/a	(153)	n/a	664
Total	747	392	—	(245)	—	894
Proposed I/M system						
Enhanced areas	451	489	221	(617)	(208)	336
Basic areas						
Centralized	67	60	n/a	(39)	n/a	88
Decentralized	95	53	n/a	(31)	n/a	117
Basic total	162	113	—	(70)	—	205
New system total	613	602	221	(687)	(208)	541

Source: U.S. EPA 1992b.

total emission reductions of a no-change option and of the proposed model program compared to a no-I/M baseline. The overall emission reductions (as a percent of total highway vehicle emissions) for each I/M program option can be calculated based on changes to the vehicle fleet emission factors.

As discussed previously, estimating emissions reductions depends on the ability of a test program to detect failing vehicles (the failure rate), the type of repair required, and the effectiveness of repairs. Using the MOBILE4.1 emissions model which included estimates for these items, EPA estimated the VOC and CO emissions factors for the vehicle fleet (in the year 2000) subject to different I/M program options (see Table 3). For instance, a vehicle operating in an area subject to the model enhanced I/M program would emit, on average, 1.5 grams of VOCs for each mile it is driven. In contrast, a vehicle that is not subjected to an inspection/maintenance program would emit an average of 2.1 grams of VOCs per mile driven. The changes to the vehicle fleet emissions factors indicates that the model enhanced I/M program would reduce on-highway mobile source VOC emissions by 28% and CO emissions by 31%.

Based on these vehicle fleet emissions factors, EPA calculated the nationwide emission reductions resulting from continuing I/M programs unchanged and from making the recommended changes to basic and enhanced I/M program areas. By assuming that all enhanced I/M areas nationwide would implement the recommended biennial, high-option program and that all basic areas would implement the new basic I/M program, EPA compared the nationwide emission reductions in the year 2000 of implementing I/M programs to a no-I/M case (see Table 4). For instance, in enhanced areas, implementation of the biennial, high-option I/M program reduces the vehicle VOC emission factor 0.58 gram/mile to 1.495 grams/mile; with approximately 62 million vehicles in enhanced I/M

Table 3. Benefits of I/M Program Options. (Based on total highway mobile source emissions in 2000)

	VOC emission effects		*CO emission effects*	
Scenario	*Emission factor (g/mi.)*	*Percentage reduction*	*Emission factor (g/mi.)*	*Percentage reduction*
Baseline—No I/M	2.084	—	11.874	—
Basic I/M	1.971	5.4	10.021	15.6
Low option enhanced	1.870	10.3	8.927	24.8
Medium option enhanced	1.661	20.3	8.927	24.8
Biennial high option enhanced	1.495	28.3	8.223	30.7
Model enhanced performance standard	1.503	27.9	8.230	30.7

Sources: U.S. EPA 1992a, 1992b

Table 4. National Benefits of I/M. (Annual tons of emissions reductions in year 2000 compared to no-I/M case)

Scenario	*VOC reductions*	*CO reductions*
Reductions from continuing I/M unchanged		
Centralized areas	55,540	775,288
Decentralized areas	60,476	791,167
Total	116,016	1,566,395
Expected reductions from I/M proposal		
Enhanced areas	384,130	2,345,278
Basic areas		
Centralized	23,289	326,290
Decentralized	12,996	174,186
Basic area total	36,285	500,476
Total annual benefits	420,415	2,845,754

Source: U.S. EPA 1992b.

areas, each driven slightly less than 10,000 miles per year, over 384,000 tons of VOCs are eliminated because of the enhanced I/M program.

Cost-Effectiveness Analysis

Based on its cost and emissions reductions analyses, EPA conducted a cost-effectiveness analysis of the different I/M program options. To do this, EPA estimated the costs (excluding inconvenience costs) and emissions reductions associated with testing one million vehicles in each different program type. Based on these figures, EPA calculated the average cost-effectiveness (on a cost per ton basis) of reducing emissions through each program option, both with all program costs attributed to VOC reduction and with costs spread across the reductions of VOCs, NO_x, and CO. (See Table 5.) For example, the annual model I/M program, which is expected to reduce 6,724 tons of VOCs per million vehicles at a cost of $11.4 million, is found to reduce VOCs for about $1,700/ton (excluding inconvenience costs). If the model program is operated biennially, cost-effectiveness is almost halved to about $880/ton of VOCs avoided.

To determine the cost-effectiveness of the program options considering the emissions reductions of all three pollutants, EPA needed to determine how to allocate program costs among the pollutants. To do this, EPA first assumed that the value of NO_x reductions could be estimated using the cost to reduce NO_x through alternative stationary source controls, namely low-NO_x burners for utility boilers that reduce NO_x for $300/ton. Based on this $300/ton figure and the number of tons of NO_x reduced for each I/M option (that is, 1,800 tons of NO_x for the high option and none for the medium and low options), EPA reduced the total cost of each pro-

Table 5. Cost-Effectiveness (C/E) of I/M Program Options. (Dollars per ton of pollutant removed)

	Total annual program cost	*C/E with all costs allocated to VOC reduction*		*C/E with all costs allocated to VOC, NO_x, CO reduction*	
Scenario	*Without I.C.*	*Without I.C.*	*With I.C.*	*Without I.C.*	*With I.C.*
Basic I/M	6,412,000	5,410	n/a	4,518	n/a
Enhanced I/M					
Low option	9,885,000	4,404	10,050*	3,655	9,300*
Medium option	11,628,000	2,621	5,625*	2,242	3,500*
High option (annual)	11,390,000	1,694	3,925*	1,271	1,675*
High option (biennial)	5,429,000	879	1,983	461	1,566

Note: Author's estimate is based on EPA data. I.C. is inconvenience costs. (* indicates author's estimate.)

Sources: U.S. EPA 1992a, 1992b.

gram option. Similarly, EPA assumed that carbon monoxide could be reduced through other control programs at a cost of $125/ton. Since only 44% of the vehicles subject to enhanced I/M were in CO nonattainment areas, EPA considered 44% of the CO reductions resulting from the different I/M program options. After reducing the total costs of the various I/M program options to account for the CO reduction benefit, EPA recalculated the cost-effectiveness figures for the various program options to determine the cost-effectiveness of VOC reduction accounting for the NO_x and CO reduction benefits. For instance, EPA estimated that the cost-effectiveness of reducing VOCs using the recommended biennial, high-option testing program, accounting for CO and NO_x reductions, is $461/ton, over $3,000/ton less than the cost-effectiveness of the low I/M option (estimated to be $3,655/ton).

Although EPA often touted the ability of its preferred biennial inspection/ maintenance program of transient IM240 testing coupled with pressure and purge testing to reduce pollution at a cost under $500 per ton, this cost-effectiveness figure did not include the inconvenience cost to motorists of inspecting their vehicles. While EPA did not determine the cost-effectiveness including inconvenience costs for all the I/M program options, it did estimate the cost-effectiveness (including inconvenience costs) of operating a high-option I/M program biennially. Assuming that driving to the test center, waiting for the test to be completed, and driving home would take 45 minutes total and that motorists' time was worth $20 per hour, EPA estimated that the total cost-effectiveness of a biennial high-option I/M program would be about $2000/ton with all costs attributed to VOC reduction and about $1600/ton if costs were allocated across the three pollutants. Using EPA's estimates for the program costs and emission reduction benefits of the various other I/M test options, I calculated similar cost-effectiveness numbers for the annual high option, the medium option, and the low option (the results can be found in Table 5).[9] As can be seen, when inconvenience costs are considered, the cost of reducing VOCs through low-option I/M testing is about $10,000/ton, over twice the cost of reducing VOCs under the model I/M program.

Using its estimates of cost-effectiveness, EPA calculated the national cost of choosing the low-option I/M program. EPA determined that implementation of a low-option I/M program in nonattainment areas nationwide would have a direct and an indirect cost. According to EPA's estimates, the total cost of a low-option I/M program would be more than the costs of running a high-option I/M program (mainly due to the higher fuel economy benefits that offset the higher inspection and repair costs of a high-option I/M program). Thus, there is a direct cost of choosing to implement a low-option I/M system rather than a high-option system. EPA found this cost to be $350 million.

EPA further claimed that there is an indirect cost associated with achieving the emission reductions foregone by choosing the less effective low I/M option. Since, as discussed above, ozone nonattainment areas are required to reduce VOC emissions 15% by 1996 and 3% per year thereafter, EPA assumed that any VOC reductions that were not obtained by vehicle inspection/maintenance would be obtained through additional stationary source controls which cost $5,000/ton on the margin. According to EPA estimates, a low-option I/M system would reduce 250,000 fewer tons of VOCs than high-option I/M, yielding an indirect cost of $1.25 billion. Therefore, EPA estimated that the total cost of implementing the less stringent low-option I/M program nationwide would be $1.6 billion.

Inspection and Repair Industry Impact Analysis

As part of its responsibility to perform an RIA,[10] EPA estimated the impact the I/M rulemaking would have on small entities. For this analysis, EPA concentrated on the effects of the I/M rule on inspection and repair facilities.[11] For this analysis, EPA considered three basic elements: the impact of the rule on the inspection business, the impact on the repair industry, and the impact on jobs. To evaluate the effect of the proposed I/M rule on the inspection industry, EPA needed to assess to what extent inspection business and profits would shift due to the new rule. EPA also needed to investigate how these changes would be distributed throughout the existing inspection industry and whether the loss of inspection business would affect businesses that chose to perform vehicle repairs only.

Similarly, EPA sought to evaluate how much business in the repair segment would grow due to the new I/M rule and the extent to which this repair business might mitigate loss of test revenues for individual businesses. Although EPA found that total test revenues for a high-tech I/M system would be less than if no changes were made to the I/M program, the adoption of IM240 and the evaporative emission tests would increase repair business by about $430 million.[12] EPA also estimated the changes in employment resulting from the I/M rule. Based on the need for additional inspectors and repair mechanics, EPA stipulated that the enhanced I/M rule would result in the net gain of 3,800 to 11,600 positions. In general, EPA found that, for most parties involved, the effect of instituting an IM240-based, centralized I/M testing program would be a shift of business revenues but not a loss of revenues. Numerous repair stations that had formerly supplemented their incomes with emissions testing would no longer conduct tests in a centralized I/M system. However, due to the superior ability of IM240 to detect failing vehicles, these stations would see an increase in repair activity.

REFLECTIONS ON THE I/M RULE DEVELOPMENT

EPA faced quite a bit of criticism regarding the assumptions it made in the regulatory impact analysis. Debate about what numbers to use occurred both within the agency and from the outside, coming from the Office of Management and Budget, environmental groups, states, and inspection and repair industry groups. Critics were concerned both that the numbers EPA used did not accurately reflect the real world and also that EPA's assumptions were not made clear. Among the specific items of concern were: treatment of inconvenience costs and the extent to which failing motorists would ping-pong between separated test and repair facilities in centralized systems; test costs in centralized systems compared to the test costs in decentralized systems, including the related assumptions regarding labor costs and vehicle throughput in these two different types of inspection systems; the effect of separating the test and repair functions of inspection/maintenance on small businesses; and the effectiveness of transient, loaded-mode testing compared to other test methods.

Many of these criticisms were made by commentors after EPA published its proposed I/M rule and released its preliminary regulatory impact analysis. Yet, even though many of the criticisms were well founded, as discussed above, the final RIA had nearly identical information regarding costs, emission reductions, and cost-effectiveness as did the RIA that accompanied the proposed rule. Furthermore, EPA's final rule for I/M programs closely followed the I/M system advocated in the proposed rule. To better understand the relation of the RIA to EPA's I/M rule, these criticisms must be explored further.

Critique of EPA's Regulatory Impact Analysis

In conducting its regulatory impact analysis, EPA was often hampered by a lack of sufficient data. The proposed model program had only been tested in small, prototype test lanes and many of the data available from current programs were not directly applicable to new inspection system alternatives. EPA often had to draw conclusions from limited test data and from existing, yet potentially inappropriate, data drawn from current I/M installations. Because of the lack of empirical evidence, EPA often used its best judgment in assessing costs and emission reduction benefits. While EPA has a great deal of experience to draw on when making these judgments, the process necessarily introduces some uncertainty into EPA's analyses.

The difficulties that arise due to such uncertainty can readily be seen in EPA's efforts to estimate the costs of I/M testing. The uncertainties involved in predicting the value of specific key variables can substantially change the overall assessment of the cost of performing vehicle inspections. Some of the key factors where uncertainty played the greatest role include: test throughput when conducting a high-tech IM240 test with pressure and purge testing (for both centralized and decentralized systems), labor costs in centralized and decentralized systems performing the IM240 test, inconvenience costs and the extent to which ping-ponging between test and repair facilities is a concern for centralized systems, repair types and costs associated with different test elements, and test equipment costs. EPA's estimation of I/M cost-effectiveness, which relied on its cost and emissions reductions analyses, likewise was affected by uncertainties arising from insufficient data. Unfortunately, given these uncertainties, EPA did not conduct any sensitivity analysis of its cost assessment (such as comparing the change in net costs with different assumptions for equipment and repair costs, test throughput, inconvenience time, and so forth).

To try to shed light on this issue, EPA's data can be used to estimate the cost-effectiveness of an annual, high-option I/M program with different assumptions for vehicle throughput in inspection lanes, cost of labor, and motorist inconvenience cost. If centralized IM240 inspection lanes are assumed to have a maximum throughput of ten vehicles per hour rather than the fifteen vehicles per hour assumed by EPA, cost-effectiveness (including inconvenience costs) of annual, high-option I/M rises to $4,500 per ton of VOCs avoided (with no allocation of costs to CO and NO_x reduction). If, along with this lower throughput, labor is assumed to cost $10 per hour rather than $6 per hour, cost-effectiveness of the high option rises to $5,400/ton of VOCs avoided. After allocating some of the program costs to CO and NO_x reductions (as described above), this new set of assumptions leads to a cost-effectiveness of $4,900/ton.

The cost-effectiveness ratios cited include an inconvenience cost of $15 per test. Although EPA did not consider the additional inconvenience to motorists who failed the inspection and needed to have their vehicles repaired and retested, EPA's estimate of the value of motorists' time ($20 per hour) can be considered generous. Using an entirely different set of assumptions, researchers at Resources for the Future estimated that the inconvenience cost to motorists of testing a vehicle in a high-option, centralized inspection facility is $17 per test (McConnell and Harrington 1992). Based on this higher inconvenience value (and assuming the lower throughput value and higher labor costs as above), the cost-effectiveness of annual, high-option I/M is $5,200 per ton with costs allocated across the three pollutants.

These sensitivity calculations appear to support EPA's choice of centralized, high-option I/M. Under all of the scenarios considered above (which all increase cost-effectiveness), high-option I/M is found to reduce pollution at a lower cost than low-option I/M, which is estimated to cost nearly $10,000 per ton avoided (even without changing assumptions regarding labor and throughput). In most cases, high-option I/M is estimated to reduce VOCs as cost-effectively as currently operating basic I/M programs (even when no inconvenience costs are attributed to basic I/M). Furthermore, under almost all of the scenarios considered, the cost of reducing VOCs is less than the marginal cost of reducing VOCs through alternative stationary source programs ($5,000/ton) when costs are allocated across the three pollutants. If biennial testing is chosen, as EPA recommended, cost-effectiveness of high-option I/M would be substantially less than $5,000/ton of pollution avoided.

Another concern regarding EPA's regulatory impact analysis is the agency's estimation of costs and emissions reductions benefits compared to a no-I/M baseline and the subsequent consideration of average, rather than marginal, cost-effectiveness values. Generally, it is more useful to consider marginal cost-effectiveness rather average cost-effectiveness when making decisions. However, EPA's consideration of the cost to motorists when assessing I/M programs leads to an interesting situation. Since more stringent I/M tests are better at discovering engine and evaporative systems failures (leading to vehicle repairs and subsequent fuel economy improvements), the savings resulting from reduced expenditures for fuel significantly offset the higher inspection and repair costs associated with more stringent I/M testing. As a result, the marginal cost-effectiveness of EPA's proposed I/M options declines as tests become more stringent. In fact, if inconvenience costs are ignored, the marginal costs of high-option IM240 are actually negative. Based on EPA's estimates of program costs and emission reduction benefits together with the inconvenience costs calculated above, the marginal cost-effectiveness of the medium option and the annual high option can be estimated to be $1,100/ton and $600/ton, respectively (all costs attributed to VOC reduction).[13]

Exploring EPA's analysis of the emission reductions resulting from various I/M options might also elucidate some of the discontentment with the final I/M rule. The uncertainties and data problems EPA faced in performing the analysis of emissions reductions mirror those inherent in the cost analysis. As with the cost analysis, EPA was forced to make assumptions and judgments based on limited test data and data from existing programs that were not perfectly applicable to the model program case. Once again, EPA did not conduct any sensitivity analysis in its benefits evaluation. Furthermore, EPA's benefits calculations were far removed from public scrutiny. Information regarding the ability of an I/M program to detect and repair failing vehicles was embedded in EPA's vehicle emissions model, MOBILE4.1. Thus, those interested in how credits were assigned simply had to trust the accuracy of the outputs of the "black box" MOBILE4.1 model. At the very least, while EPA's estimated emissions reductions credits may very well be accurate, its assumptions leading to these credits were not transparent to those interested in evaluating EPA's analysis.

It may be, however, that the emission reduction credits assumed in MOBILE4.1 were not correct, especially for I/M options not tested extensively in Maryland and Indiana. It is quite possible, considering the amount of data EPA had collected on IM240 and pressure and purge testing, that EPA's estimates of the emission reduction capabilities of the low, medium, and high I/M options were fairly accurate. Therefore, EPA's estimates of the ability of these options to reduce vehicle emissions and the cost-effectiveness of these options may also be accurate. However, less data were available for new I/M test procedures (that were not yet in place nor tested by EPA), and EPA's estimates of the emission reduction potential of these procedures might not be accurate. If this were the case, some of these alternative inspection procedures might be more cost-effective than projected using the MOBILE model. Thus, although EPA's preferred high-option I/M program might be more cost-effective than alternative stationary source programs, it could be less cost-effective than an alternative I/M program preferred by a given state.

Finally, in its regulatory impact analysis, EPA did not explicitly calculate benefit-cost ratios for the various program options. Since EPA did not assign monetary value to the emission reductions that would result from IM240 inspection programs, it could not show that its recommended I/M program yielded positive net benefits. However, an estimate of the benefit-cost ratio may be made using EPA's stated rules of thumb (namely, that reducing VOCs has a value of $5,000/ton removed; NO_x reductions are worth \$300/ton; and CO reductions are worth \$125/ton). Based on data provided in the RIA regarding the amount of VOCs, NO_x, and CO reduced by implementing I/M nationwide and the cost of administering these programs, EPA implies that adoption of I/M programs yields over \$1.7 billion in net benefits (see Table 6). This net benefits figure is based on rules of thumb for the value of reducing air emissions and does not include the cost of motorist inconvenience. In a study for Southern California Edison, an economic consulting firm estimated the damages caused per ton of VOCs and NO_x emitted to be \$2,700 and \$4,700, respectively (Harrison and others 1992). Using these figures and including the cost of motorist inconvenience, one could estimate that the benefits of implementing EPA's recommended biennial I/M program would exceed the costs by over \$500 million.

Relation of the RIA to the Policy Decisionmaking Process

In large measure, the final rule EPA promulgated regarding enhanced inspection/maintenance was consistent with the findings of the regulatory impact analysis. Specifically, the final rule forwarded a model I/M program based on transient, loaded-mode testing (IM240) coupled with pressure and purge evaporative testing, to be performed in a centralized inspection network. In part, this occurred because the RIA showed the model program to be an effective and cost-effective method of addressing in-use motor vehicle emissions without a significant net impact on jobs, small businesses, or motorists.

Yet it is worth delving deeper into this relationship to see what lies below the surface. Certainly, the fact that the chosen model program was

Table 6. Benefit-Cost Estimates of I/M Programs. (millions of dollars)

Nationwide I/M (basic and enhanced)	*EPA's implied estimate*	*Author's estimate*
Benefits		
VOCs (420,415 tons)	2,100	1,100
CO (1,252,000 tons)	157	157
NO_x (111,600 tons)	3.4	525
Total	2,260	1,780
Costs	541	1,200
Benefit/cost ratio	4.2	1.5

Notes: This table refers to EPA's recommended biennial enhanced and basic programs. The EPA benefit estimate based on nationwide emission reductions due to I/M and stated rules of thumb (that is, VOCs worth 5,000/ton; CO worth 125/ton; NO_x worth 300/ton). Cost estimate includes inspection cost, repair cost, and fuel economy benefits; however, motorist inconvenience is not included.The author's benefit estimate uses EPA's estimate of emission reductions but values them as follows: VOCs worth 2,700/ton; NO_x worth 4,700/ton; CO worth 125/ton. Cost estimate includes an additional 675 million for motorist inconvenience.

shown to be quite cost-effective had a bearing on the decisionmaking process, but other factors were at work. First, the constraints placed by the requirements of the Clean Air Act severely limited available policy choices. Thus, the RIA did not need to weigh I/M programs in general against other in-use vehicle emission programs. It only needed to compare various model I/M programs. Furthermore, lobbying by interest groups and institutional preferences probably had a greater effect on EPA's policy choices than did the findings of the RIA. For instance, under pressure from the Natural Resources Defense Council and the American Lung Association, and in sync with its own predilections, EPA dropped the provisional equivalency of decentralized programs in its final rule.

Another indication that the RIA was subject to institutional pressures was that the criteria by which the RIA judged alternative programs was buried in the MOBILE emission factor model. The RIA failed to uncover many of the issues involved in evaluating different I/M program options. This resulted from the fact that, in many instances, the RIA relied on the use of EPA's mobile source emission factor model, MOBILE4.1, to evaluate different program choices. It was in the MOBILE model that EPA embedded its assumptions regarding test, failure, and repair effectiveness, not in the regulatory impact analysis. Thus, the RIA did not directly address many of the most contentious issues, including the effectiveness of decentralized (test-and-repair) networks versus the effectiveness of centralized (test-only) networks; the effectiveness of IM240 compared to the effectiveness of other loaded-mode, steady state tests; and the credit given to I/M programs that differed from the model program. Viewed perhaps cynically, this indicates that the policy choice selected the form of the RIA, not the other way around.

Recent events highlight other limitations of the regulatory impact analysis regarding the development of enhanced I/M policy. In September 1995, under intense pressure from numerous states and federal legislators, EPA promulgated its Inspection/Maintenance Flexibility Amendments rulemaking (U.S. EPA 1995). The amendments were designed "to provide greater flexibility to states ... that may not need an enhanced I/M program as effective as the one EPA adopted in 1992 to meet the Act's reasonable further progress and attainment demonstration requirements." Under the amendments, EPA would allow states to implement, with certain restrictions, an alternative low-enhanced I/M program that did not include transient, loaded-mode testing or evaporative systems testing. (States could approve any program falling between the low- and high-enhanced I/M program effectiveness targets as long as they could show that they would meet the CAA-required reasonable further progress goals and attainment dates.)

Even though these amendments constituted a substantial change in I/M policy, EPA found that the change would not create an annual effect on the economy of $100 million or more, and therefore a new regulatory impact analysis did not need to be performed. Presumably, EPA still considers the findings of the 1992 analysis to be valid. If this is the case, states either believe that they can implement more cost-effective programs (than high, option IM240) to meet reasonable further progress goals and attainment dates or they would rather implement less cost-effective programs that are not as visible to the public at large. At any rate, it seems that EPA's policy change reflects the agency's understanding that many states strongly desire to have the flexibility to adopt programs whose emission benefits reflect the individual needs and concerns of those states.

In late 1995, President Clinton signed the National Highway System Designation Act of 1995.[14] Section 348 of this Act ("Moratorium on Certain Emissions Testing Requirements") forbade the administrator of EPA from automatically discounting the emission reduction credits of a test-and-repair enhanced I/M program. Section 348 allowed states 120 days to revise their state implementation plans regarding enhanced inspection/maintenance. The Act required that the EPA administrator "approve the [revised SIP] based on the full amount of credits proposed by the State for each element of the program if the proposed credits reflect good faith estimates by the State and the revision is otherwise in compliance with such Act."

Due to the flexibility provided by EPA's I/M amendments and the National Highway System Designation Act rider, many states have submitted revised SIPs to EPA to change their I/M programs. As a whole, the I/M programs chosen by the states represent a wide array of I/M options (U.S. EPA 1996). Submitted programs include: centralized IM240 testing with pressure and purge testing, centralized IM240 with no evaporative system testing, IM240 testing with evaporative pressure testing but no purge testing, IM240 testing in a decentralized network, ASM testing in a decentralized network, and BAR90 two-speed idle testing. Three explanations might explain why states have chosen such a wide variety of I/M programs: states have widely divergent methods of weighing the costs and benefits of different I/M program options; the relative costs and benefits of a particular I/M program vary from state to state; or states, for whatever reason, do not always choose the most cost-effective solution.

Lessons Learned

The I/M case sheds light on both the potential usefulness and the limitations of economic analysis. Economic analysis can help policymakers determine if a particular policy choice is cost-beneficial and if it can help policymakers make comparisons among different options. However, it cannot ensure that the widest and most relevant set of policy options are evaluated, and it cannot ensure that all elements of the policy debate are given attention. In contrast to the general case, a particular economic analysis may be restricted by statute in the choice of policy options considered or the analysis may be conducted reactively rather than proactively, as a tool to defend a prior policy choice rather than as an integral part of the policy development process.

In the RIA performed for the I/M rule, the centralized loaded-mode transient test with pressure and purge testing was determined to be both effective at reducing pollution and quite cost-effective. Even if all the technical criticisms were heeded, the RIA likely would have pointed toward selection of this high-technology IM240 as the preferred alternative. Yet, as indicated by the strong negative reaction of many states, centralized IM240-type testing was not universally seen as the best policy choice. Concerns about intangible costs not considered in the RIA, the potentially adverse impact on small business, and different public preferences across states all may have contributed to the backlash. Whatever the cause, EPA's chosen I/M policy was so disliked that it became a banner for states' rights, unfunded mandates, and other related concerns. The I/M rule also became the topic of congressional hearings and constitutional cases brought by states against the federal government.

Why then did the RIA for the IM240 program find it to be such a winner? The short answer: it failed to include critical issues and to capture key elements of the policy debate. States apparently desired the flexibility to choose the type of I/M program they felt was appropriate for their own nonattainment areas. In the inspection/maintenance case, states may have wanted to choose an I/M system that was not as cost-effective if it had less of an impact on local businesses or was not as visible or inconvenient to the driving electorate. Additionally, the RIA did not reflect the political issues regarding the I/M rulemaking. For example, it does not reflect the growing concern of state governments regarding unfunded federal mandates or that state legislators may be under pressure from the public to reduce government intrusion into individuals' lives.

ACKNOWLEDGMENTS

I would like to thank Richard Morgenstern for his help and guidance, both with this paper and during my years at the U.S. EPA. Additionally, I wish to thank David Harrison and Jonathan Wiener for their helpful comments on an earlier draft of this paper. Finally, I would like to extend special thanks to Liz Farber for her assistance and patience throughout this project.

ENDNOTES

[1]Section numbers refer to the Clean Air Act (42 U.S.C. 7401-7626).

[2]Section 109, The Clean Air Act (42 U.S.C. 7401-7626).

[3]Section 184 of the Clean Air Act also required metropolitan areas in the northeast Ozone Transport Region with populations of 100,000 or more to implement enhanced I/M programs.

[4]Although Congress required EPA to base its guidance on a performance standard, the specific program elements prescribed by the Act (computerized emissions analyzers, registration denial, a $450 minimum expenditure to be granted a waiver, and so forth) limited the flexibility afforded by such a standard. Furthermore, as described in more detail in the text, EPA's choice of a model program used to set a performance standard creates a *de facto* design standard for I/M programs. When reference is made in the text to a performance standard or a model program, these considerations should be kept in mind.

[5]Modern vehicles are equipped with a closed fuel system that includes a charcoal canister to collect gasoline vapors from the fuel tank which can then be delivered to the engine to be burned as part of normal engine operation.

[6]The reason EPA placed such an emphasis on test-only inspection networks is that the data it had collected indicated that, historically, decentralized test-and-repair networks had suffered from quality problems and poor performance. EPA contended that this lower performance was a result of the conflict of interest inherent in stations that perform both the testing and repair functions.

[7]The figures cited in the discusion of the cost analysis regarding EPA's estimates of I/M costs and benefits come from U.S. EPA 1992c. Data regarding EPA's medium- and low-program options come from an earlier draft of this document, U.S. EPA 1992a.

[8]Thus, EPA did not conduct a true benefit-cost analysis where the costs of regulating could be compared to the benefits associated with the regulation. Instead, EPA's estimated costs and "benefits" could be used to determine the cost-effectiveness of reducing pollution through the I/M program which could then be compared to other air pollution regulatory programs.

[9]For this analysis, I assumed that testing under the medium option would require 40 minutes of the motorist's time, five minutes less than high-option testing (since EPA estimated that purge and transient testing will require five additional minutes of test time). I assumed low-option I/M testing would require 38 minutes total of a motorist's time, since EPA estimates that pressure testing will require two minutes.

[10]The Regulatory Flexibility Act, 5 U.S.C. 605(b).

[11]For this analysis, EPA did not consider the impact of the I/M rule on emissions equipment manufacturers since these businesses are large enough that they do not fall under the requirements of the Regulatory Flexibility Act.

[12]Figures for the increases in repair business and additional jobs required as a result of the enhanced I/M rule can be found in U.S. EPA 1992d.

[13]As no figures were available for the inconvenience of basic I/M testing, I did not calculate the marginal cost-effectiveness of moving to a low-option inspection program.

[14]P.L. 104-59. 42 U.S.C. 7511a.

REFERENCES

Harrison, David Jr., Albert Nichols, John Evans, and J. Douglas Zona. 1992. *Valuation of Air Pollution Damages*. March. Cambridge, Massachusetts: National Economic Research Associates.

McConnell, Virginia and Winston Harrington. 1992. Cost-Effectiveness of Enhanced Motor Vehicle Inspection and Maintenance Programs. Discussion Paper QE92-18-REV. September. Washington, D.C.: Resources for the Future.

U.S. EPA (Environmental Protection Agency). 1992a. *I/M Costs, Benefits, and Impacts Analysis*. Technical Support Document [Draft]. February. Washington, D.C.: U.S. EPA. Office of Air and Radiation.

———. 1992b. Notice of Proposed Rulemaking: Vehicle Inspection and Maintenance Requirements for State Implementation Plans. *Federal Register* vol. 57, no. 134, July 13, 31058-31087.

———. 1992c. *I/M Costs, Benefits, and Impacts*. Technical Support Document. November. Washington, D.C.: U.S. EPA. Office of Air and Radiation.

———. 1992d. Inspection/Maintenance Program Requirements; Final Rule. *Federal Register*, vol. 57, no. 215. November 5. Washington, D.C.: U.S. EPA. Office of Air and Radiation.

———. 1994. *National Air Pollutant Emission Trends, 1900–1993*. EPA-454/R-94-027. October. Research Triangle Park, North Carolina: U.S. EPA. Office of Air and Radiation.

———. 1995. Final Rule, Inspection/Maintenance Flexibility Amendments. *Federal Register*. September 18. vol. 60, no. 180, 48029-48037.

———. 1996. I/M Roundup: State Programs Reflect Major Changes. *Inside EPA's Mobile Source Report* 4(9) May 3. Washington, D.C.: U.S. EPA.

13

Municipal Sewage Sludge Management

Mahesh Podar, Susan Burris, and Robert S. Raucher

On February 19, 1993, EPA published its final rule for regulating the use and disposal of sewage sludge. The RIA accompanying the proposed rule assessed the costs and some of the benefits of four alternative regulatory options. The annual compliance cost of the regulatory option preferred by EPA was estimated to be over $150 million. The revised RIA accompanying the final rule similarly presented an assessment of the costs and portions of the potential benefits of regulatory options. However, the estimated cost of compliance for the preferred option in this case was less than $50 million annually—a difference of well over $100 million per year. Such a dramatic reduction in costs from proposal to promulgation resulted both from an improvement in the quality of data used in the analysis and from a lessening in regulatory stringency. Benefits, while not monetized, were probably underestimated in both the initial and revised RIAs because of a failure to recognize potential cost-savings from expanded reuse opportunities. Additionally, in the rulemaking process, other cost-saving opportunities were disregarded, due to concerns about adverse public reaction. Despite these omissions, the total benefits of the scaled-back rule likely still exceeded total costs. In this case study, Mahesh

---

MAHESH PODAR is the director of the Policy and Budget Staff of the Office of Water, U.S. EPA. SUSAN BURRIS is a senior economist in the Office of Science and Technology in the Office of Water, U.S. EPA. ROBERT RAUCHER is a director of Hagler Bailly Consulting, Inc., a private firm that provides technical services in environmental economics, sciences, and policy. The views expressed here are those of the authors and should not be construed as the official position of the U.S. EPA or Hagler Bailly, Inc.

Podar, Susan Burris, and Robert Raucher introduce the complexities of the sewage sludge management task (including the wide range of multimedia issues it poses), present an overview and critique of both RIAs, and discuss the evolution of the rulemaking from proposal through promulgation.

INTRODUCTION

Municipal sewage sludge is generated at *publicly owned treatment works* (POTWs) as a by-product of domestic sewage wastewater treatment. These sludges are a combination of water, solids, and contaminants: dissolved substances, including nutrients, metals, inorganic and organic chemicals, pathogens, pesticides, and household chemical wastes. The exact composition and concentrations of these sludge contaminants depend on several factors, including the types and levels of discharges the POTW receives from industrial sources ("indirect dischargers" who are subject to "pretreatment requirements"), and the level of treatment provided at the POTW. Because of the volume and potential toxicity of municipal sewage sludge, public health and environmental concerns about disposal and potential reuse have developed. In 1987, as a result of such concerns, Section 405 of the Clean Water Act was amended, requiring EPA to regulate the use and disposal of sewage sludge.[1]

Managing Municipal Sewage Sludge

At least five important characteristics of the municipal sewage sludge problem make managing municipal sewage sludge a complex and challenging task:

First, the regulated substance arises as a *by-product generated by wastewater pollution control* rather than as a waste generated by industrial production or other economic activities. Thus, it is not a straightforward case of balancing environmental and public health benefits against regulatory compliance costs. Rather, it raises the issue of risk balancing—of how to account for and evaluate trade-offs in the benefits and costs associated with wastewater pollution abatement, relative to the risks and costs associated with managing its by-product, to maximize net social benefits (or minimize total social costs).[2]

Second, the problem is one of *multimedia pollution*. Municipal sewage sludge represents a problem whereby environmental protection efforts in one media (water) create potential environmental problems in others. In this instance, sludges become a solid waste problem with potential management options that include surface disposal (landfilling), incineration, and reuse (such as land applications). Evaluating the risks and benefits of the rule therefore raises challenging problems in terms of assessing multiple exposure pathways, through different media, that potentially affect different receptor populations (both human and ecologic). On the cost side, the rule may directly or indirectly affect a range of different economic sectors (including farmers, industries, and municipalities), depending on the type of management option selected.

Third, because it is a multimedia issue, sludge controls also raise the problem of *cross-programmatic regulation*—of how to design and evaluate environmental protection programs where efforts to control one problem (in this instance, water pollution) may simply shift the problem to one or more other program areas (such as regulatory programs related to surface disposal and incineration, and agricultural practices). In designing and implementing a regulatory approach that strives to maximize net social benefits regardless of where the contaminants reside, it is important to recognize that each medium is governed by its own statutory provisions, regulatory frameworks, precedents, and policy priorities. Thus, the sludge issue raises the specter of developing a single rule that can be logically consistent in how it addresses risk across media, yet also consistent within each of the separate regulatory programs and their governing statutes.

Fourth, municipal sewage sludge can be *a valuable resource* for agricultural productivity and land stabilization because of its nutrient content and soil-conditioning properties. Therefore, the regulation of sludges needs to balance a desire to promote beneficial reuse (rather than costly and wasteful disposal) against the desire to protect public health and the environment in reuse applications. Overly stringent controls on land applications could be protective with a margin of safety but might also unduly inhibit wise resource use by creating financial incentives to dispose of rather than reuse sludges.

Fifth, the rule imposes *costs directly on municipalities*. However, the rule also has implications for other sectors (for instance, by promoting industrial wastewater pretreatment). Therefore, the rule must be evaluated based not only on its cost implications to local governments (raising the issue of unfunded mandates) but also on cost implications that could potentially extend to numerous other sectors.

In sum, good municipal sewage sludge management requires weighing risks against other risks, and balancing benefits and costs across different media and regulatory programs—no easy task.

EPA's Approach to Managing Sewage Sludge

In developing its sludge regulation, EPA had to appreciate and deal with a wide range of data and analytical issues, including:

- the volume of sludges generated and methods available for their reuse and disposal;
- the types and concentrations of pollutants potentially found in sludges (metals, pesticides and other organic contaminants, and pathogens);
- the range of environmental media (air, water, soil) into which pollutants in sewage sludge may be placed, and through which the contaminants may be transported to human or ecological receptors;
- the potential types of adverse effects on individuals, on groups of individuals, and on plants or animals (that is, the human health and environmental risks posed by the use and disposal of sewage sludge);
- definition and assessment of risk levels that are "adequately protective" of human health and the environment; and
- estimation of the anticipated costs and benefits of various options.

EPA took several simplifying steps in an attempt to cope with these complexities. In essence, the agency focused primarily on whether the regulations should be based on health risk levels. Then between proposal and promulgation, EPA proceeded to grapple with the methodological issues associated with trying to assess health risks for each contaminant and reuse or disposal option, and then the policy issues of defining contaminant concentrations (performance standards) that would conform with a notion of providing "adequate protection" of human health. Concurrently, in an attempt to develop regulations that encouraged sludge reuse (land application) over disposal (landfilling or incineration), the agency strove to balance these risk-based objectives with economic objectives.

The final sludge rule was not a product of an explicit and broadly encompassing benefit-cost analysis. However, the work performed in the RIA assisted the agency as it developed a regulatory approach guided by the concerns of reducing public health risks (that is, generating benefits) and containing costs, and that was sensitive to the potential impact that regulation-driven compliance costs might have on the financial incentives and behavior of the regulated entities.

The Proposed Rule and Its Regulatory Options

The proposed rule, published in February 1989, included four regulatory options, applicable to five use and disposal practices (incineration, sludge monofills, land application, distribution and marketing, and surface impoundments). For each regulatory option, the proposed rule defined maximum allowable pollutant concentration in sewage sludge (performance standards) and/or minimum acceptable management practices (such as sludge testing) to ensure that the "desired level of protection" was maintained (technology or management standards). Specific numeric limits were proposed for seventeen pollutants. Incremental impacts of new or additional requirements were estimated after a baseline was established via an assessment of current use and disposal methods. The four options developed and considered at proposal are discussed below. The costs and benefits associated with the four options included in the proposed rule are shown in Table 1.

Option 1: Media Specificity (Statutory). Under this option, sewage sludges would be regulated in accordance with whatever media-specific laws and regulations already applied, regardless of whether or how human health risks were embodied in these media-specific rules. For instance, sewage sludge failing the Toxicity Characteristic Leaching Procedure—a test to determine whether the sewage sludge is a "hazardous waste"—would be regulated as a hazardous waste under Subtitle C of the Resource Conservation and Recovery Act, and all other sewage sludges would be managed under existing media-specific regulations *except* when limits for lead are exceeded in sludge that is incinerated. In addition, sludge would be tested periodically and sludge products that are marketed would have to be labeled appropriately.

Option 2: Pollutant Concentration Limits. In addition to the testing and labeling requirements in Option 1, limits on allowable pollutant concentrations in five sewage sludge use and disposal practices were proposed. These limits were based on the 98th percentile concentration as derived from the agency's national database on sewage sludge for each contaminant of concern. This option would have restricted the use and disposal of sewage sludges that have concentrations considerably higher than those found in "typical" sewage sludge. Based solely on the distribution of observed concentration levels, these limits would have no direct relationship to health risks or environmental benefits.

Option 3: Human Health Risk Assessment. This option, which was ultimately selected to serve as the basis for the proposed regulation, used human health risk assessment to set maximum allowable pollutant concentrations for each use and disposal practice. The maximum allowable concentrations were designed conservatively (stringently) in order to limit risks to individuals with high levels of pollutant exposure, or to compensate for significant scientific uncertainties. For carcinogens, the risk of cancer to the *most exposed individual* (MEI) was the basis for concentration limits; for noncarcinogens and endpoints other than human health, comparable MEI exposure scenarios, human health criteria, or plant/animal toxicity values were selected as a basis for concentration limits. The 98th

Table 1. Summary of Costs and Benefits of the Proposed Regulatory Options.

	Incineration	*Ocean disposal*	*Monofill*	*Land application*	*Distribution and marketing*	*Surface impoundments*	*Total*
Number of facilities using disposal practice	169	25	49	2,623	106	2,395	5,367
Baseline aggregate risk							
Cancer (cases/year)	13[a]	NA	0.02	0.18	0.02	NA	13
Other health effects (cases/year)	5,976[a]	NA	26	41.5	95	NA	6,138.5
Number of facilities in noncompliance							
Option 1[b]	96	NA	0	0	0	NA	96
Option 2	97	NA	7	54	0	NA	158
Option 3	122	NA	49	278	35	25	509
Option 4	122	NA	49	2,623	35	NA	2,829
Annual incremental cost of compliance ($million)[c]							
Option 1	$22.8	NA	$0.2	$4.1	$0.5	NA	$27.6
Option 2	22.8	NA	0.9	9.9	0.6	NA	34.3
Option 3	103.8	NA	25.5	15.2	7.8	5.3	157.6
Option 4	103.8	NA	25.5	244.3	25.4	NA	399.0
Risk level to be ensured by limitations, in terms of MEI							
Option 3	10^{-5}	NA	10^{-4}	10^{-4}	10^{-4}	NA	NA
Option 4	10^{-5}	NA	10^{-5}	10^{-5}	10^{-5}	NA	NA

Estimated Reduction In Aggregate Risk[b] - Cancer (cases avoided/year)							
Option 1	2.8	NA	0.00	0.00	0.00	0.00	2.8
Option 2	2.8	NA	0.01	0.06	0.00	0.00	2.9
Option 3	9.4	NA	0.02	0.06	0.02	NA	9.5
Option 4	.3	NA	0.02	0.18	0.02	0.00	7.5
Other health effects (cases avoided/year)							
Option 1	5,155	NA	0	0	0	0	5,155
Option 2	5,155	NA	0	4	0	0	5,159
Option 3	5,163	NA	26	21	56	NA	5,266
Option 4	4,769	NA	26	42	62	0	4,889

Notes: NA = Not applicable

[a]Baseline risk includes the risks of incineration by POTWs that are required to cease ocean disposal.

[b]The four regulatory options noted here were applicable to five use and disposal practices (incineration, sludge monofills, land application, distribution and marketing, and surface impoundments). Option 1: media specificity (statutory); Option 2: pollutant concentration limits; Option 3: human health risk assessment; Option 4: more stringent human health risk assessment.

[c]Incremental costs and benefits of the regulation associated with the twenty-five POTWs affected by the ocean ban are included in the costs and benefits for incineration.

Source: U.S. EPA 1989, Table 1-2.

percentile sludge concentrations were also used under this option for setting pollutant limits for nonagricultural land application and surface disposal sites.

Also included in this option were a number of minimal acceptable management practices, such as testing; continuous monitoring and recording for incineration; runoff and run-on controls for monofilling and surface impoundments; access and use controls; pathogen reduction and testing for land application; sludge product labeling for distribution and marketing; and record-keeping and reporting requirements for all disposal methods.

Option 4: More Stringent Human Health Risk Assessment. As in Option 3, maximum pollutant concentrations were proposed on the basis of risk assessments; however, the MEI-based human health risk levels were generally more stringent than the risk levels defined in Option 3. Surface impoundments were not covered in this option. Management practices required under Option 3 were also required in this option.

EPA Preferred Option. EPA proposed Option 3 as its preferred option, with an estimated annual incremental compliance cost of $157.6 million. Option 3 was also associated with health risk benefits from reduced aggregate risk, with benefits consisting of an estimated excess of 9.5 cancer cases and 5,266 other health effects cases avoided annually.

The Final Rule

The final rule changed substantially from the proposal in response to new data, public comments, and additional analyses. The number of use and disposal practices was reduced from five to three, and the size of the community to be regulated was expanded. Domestic septage haulers—operators that service and empty septic tanks and take the material to POTWs or apply it to land—were added to the number of facilities affected by the final rule. The total number of explicitly regulated contaminants was also reduced from seventeen to eleven, as shown in Table 2 for the relevant use and disposal practices.

Distribution and marketing was incorporated into land application, and sludge monofills were combined with surface impoundments to create the "surface disposal" route. The predominant reuse and disposal options for sludge management were thus defined as follows:

- *Beneficial Reuse for Soil Conditioning and Nutrients*—The organic and nutrient content of sewage sludge makes it a valuable resource for use both in improving marginal lands and as a supplement to fertilizers. The organic content in sewage sludge contains $30 to $60 per dry

Table 2. Pollutants with Numerical Limits under the Promulgated Rule.

Pollutants	*Land appliation*	*Surface disposal*	*Incineration*
Arsenic	X	X	X
Cadmium	X		X
Chromium	X	X	X
Copper	X		
Lead	X		X
Mercury	X		
Molybdenum	X		
Nickel	X	X	X
Selenium	X		
Total hydrocarbons			X
Zinc	X		
Totals	10	3	6

Source: U.S. EPA 1993b, Table 1X-1, p. 9,319.

ton's worth of organic nitrogen and phosphors. Sewage sludge can be used beneficially to increase production of agricultural commodities; increase forest productivity; revegetate and stabilize forest land after harvest, fire, landslides, or other natural disasters; and stabilize and revegetate areas destroyed by mining, dredging, and construction activities.

- *Disposal Methods*— Sewage sludge is disposed of using one of two different practices: land application at dedicated sites (sites set aside specifically for sewage sludge disposal) and landfilling (deposited in a dedicated area, alone or with solid waste), and incineration (destruction of organic pollutants and reduction in volume).

Alternative Regulatory Approaches and the Final Rule. A central feature carried over from the proposed rule to the final rule was the risk-based approach. Four regulatory options were developed and considered for promulgation. The first two of these "alternative regulatory approaches" replicated Options 1 and 2 from the proposal stage; that is, *Approach I* was based on using existing regulations, and *Approach II* was based on using the 98th percentile pollutant concentrations where existing regulations provided no numerical limits. Both of these options (Approaches I and II) were again rejected by the agency. The two remaining alternatives were given more serious consideration, and the final rule was a hybrid variation of these two:

Under *Approach III*, the agency would establish concentration limits based exclusively on the agency's exposure assessment modeling, with the limits based on regulating exposures to "protective" levels for human or environmental MEIs. Protective levels were defined for carcinogens as incremental lifetime excess risk levels of 1 x 10^{-5} (and 1 x 10^{-6} for reuse in a distribution and marketing mode) and oral reference doses (which anticipated no effect exposure levels) were used for noncarcinogens. Ecological MEI endpoints were constructed using the most sensitive species with steady state duration and concentration of exposure over a critical life period. Such exposures were benchmarked against established plant and animal toxicity values published by the agency or in the peer reviewed literature.

Approach IV replicated the option selected at proposal (Option 3, above) in that it used a combination of MEI risk levels and 98th percentile concentrations as the basis for setting concentration limits in sludges. This option differs from Approach III in that it used a less stringent 1 x 10^{-4} target incremental cancer risk level for sludge reuse and disposal options (except that incineration-based limits remained based on 1 x 10^{-5} limits), but was more stringent in some practices in that the 98th percentile levels were applied in cases where humans were unlikely to be exposed (that is, forest and reclaimed lands applications, and disposal at surface application sites).

Based on the comments received and its own evaluations, EPA ultimately promulgated a final rule that resembles a hybrid of Approaches III and IV, with some new dimensions added as well. The final rule establishes pollutant limits solely on the basis of exposure and risk assessments (as in Approach III), but the regulatory limits set for pollutants are more lenient in that the numerical limits are based on a 1 x 10^{-4} incremental cancer risk level for all disposal and reuse modes (similar to Approach IV). The novel element of the final rule is that the pollutant concentration limits are based on the construct of *highly exposed individuals* (HEIs) rather than on MEIs (which the agency labeled as an "unrealistic worst-case approach"). The agency appears to have selected the HEI-based, 1 x 10^{-4} target-level risk assessment option based on various logical and practical considerations rather than on any empirical assessment such as a regulatory impact analysis (RIA). However, the selection of the final numerical limits appears to be based on empirical assessments of incremental risk-level changes under alternative risk targets, with limits set to prevent undue transfer of sludges to incinerators, which are associated with high population risks relative to the other disposal or reuse options.

Economic Sectors Affected by the Final Rule. The final rule covers nearly 35,000 entities. These entities include about 17,600 primary treatment POTWs, secondary and advanced treatment POTWs, privately owned treatment works, and federally owned treatment works and about 17,000 domestic septage haulers (U.S. EPA 1993b, Table 1-8 and page 1-20).

However, only about 15,300 entities incur compliance costs due to the promulgated rule.

Approximately 12,750 POTWs use and dispose of about 5.4 million dry tons of sewage sludge annually, or 47 pounds of sewage sludge (dry weight basis) for every individual in the United States. Table 3 shows the amount of sewage sludge used and disposed of by the size of a POTW, the amount of sewage sludge that is used and disposed of using a particular practice, and the number of POTWs using each practice. This volume is expected to double by the year 2000 due to population growth, stricter wastewater treatment requirements, and a greater number of better operated POTWs.[3]

THE REGULATORY IMPACT ANALYSIS

The "Forty City Study"[4] was the primary source of information used to develop the proposed RIA. At the time of the proposed regulation, this study provided the most comprehensive and best documented database on the occurrence and concentration of pollutants in sewage sludge of forty-three to forty-five POTWs.

While developing the proposed regulation, EPA recognized several deficiencies in the Forty City Study as it relates to sewage sludge quality.

Table 3. Sewage Sludge Use and Disposal. (thousands of dry metric tons)

	Reported flow rate (mgd)					*POTWs*	
Use or disposalpractice	*>100*	*>10 to 100*	*>1 to 10*	*<1*	*Total (% of total)*	*using a practice number*	*Percent of POTWs*
Incineration	384	348	126	11	869 (16%)	381	2.8
Land application	390	667	540	196	1,793 (33%)	4,657	34.6
Codisposal (landfill)	521	676	496	110	1,803 (34%)	2,991	22.2
Surface disposal	80	266	123	87	556 (10%)	1,351	10.0
Unknown	167	158	8	3	336 (6%)	4,078	30.3
Total (% of total)	1,542 (29%)	2,115 (39%)	1,293 (24%)	407 (8%)	5,357 (100%)	13,458	100.0

Note: The total number of POTWs does not equal the number of POTWs in the text because some POTWs use more than one practice.

Source: U.S. EPA 1993a, Tables I-1 and I-2, pp. 9,256–57.

Thus, to supplement the existing information as well as to fill in the data gaps, EPA conducted the National Sewage Sludge Survey (NSSS; U.S. EPA 1990) to obtain a current and reliable database for developing the final regulation. The NSSS data collection effort began in August 1988 and was completed in September 1989. EPA sampled 180 POTWs and analyzed their sludges for more than 400 pollutants. In addition, through the use of a detailed questionnaire, information was collected on sewage sludge use and disposal for approximately 460 POTWs with at least secondary wastewater treatment.

The NSSS solved many of the data problems associated with the RIA for the proposed regulation; however, some data limitations persisted. Data for privately and federally owned treatment works were limited. Thus, numbers of privately owned treatment works were estimated from permits. Since only a few states collect data on use and disposal practices for privately owned treatment works, the agency had to extrapolate up to the total population. Secondly, information on federally owned treatment works was unavailable. For this reason, the distribution of use and disposal practices for privately owned treatment works was applied to federally owned treatment works. Information on domestic septage haulers was also very limited. EPA's estimates of the number of septage haulers and typical use and disposal practices were based on data from a limited number of states.

Cost Analysis

The final RIA included compliance costs for management practices and monitoring, record keeping, and reporting, as well as the potential cost a POTW or other affected entity might incur if it were required to alter its current disposal practice. These changes could include shifts to a new or different combination of use and disposal practices or increased reliance on pretreatment. The costs of these changes were evaluated for each use and disposal practice that was expected to be altered by the regulation.

It is important to note that the EPA analyses of neither costs nor benefits explicitly incorporated the cost-savings (and risk changes) that would accompany a shift from disposal practices to reuse applications. This static aspect of the analysis in relation to the regulation's potential for enabling a higher volume of sludges to be placed into reuse applications (rather than more costly disposal routes) is important due to the large potential cost-savings that may be realized. This issue is discussed further later in the chapter in relation to the potential benefits of the rule.

Most of the EPA cost analysis focused on secondary and advanced treatment POTWs because the agency had the most information on this subject, as furnished by the NSSS.[5] NSSS data on sewage sludge quality were compared with the applicable pollutant limits in the final regulation to determine whether the sewage sludge used and disposed of at each POTW could meet those limits. Results were then extrapolated to the population of regulated POTWs using the NSSS statistical weighing factors.

Data from surveyed POTWs that exceeded the potential pollutant limits at their current operating conditions (given the POTWs' existing sewage sludge feed or land application rates) were further evaluated to develop the mix of anticipated compliance strategies. The costs of these compliance strategies were then estimated. A failure could be either economic or regulatory. For example, in the case of land application, if either application rates or site life had to be reduced, then a cost associated with these changes was estimated.

The costs required to make changes to current use and disposal practices in order to comply with a potential regulatory limit were added to the costs estimated for management practices, monitoring, record keeping, and reporting. These results from the sampled POTWs were then extrapolated to the population of regulated entities.

For each use and disposal practice, EPA conducted a "pass/fail" analysis. First, the quality of sewage sludge in the NSSS was compared with either the pollutant concentration limits or the pollutant loading rate limits. Second, site-specific risk assessment-based pollutant limits were developed for those POTWs that "failed" pollutant concentration limits. Third, individual compliance strategies for each failing POTW were developed. These compliance scenarios included such options as improved pretreatment at industrial dischargers, best management practices, and better operations to minimize costs. Finally, these individual compliance strategies served as the basis for estimating costs.

Several sensitivity analyses were conducted based on a number of different pollutant limits for each of the three use and disposal practices. For example, numeric pollutant limits vary depending on the use and disposal practice, and the agency ran a pass/fail analysis to determine how many POTWs did and did not meet the limit based on their current disposal practices. One key sensitivity analysis evaluated the cost of setting the *total hydrocarbons* (THC) limit for incineration at various levels ranging from 30 to 100 parts per million (ppm).

Table 4 shows the agency's estimated total costs of meeting requirements and limits for land application, including pathogen reduction, surface disposal, and incineration. Added to these $44.2 million is the cost of reading and interpreting the regulation, which adds $1.7 million annually. Thus, the EPA estimate of the total costs of compliance with the promulgated regulation is $45.9 million annually. These estimated compliance costs are substantially less than the proposed rule's estimated incremental annual costs of $157.6 million.

Table 4. Summary of Total Annual Costs. ($1992)

Affected entity	*Number of affected entities*	*Land application*	*Surface disposal*	*Incineration*	*Total*
Primary treatment POTWs	1,296	$1,854,545	$1,775,478	$489,000	$4,119,023
Secondary and Advanced Treatment POTWs	6,249	11,639,934	12,107,469	11,214,000	34,961,403
Privately owned treatment works	1,580	438,354	2,162,675	0	2,601,029
Federally owned treatment works	81	22,578	109,900	0	132,478
Domestic septage haulers	6,120	226,994	2,179,062	0	2,406,059
Total	15,326	$14,182,405	$18,334,584	$11,703,000	$44,219,989

Note: The final total of all annual costs does not include $1.6 million to read and interpret the regulation. All numbers may not add as indicated in the totals because of rounding.

Source: U.S. EPA 1993a, Table I-7.

There are several reasons for the large decrease in the expected costs of the regulation.

First, quality of sewage sludge was much better than anticipated. EPA used the Forty City Study to develop the proposal; however, data subsequently collected as part of the NSSS show that concentrations of pollutants of concern were much lower than anticipated. As a result, more POTWs were able to meet the limits in the final rule without incurring additional treatment costs. Second, monitoring and record-keeping requirements are reduced in the final rule. Rather than requiring uniform monitoring and record keeping from all POTWs, the final rule specifies minimum requirements and allows permit writers discretion to impose more stringent requirements, if warranted. In addition, the sensitivity analyses conducted in the RIA showed that setting the THC standard at 100 ppm would substantially lower costs without adverse effect; hence, the THC monitoring requirement was set at 100 ppm rather than 30 ppm as proposed. And finally, thresholds for pass/fail analysis were raised based on revised risk assessments. As a result, a smaller number of POTWs failed to meet the applicable limits and were projected to incur compliance costs.

The largest cost decrease was concentrated in incineration, where estimated annual costs were reduced from $103.8 million to $11.7 million. These cost reductions were attributed to the fact that 171 out of 185 POTWs with incinerators are expected to meet the final rule's limits with minimum additional expenditures, primarily for monitoring and record keeping, and because monitoring requirements in the final rule were substantially reduced—both in terms of frequency and level—relative to the proposed rule.

Estimated annual costs for surface disposal declined from $30.8 million to $18.3 million, and for land application from $23.0 million to $14.2 million. For land application, only 14 of 4,328 POTWs were projected to fail to meet the applicable limits. The failing POTWs are projected to use more stringent pretreatment requirements—requiring industrial dischargers to improve their treatment and reduce the quantities of pollutants discharged—and shift to co-disposal. For surface disposal, only 8 of 1,410 POTWs that use surface disposal fail to meet the applicable limits. However, these POTWs would be able to obtain site-specific permit limits to enable them to minimize compliance costs.

Benefits Analysis

As mentioned previously, the RIA discusses both the benefits and the costs of the sewage sludge regulation. However, the benefits analysis is limited to quantitative estimates of cancer and noncancer cases avoided, and limited qualitative discussion of other potential benefits. No monetization of benefits is attempted. Potential cost-savings associated with switching from disposal to beneficial reuse are also estimated in the RIA, although these cost-savings were not explicitly added to the benefits of the rule (or netted out of the estimated compliance costs).

Health Risk Reduction Benefits. The risk assessment for the benefits analysis in the RIA follows the process outlined by the National Academy of Sciences (NAS 1983). The agency also followed several other sets of guidelines (U.S. EPA 1986) for health and risk assessment in developing the regulations. The assessment begins with a hazard identification and source assessment and follows with fate and transport estimates, exposure assessments, pharmacokinetics analysis, and dose-response assessment. These factors are used to estimate changes in public health, measured as morbidity and mortality.

The first step in estimating benefits involves estimating the baseline public health risks of sewage sludge use and disposal. These risks are presented as cases of cancer and other adverse health effects. The key inputs for estimating baseline risks include POTW information, sewage sludge pollutants, and ultimate use and disposal site characteristics. Baseline risk from sewage sludge use and disposal practices is characterized using sewage sludge quality based on NSSS data, including the amount of sewage sludge used and disposed of by the POTWs, and the fate and transport of the pollutants subsequent to use and disposal. Using the inputs described above, the EPA risk analysis estimates the potential pathways of human exposure and models the fate and transport of key sewage sludge pollutants for these pathways. Next, the potential population exposed is estimated. This information, in concert with dose-response data for each of the sewage sludge pollutants of concern, is used to characterize baseline public health risks.

After baseline risks were estimated, risk estimates were developed assuming that the alternative sewage sludge requirements were met. The regulatory compliance strategies for the public health risk analysis assessment paralleled the strategies used to estimate the compliance costs. Furthermore, the same risk assessment process is used to derive the change in the baseline risk as a result of the requirements for each use and disposal practice. This change in health risks (such as the number of excess cancer cases) relative to the baseline is used as the measure of benefits.

A number of limitations affect the estimates of risk reduction benefits in this regulation. These limitations include the exclusion of certain exposure pathways, pollutants, and health effects, and the inability to account for population growth and mobility.

Table 5 highlights the anticipated health benefits and total cost of compliance for each of the three use and disposal practices regulated by the final rule. The benefits of complying with this regulation are expressed as reductions in risk—the number of baseline cases that are avoided. In total, the benefits are expected to consist of between 0.09 and 0.7 excess cancer cases avoided, and between 90 and 600 noncancer health effects avoided each year. For land application, the benefits are estimated to be less than 1 cancer case avoided and 0 to 500 cases of other adverse health effects avoided. The benefits of complying with the surface disposal requirements are estimated to be 0 to 0.07 cancer cases

Table 5. Estimated Annual Costs and Benefits for All Treatment Works ($1992).

		Benefits	
Regulatory options	*Annual costs (millions)*	*Cancer (annual excess cases avoided)*	*Noncancer (annual excess cases avoided)*
Land application	$14.18	0–0.5	0–500
Surface disposal	$18.34	0–0.07	<1
Incineration	$11.70	0.09	90
Other costs	$ 1.68	NA	NA
Total costs/benefits	$45.90	0.09–0.7	90–600

Note: NA = Not applicable.

Source: U.S. EPA 1993a, Table XIII-5, p. 9,371.

avoided and less than 1 other health effect case avoided. Complying with the incineration requirement results in avoidance of an estimated 0.09 cancer case and 90 other adverse health effects cases.

Compared to the proposed rule, the reduction in aggregate risk is substantially less for the final rule. In the RIA for the proposed rule, the preferred option was projected to avoid 9.5 cancer cases and 5,266 other health effects cases (incineration accounted for 9.4 of the cancer cases and 5,163 of the other health effects cases avoided). The primary reason for reduced health effects benefits is that the sewage sludge quality was better than anticipated (as discussed earlier); that is, the baseline health risks are much lower than estimated at proposal.

Cost-Savings. Sewage sludge of good quality—that which is shown to have few or no adverse health effects —has the potential to generate revenue for POTWs that can be used to offset the cost of improving sewage sludge quality. In other words, POTWs can generate revenue from sewage sludge if they can assure the potential users that the sewage sludge has no adverse health effects. The logic is that as misapprehensions about the quality of sewage sludge are cleared—in large measure due to the promulgated standards—high-quality sludge that would otherwise be disposed of through incineration and landfill could be shifted to land application. Therefore, EPA analyzed the potential cost-savings associated with the final rule that some POTWs would likely experience.[6]

An analysis of the NSSS data showed that sewage sludge of about 2,927 POTWs can meet pollutant concentration limits and is therefore eligible for land application. These 2,927 POTWs dispose of 1,286,000 dry metric tons (dmt) annually—525,000 dmt to incineration, 621,000 dmt to co-disposal, and 140,000 dmt to surface disposal (see Table 6). Cost-savings to shift from disposal options to land application vary depending on the size of the POTW and are also shown in Table 6.

It is not possible to predict how many POTWs would be able to, and elect to, shift to land application. Therefore, EPA estimated potential cost-savings for three scenarios under which either 30%, 60%, or 90% of sewage sludge is shifted from disposal to land application. The estimated costs savings would range from $69 million with a 30% shift to $141 million at 60% up to $211 million at 90% (see Table 7). Even a 30% shift from disposal to land application (a beneficial reuse), with an associated potential cost-savings of $68.6 million, would exceed the estimated compliance costs of $44.2 million and generate a net savings of $24.4 million annually.

In addition to these direct cost-savings, additional cost-savings are also possible. Commercial fertilizers are a significant portion of a farmer's cost of growing crops. Increased land application will enable farmers to meet the agronomic need of crops for nitrogen from sewage sludge at

Table 6. Sewage Sludge Disposed* and Potential Cost Savings. ($1992)

Reported	*Incineration*		*Co-disposal*		*Surface disposal*	
flow group	*Dry metric tons*	*$/ton*	*Dry metric tons*	*$/ton*	*Dry metric tons*	*$/ton*
>100 mgd	163,650	$105	139,890	$58	0	$111
10–100 mgd	294,470	$117	259,470	$54	47,920	$237
1–10 mgd	64,330	$150	162,700	0	42,550	$596
<1 mgd	2,500	$457	58,640	0	49,420	$2,243
Total	524,950		620,700		139,890	

*These quantities of sewage sludge disposed are those that met pollutant concentration limits.

Source: Jones 1992.

Table 7. Potential Cost Savings from Switching from Disposal to Reuse/Land Application. ($1992)

	Scenario (percentage of sludges shifted from disposal to reuse)		
Type of disposal	*30%*	*60%*	*90%*
Surface disposal	$17,670,000	$35,340,000	$53,000,000
Co-disposal	$6,640,000	$17,070,000	$25,610,000
Incineration	$44,270,000	$88,540,000	$132,810,000
Total	$68,580,000	$140,950,000	$211,240,000

Source: Jones 1992.

least partially, thus reducing the need to purchase commercial fertilizer. This shift could result in substantial cost-savings to farmers that have thus far not been quantified.

Other Potential Benefits. The sewage sludge regulation has the potential to yield other environmental benefits in addition to human health risk reductions and cost-savings for POTWs. For example, a reduction in the toxicity of sewage sludge would benefit the environment, especially when the sludge is applied in environmentally sensitive areas. Reductions in particulate emissions would also minimize the potential adverse effect on nearby buildings, automobiles, other structures, and nearby farms. These benefits were not quantified by EPA.

Other Analysis Conducted

In addition to the assessment of overall costs and benefits, EPA also assessed impacts on small entities under the Regulatory Flexibility Act. EPA determined that about 90% of all entities affected by this regulation are small entities; however, only about 40% of the entities affected employ a use or disposal practice covered by this regulation. The total estimated compliance costs for small entities are $14.1 million, mostly for land application and surface disposal. Small POTWs account for about $11.0 million and domestic septage haulers account for about $2.4 million. EPA's assessment shows that these costs are not estimated to adversely affect the small entities.

CRITIQUE OF THE RIA

EPA devoted considerable effort to develop the risk assessments and cost estimates for both the proposed and final rules as reflected in the accompanying RIAs. However, EPA did not attempt to monetize cancer and noncancer benefits, nor to quantify nonhealth benefits to human welfare or ecologic systems. Therefore, the agency apparently had little interest in developing, much less using, benefit-cost information as a basis for designing and evaluating its regulatory options with respect to municipal sewage sludge.

From a review of how the rule evolved (proposal through final promulgation), it appears the agency made several implicit benefit-cost-type judgments to steer the rule toward an endpoint that would focus on protecting health and the environment while minimizing compliance costs and related financial disincentives to productive reuse of the sludge resource. Thus, while it appears the agency fell short of the desired mark in terms of developing a useful benefit-cost analysis, EPA's final rule appears to be implicitly influenced by—and may well be consistent with—the underlying rationale of the benefit-cost paradigm. To examine this hypothesis more closely, three economic interpretations are provided below.

Cost-Effectiveness Perspective

If cost-effectiveness is used as a decision criterion, one would deduce that EPA's decision to promote beneficial use, rather than disposal, of sewage sludge was a logical choice. Given the large volume of sewage sludge that must be used and disposed of in an environmentally sound manner (without causing harm to human health or the environment), the cost-effectiveness of each practice must be considered in making decisions. Table 8 displays the cost-effectiveness of each use and disposal option. Surface disposal at $8.11/ton appears to be most cost-effective, land application at $8.31/ton is next, and incineration at $13.68/ton is least cost-effective.[7] However, such measures of cost-effectiveness are of limited value from a social welfare maximization perspective because the compliance cost per metric ton does not indicate how much risk exists, or is reduced, with the different tons of sludge. This is because each unit of sludge mass can generate different levels of environmental risks and other social costs

Table 8. Cost-Effectiveness of Sewage Sludge Use and Disposal. ($1992)

Regulatory options	*Dry metric tons/year*	*Annual cost*	*$/ton*
Land application	1,793,000	$14,900,000	$ 8.31
Surface disposal	2,359,000	$19,120,000	$ 8.11
Incineration	869,000	$11,890.00	$13.68

Notes: Other costs of $1.7 million are redistributed to other practices based on the percentage of sewage sludge used or disposed. As a result, annual costs increase slightly for each practice. Quantities are from Table 3, Sewage Sludge Use and Disposal. Surface disposal includes both co-disposal and surface disposal.

depending on several factors: what contaminants (and levels) the sludge contains, where it is located, how it is being managed, and what receptors might be exposed to its contaminants in different settings and through different disposal or reuse patterns.

Benefit-Cost Perspective (Absent Cost-Savings)

If the estimated benefits were monetized and decisions were made based on net positive benefits, welfare-based policy conclusions would differ from the cost-effectiveness results. The available benefit and cost information is recast in Table 9, using the static analysis presented by EPA's RIA in which cost-savings associated with potential shifting of sludges from disposal to reuse are ignored. In Table 9, the cancer case reductions from Table 5 have been valued, for illustrative purposes, at $5 million per case avoided (based on the EPA-accepted range of values for a "statistical life" of $2 million to $10 million). After the monetized health risk reduction benefits are netted out of costs, it is clear that there are still appreciable levels of net costs to weigh against the remaining potential benefits.

For example, for the rule as a whole, the monetized benefits range from less than $0.5 million to $3.3 million per year. These benefits are more than an order of magnitude less than the annualized costs of $45.9 million. (Even if the cancer cases were valued at twice the amount—$10 million per case—substantial disparity between costs and benefits remains.) The remaining, nonmonetized health benefits would have to be valued at between $71,000 and $506,000 per potential noncancer health effect avoided in order for benefits to be roughly equivalent to costs. The monetary value of morbidity effects depends greatly on the health endpoint that is relevant, and the RIA does not indicate what illnesses are potentially associated with this regulation. Nonetheless, these values greatly exceed accepted estimated willingness-to-pay values for most nonfatal, noncancer health effects (such as the estimates by Tolley and others 1986 of approximately $10 per day for acute respiratory symptoms). Thus, while this monetized comparison fails to include potential

Table 9. Interpretation of Benefit-Cost Findings. ($1992)

Regulatory option	*Annual costs (millions)*	*Monetized benefits (cancer risk reductions)*[a]	*Nonmonetized benefits (noncancer cases avoided annually)*	*Implicit value for nonmonetized benefits (cost per case avoided)*
Land application	$14.2	$0.0–$2.50	0–500	>$0.02
Surface disposal	$18.3	$0.0–$0.35	<1	>$17.9
Incineration	$11.7	$0.45	90	$0.13
Other costs	$1.7	NA	NA	NA
Totals[b]	$45.9	$0.5–$3.3	90–590	$0.71–$0.51

[a]Annual cancer cases avoided are valued at $5 million per case, based on published values (Tolley and others 1986) for risk reductions (value per "statistical life").

[b]All numbers may not add as indicated in totals due to rounding. NA = Not applicable.

nonhealth benefits of the rule, it does help to illustrate potential concerns regarding whether the net benefits are positive.

Of additional policy interest is the evaluation of benefits and costs across the disposal and reuse options. In particular, the results raise doubts about the efficacy of the surface disposal portion of the rule. Although it is the costliest portion of the rule, it appears to have, by far, the smallest benefits. For land application, the annual costs ($14.2 million) less the monetized value of reductions in carcinogenic risks ($0 to $2.5 million) implies that the value of each noncarcinogenic case avoided would need to be at least $23,400 per case in order to justify the rule. Likewise, for incineration, costs minus monetized benefits implies that the nonmonetized, noncarcinogenic risks would need to be valued on the order of $125,000 per case avoided. While it is difficult to ascertain whether these implicit values are reasonable, the comparable value for surface disposal appears more clearly to be out of the likely accepted range. For surface disposal, a value of noncarcinogenic health risks avoided of over $17.9 million per case avoided is required to balance health benefits with costs.[8] This implies that the rule is overly stringent for surface disposal, potentially shifting an undesirable quantity of sludges to other, riskier disposal or reuse options.

If the surface disposal option were not changed from baseline, and the new rules applied only to incineration and land application, the costs of the rule would be reduced by 60% (from $45.9 million to $27.6 million per year). Concurrently, the rule would still generate between 89% and 100% of the anticipated carcinogenic risk reduction benefits, and 100% of the noncarcinogenic risk reduction benefits. Therefore, using the criteria of making decisions based on net positive benefits, this review of the RIA raises the question of why EPA opted to promulgate the rule in the manner that it did.

Benefit-Cost Comparison (Including Potential Cost-Savings)

While EPA failed to include its assessment of potential cost-savings from expanded reuse opportunities within its benefit-cost analysis, the potential size of these cost-savings is sufficiently large to alter how one interprets the benefit-cost findings portrayed above. Insufficient information is available to ascertain how accurate these potential cost-savings estimates are, or what percentage of the sludges might be subject to these savings due to the promulgated rule. Therefore, we assume that the cost-savings estimates are reasonably accurate, and that the 30% scenario developed by EPA applies.

Under this scenario, the cost-savings benefits alone ($68.6 million per year) easily outweigh the overall estimate of compliance costs ($45.9 million per year), and the other benefits, such as health risk reductions, become of secondary importance. If these conditions hold, EPA made a reasonably wise policy decision in crafting its municipal sewage sludge rule (although absent better benefit-cost information, one cannot ascertain if even better options may have been identified).

Impact of the RIA on the Rulemaking

Several of the analyses portrayed in the RIA played an important role in the decision process. The RIA also provided a forum for incorporating new data and making adjustments in response to public comments, so that the analysis supporting promulgation could be much refined and improved over the data presented at the proposal stage. For example, at proposal, the RIA demonstrated substantial reductions in risk from incineration as a result of the proposed requirements. However, subsequent data and analysis showed the baseline risks to be much lower than anticipated at proposal; thus, the final requirements for incineration were appropriately modified to make them less stringent. In addition, risk and economic impact considerations played a major role in determining the numeric limits for THC. An in-depth sensitivity analysis, included in the final RIA, resulted in reducing the monitoring requirements for THC in the final rule.

The analysis contained in the final RIA also showed that it is better to combine various use and disposal practices into broad categories because finer breakdowns are not warranted. As a result, the final regulation is much simpler than the proposed rule and is easier to implement. The final RIA also included potential cost-savings associated with improvements in sludge quality. Not only did information about improved sludge quality help alleviate misapprehensions about its impacts, it also had a substantial impact on the growth of a new industry that promotes beneficial use of sewage sludge.[9] Thus, the RIA supported a final sewage sludge regulation that embodies the agency's priority in promoting pollution prevention and recycling/reuse.

The risk-based pass/fail analyses played an important role in determining which pollutants should be regulated, and at what level, for each use and disposal practice (at desired levels of protection). As a result, compliance costs estimated for the final rule were substantially lower when compared with the costs of the proposed rule—$46 million ($1992) for the final rule as opposed to $158 million (in $1987; this is equivalent to $195 million in $1992) for the proposed rule.

CONCLUSIONS

EPA's benefit-cost analyses for the municipal sewage sludge rule were fairly limited in their attempt to quantify or monetize beneficial outcomes or to incorporate a better analysis of the potential cost-savings to be realized where reuse was facilitated by the regulations. Furthermore, the benefit-cost portions of the RIA did not have a significant explicit impact on the development of the final rule. For example, on a benefit-cost basis, the agency might only have regulated incineration and land application at or near the levels imposed by the final rule, whereas surface disposal might have been regulated much more leniently.

Nonetheless, the basic paradigm of benefit-cost analysis did have a significant influence on the sewage sludge regulation as it evolved from proposal through its final promulgation. Although benefit-cost analysis did not serve as an overt basis for developing and evaluating regulatory options, the agency's logic in developing its overall risk-based approach to the rule, and the specific numerical limits that the rule embodies, were clearly based on risk assessments, implicit trade-offs between costs and risk reductions, as well as a desire to promote beneficial reuse of sludge by reducing misapprehensions about sludge quality. However, the decisionmakers might have been better served if the next step were taken and decisions were made based on better developed and more explicit risk, benefit, and cost considerations.

ENDNOTES

[1]The proposed rule, *Standards for the Disposal of Sewage Sludge: Proposed Rule,* resulting from Section 405 was published on February 6, 1989 (*Federal Register,* vol. 54, no. 23, pp. 5,746–902). The final rule, *Standards for the Use or Disposal of Sewage Sludge,* was published on February 19, 1993 (U.S. EPA 1993a). The related National

Sewage Sludge Survey (U.S. EPA 1990), published on November 9, 1990, was conducted to provide data to support the development of the sewage sludge regulation.

The proposed rule contained requirements for land application, distribution and marketing, municipal landfills, surface disposal, and incineration. It included limits for a total of seventeen pollutants, although the number of pollutants regulated under each use and disposal practice varied.

Based on public comments, the five proposed categories of disposal practices were consolidated into three categories: land application, surface disposal, and incineration. The final rule sets limits for a total of eleven pollutants, although the number regulated under each practice varies.

A number of EPA staff (one full-time equivalent) monitored the contractor, reviewed the analysis, and provided technical guidance during the proposal, the notice of data availability, and the final regulation.

[2]The issue of balancing the benefits and costs related to water quality improvement resulting from wastewater treatment with the risks and costs associated with wastewater sludges is, in and of itself, an interesting analytic issue worthy of further consideration. However, the trade-offs between water quality improvements and any risks associated with sludge management would be extremely site specific. No empirical research appears to have been conducted in this area to date.

[3]Municipal wastewater treatment plants may use one or more levels of treatment (that is, primary, secondary, or tertiary). Each level of treatment provides cleaner water and generates greater amounts of sewage sludge. Primary treatment processes remove the solids that settle out of the wastewater by gravity and generate between 2,500 and 3,000 liters of sewage sludge per million liters of wastewater treated. Secondary treatment uses biological treatment processes (such as activated sludge systems, trickling filters, and so forth, that use microbes to break down organic matter) and generates between 15,000 and 20,000 liters of sewage sludge per million liters of wastewater treated (over primary treatment). Advanced wastewater treatment (such as chemical precipitation and filtration) generates about 10,000 liters of sewage sludge per million liters of wastewater treated (over secondary treatment).

Sewage sludge contains between 93% and 99.5% water, solids, and dissolved substances. Before reusing or disposing of sewage sludge, treatment works generally thicken, stabilize, and dewater the sludge to reduce volume and transportation cost. Before sewage sludge can be land applied, it is frequently composted or digested to reduce odors and pathogens.

[4]The "Forty City Study" is shorthand for an agency study, Fate of Priority Toxic Pollutants in Publicly Owned Treatment Works, completed in 1981. Although it was not specifically designed to support the sewage sludge regulation, it provided data from the largest available sample of nationally distributed treatment plants. EPA uses these data to develop three profiles of sludge quality to represent sludge from all POTWs.

[5]To determine the costs of compliance for primary treatment POTWs, EPA assumed that compliance costs would be similar to costs for secondary or advanced treatment POTWs because the agency determined that pollutant con-

centrations in sewage sludge from primary treatment POTWs are no worse than those in sewage sludge from secondary or advanced treatment POTWs. Compliance costs developed for each use and disposal practice and for each reported flow group were then applied to the appropriate primary treatment POTWs.

To estimate the cost of compliance for privately and federally owned treatment works, EPA assumed that the sewage sludge quality at these treatment works is similar to the sludges generated by the smallest POTWs surveyed in the NSSS, because both typically have a flow rate of less than one million gallons per day. The per-POTW costs developed for the smallest POTWs in the NSSS were applied to the estimated number of privately and federally owned treatment works that employ each use and disposal practice, in the same manner as discussed previously for primary POTWs. Finally, compliance costs for domestic septage haulers were calculated based on the requirements imposed for the use and disposal of domestic septage, with compliance costs estimated on a per-firm or per-truckload basis.

[6]During the public comment period, EPA staff met with a number of POTWs as well as entrepreneurs that were already marketing sewage sludge as a product. They projected a growing market for sludge products and argued that EPA's assurance about sludge quality would help their marketing. The relatively high costs of incineration, and the potential for generating revenue to defray the cost of operating a POTW, combined with the information about sludge meeting the requirements being harmless, served as strong incentive for POTWs to explore beneficial reuse. Potential users saw sludge products as organic materials that would result in decreased use of synthetic fertilizers.

[7]Cost-effectiveness value is derived when total annual cost for each option is divided by the total volume. These disposal costs do not account for potential cost-savings POTWs would realize as they shift from disposal to beneficial reuse. These regulations play an important role in assuring potential users that the quality of sewage sludge is protective of the public health and the environment.

[8]Benefits associated with surface disposal restrictions that have not been quantified and monetized include reduced potential contamination of streams from runoff controls and reduced potential for groundwater contamination.

[9]During the public comment period, a number of new proposals were developed and submitted as evidence of the potential market for sludge products. We have no information about the actual growth of this industry other than personal observations of the large variety of products now available through local nurseries and hardware stores in various parts of the United States.

REFERENCES

Jones. 1992. Memorandum from Anne Jones, senior economist, Eastern Research Group, Lexington, Massachusetts, to Susan Burris, U.S. EPA, September 25.

NAS (National Academy of Sciences). 1983. *Risk Assessment and Management: Framework for Decision Making.* Washington, D. C.: National Academy Press.

Tolley, G. S. and others. 1986. *Valuation of Reductions in Human Health Symptoms and Risks*. Prepared at the University of Chicago for U.S. EPA. Washington, D.C.: U.S. EPA.

U.S EPA (Environmental Protection Agency). 1986. *Guidelines for Carcinogen Assessment; Guidelines for Mutagenicity Risk Assessment; Guidelines for Health Assessment of Selected Developmental Toxicants; Guidelines for Health Risk Assessment of Chemical Mixtures. Federal Register* vol. 51, no. 185.

———. 1989. *Regulatory Impact Analysis of the Proposed Regulations for Sewage Sludges and Disposal*. January. PB89-136634. Springfield, Virginia: National Technical Information Service.

———. 1990. *National Sewage Sludge Survey: Availability of Information and Data, and Anticipated Impacts on Proposed Regulations*. November 9. Washington, D.C.: U.S EPA.

———. 1993a. *Standards for the Use or Disposal of Sewage Sludge; Final Rules*. In *Federal Register*, February 19, Vol. 58, No. 32, pp. 9,248–415.

———. 1993b. *Regulatory Impact Analysis of the Part 503 Sewage Sludge Regulations*. March. EPA 821-R-93-006. Washington, D.C.: U.S. EPA.

14

Reformulated Gasoline

Robert C. Anderson
and Richard A. Rykowski

In 1994, to satisfy requirements of the 1990 Clean Air Act Amendments (CAAA), EPA issued standards for reformulated gasoline (RFG). The RFG requirements were largely dictated by Section 211(k) of the CAAA, which established a minimum requirement for oxygen content, a maximum limit for benzene content, and emission reduction requirements for volatile organic compounds (VOCs), toxics, and nitrogen oxides (NO_x). Within the constraints of this formulaic approach, EPA initiated a regulatory negotiation to develop specific proposals for implementing Section 211(k) and conducted a regulatory impact analysis (RIA) to assess the cost-effectiveness of the limited number of options. In addition to assessing fuel parameters that could be adjusted under Section 211(k), the RIA interestingly showed that several of the fuel specifications and performance requirements of Section 211(k) probably did not meet the agency's cost-effectiveness threshold set for Phase II standards and that EPA's $5,000 cost-effectiveness cutoff for NO_x may have been too high. Although the RIA did examine in detail a series of issues associated with the use of ethanol, it did not assess many of the items called for by Office of Management and Budget (OMB) guidelines. In this case study, Robert Anderson and Richard Rykowski review and assess the RFG RIA. They argue that it is long on engineering analysis and short on economic or policy analysis. They conclude that by failing

ROBERT C. ANDERSON, formerly research manager for the American Petroleum Institute, is currently president of Resource Consulting Associates, Inc. RICHARD A. RYKOWSKI, formerly chief of the Fuel Studies and Standards Branch within EPA and responsible for preparing the RIA for reformulated gasoline, is curently an environmental consultant with Air Improvement Resource, Inc.

to conduct broadscale analysis and omitting various factors from the analysis, such as benefits, market-based alternatives, impacts on competitiveness and trade, and impacts on segments of the industry that might be adversely affected,the RIA did not provide insight into a number of important RFG policy issues.

INTRODUCTION

The notion of reformulating conventional gasoline to make a cleaner burning fuel—reformulated gasoline (RFG)—originated during the late 1980s in response to challenges posed to air quality managers and to the petroleum and automobile industries. Throughout the country, ozone was proving to be a much more difficult pollutant to control than originally anticipated. While peak ozone levels in the worst nonattainment areas—such as southern California—were coming down slowly, the number of areas that failed to attain the national ambient air quality standard for ozone showed no signs of declining. Based on the large number of exceedances of the ozone standard during the summer of 1988, it appeared that EPA would have to substantially increase the number of nonattainment areas if these trends persisted. The Office of Technology Assessment (OTA; U.S. Congress 1989) and other agencies identified vehicle emissions as the largest single contributor to urban ozone problems.

One means of reducing emissions from conventional fuels was to lower Reid vapor pressure, or RVP. Historically, RVP had averaged approximately 9 pounds per square inch (psi); however, during the phase-out of lead additives in gasoline, refiners raised RVP by more than 1 psi to maintain octane levels. By the late 1980s, EPA issued rules limiting the RVP of gasoline sold during the summer in ozone nonattainment areas in southern states to 7.8 psi by 1992 and gasoline RVP elsewhere to 9 psi. At about this same time, scientists expressed concern about the possible impact that lowering RVP might have on toxic emissions, a concern that influenced the congressional design of requirements for RFG.

Largely in response to such air quality concerns, California enacted a low emission vehicle (LEV) initiative in September 1990, which featured lower VOC and NO_x emission standards and could be satisfied through alternative fuels or cleaner conventionally fueled vehicles. California also included a requirement that fuel providers supply alternative fuels if sufficient numbers of new vehicles requiring their use were sold.

Among fuel providers, ARCO developed and then marketed in southern California a specially formulated EC-1 gasoline (for emission control), which provided significant reductions in VOC, carbon monoxide (CO), and NO_x emissions. EC-1 was an 89 octane unleaded fuel and featured a reduced benzene and butane content and included an oxygenate. At the time, ARCO was the world's leading producer of methyl tertiary butyl ether (MTBE) and could produce EC-1, at least in small batches, and remain competitive in price with conventional gasoline. Soon other refiners also began making cleaner gasoline (usually with added oxygenates or lower RVP) available in selected parts of the country.

Alternative fuels and vehicles had advantages and disadvantages relative to the reformulated gasolines being marketed by the petroleum industry. Some alternative fuels, such as methanol and electricity, had far lower emissions than the cleanest of gasolines. However, reformulated gasoline could have an immediate impact on the entire fleet if it could be supplied in sufficient quantity. For all of the alternatives, cost was also an issue, with widely varying estimates in circulation. The American Petroleum Institute (API) claimed initially that basic reformulated gasoline would cost an incremental twenty-five cents per gallon to produce (according to API President C. J. DiBona, responding to media inquiries), implying a cost-effectiveness on the order of $25,000 per ton of VOC, far higher than some other options for reducing VOC. Later, API lowered this cost estimate to a range of ten to twelve cents per gallon, a figure it still uses (communication February 1996 with T. J. Lareau, Policy Analysis and Strategic Planning Department). Other production cost estimates were far lower, based in part on the observation that ARCO was marketing its EC fuel at the same price as conventional gasoline.

The 1990 debate surrounding the Clean Air Act Amendments reflects the uncertainty at that time surrounding the cost, availability, and effectiveness of reformulated gasoline and alternative fuels (Gushee 1996). President Bush's initial proposal included a program requiring that 30% of the vehicles sold in the worst ozone nonattainment areas have emission performance characteristics so strict that alternative fuels would likely be required. Proposed amendments introduced by Senator Thomas Daschle would have specified the composition of RFG. Industry called for a performance-based fuels standard. The final amendments substituted a provision requiring that the nine cities with the worst ozone problems have reformulated gasoline for all vehicles. The final RFG requirements included both performance and specification features.

The Clean Air Act Amendments incorporated the California LEV program and offered other concessions to alternative fuels, such as the oxygenate requirement for RFG; however, not until the Energy Policy Act of 1992 were alternative fuels given a prominent place in the federal energy agenda.

The 1990 Clean Air Act Amendments

Among many new initiatives, the 1990 Clean Air Act Amendments (CAAA) required that reformulated gasoline be sold, beginning in 1995, in the nine large metropolitan areas with the most severe summer ozone problems (Deal 1992). Other ozone nonattainment areas could join the program through an "opt-in" provision. The act also required that conventional gasoline sold in the remainder of the country after 1994 could not be more polluting than it was in 1990. This was to ensure that refiners not increase harmful emissions from conventional gasoline by "dumping" into fuel constituents of conventional gasoline, such as benzene and aromatics, that are limited in reformulated gasoline.

Section 211(k)(1) directs EPA to publish regulations that result in "the greatest reduction in ozone-forming and toxic air pollutants achievable through the reformulation of conventional gasoline, taking into consideration the cost of achieving such emission reductions, any non-air-quality and other air-quality related health and environmental impacts and energy requirements." While this mandate seemingly gives the agency the complete freedom to use benefit-cost analysis in selecting among alternatives, Section 211(k)(3) specifies certain parameters for reformulated gasoline that greatly restrict the alternatives available to EPA. Reformulated fuel sold between 1995 and 1999 (so-called Phase I reformulated gasoline) must reduce emissions of both VOCs and toxics by 15% relative to baseline emissions from a 1990 model car operated on baseline gasoline or meet the VOC and toxic emissions performance of a specified "formula" fuel, whichever produces the lower emissions. Reformulated gasoline must also contain a minimum of 2% oxygen by weight and a maximum 1% benzene by weight. Section 211(k)(2)(A) also specifies that RFG not increase NO_x emissions, contain no heavy metals, and contain detergents.

The VOC and toxic emission performance standards for reformulated fuels marketed for the year 2000 (so-called Phase II reformulated gasoline) and beyond must meet the performance of the foregoing "formula" fuel or result in a 25% reduction from baseline emissions, whichever produces the lower emissions. EPA may adjust this performance standard up or down to take cost into account, but in no case may the reduction from baseline emissions be less than 20%.

The RFG and antidumping programs broke new ground by using (in part) a performance-based (that is, emissions) standard for fuels. All previous EPA fuel requirements (such as limits on lead content and RVP for gasoline, and sulfur and cetane in diesel fuel) were directed at fuel constituents and fuel properties, not at combustion products.

Section 211(k) required that RFG regulations be promulgated within twelve months of passage of the 1990 amendments, that is, by November 15, 1991. This would have been a monumental task under ideal conditions but was impossible considering the numerous practical issues associated with RFG production, distribution, and enforcement. One particular challenge was the need to develop predictive models that could approximate the emission performance of individual batches of fuel, typically 10,000 to 20,000 gallons, without actually testing emissions, whose measurement could cost as much as the fuel. Another challenge for EPA was the nascent science of actually reformulating gasoline. A few refiners had recently begun to offer in limited quantity various blends of cleaner gasoline that reduced VOC emissions by up to 15%. Although Congress had set many of the requirements for RFG, several other details had to be decided by EPA, which clearly had less information than it wanted as the foundation of a rulemaking.

The Regulatory Negotiation

Following passage of the 1990 CAAA, EPA initiated a formal regulatory negotiation (more commonly referred to as "reg neg") with interested parties to develop information and specific proposals for implementing the RFG and antidumping requirements. The RFG regulatory negotiation is one of approximately thirty-five negotiated rules issued since 1982 by various federal agencies (about one-half of them by EPA with the remainder originating with the Departments of Agriculture, Labor, Education, and Transportation; Farm Credit Administration; Federal Communications Commission; and Nuclear Regulatory Commission). Under the procedures of the Federal Advisory Committee Act, EPA was obligated to take reasonable efforts to ensure that all interest groups and other parties that would be affected by the rule were aware of the proceeding. Parties that sign off on a negotiated rulemaking agree not to subsequently challenge the rule in court if it is promulgated in accord with the negotiated agreement.

Early in 1991, EPA announced the formation of the regulatory negotiation committee of interested parties, which included representatives of petroleum and automobile manufacturers, state air pollution control agencies, vehicles owners, gasoline marketers, oxygenate producers, citizen groups, and environmental organizations (U.S. EPA 1991). Absent from the committee was any representation by foreign refiners.

Research over the course of the next two years was funded largely by the automobile and petroleum manufacturers—the so-called Auto/Oil Air Quality Improvement Research Program—with refinery modeling provided by Bonner and Moore—an engineering consulting firm with noted expertise in oil refining—and funded by EPA. The research revealed many of the details that would be pertinent to a rulemaking. EPA's regulatory impact analysis (RIA) for reformulated gasoline draws on this work.

In the process of satisfying its mandate under Section 211(k), EPA would eventually issue three proposed rules (U.S. EPA 1991, 1992, 1993a) and two final rules (U.S. EPA 1994), as well as defend its RFG rule in court. The RIA (U.S. EPA 1993b) contains information and analysis that EPA used in establishing and defending the many regulatory actions. The purpose of this chapter is to characterize the RIA and evaluate its relationship to the regulatory and legal actions on RFG. Did the RIA play a central role in these processes? Did the RIA identify new regulatory alternatives and lead to better (more socially desirable) outcomes or was it produced primarily to satisfy Executive Order 12291, which required an RIA for all rulemakings with an anticipated cost to the economy of at least $100 million? With the benefit of hindsight, should the RIA have provided more or different analysis to aid in decisionmaking?

Overview of Case Study

This chapter reviews and assesses the regulatory impact analysis that EPA developed in conjunction with its regulatory actions on RFG. The next section provides an overview of several interrelated activities: the regulatory negotiation process, EPA's treatment of ethanol, the development of simple and complex models used to predict emissions, the inclusion of NO_x emission reductions as part of the rule, and antidumping provisions and enforcement. The following section describes the RIA in some detail, the methods used to develop costs, the reasons that benefits were not quantified, the resulting emphasis on cost-effectiveness, and how the RIA influenced the final rule on RFG promulgated by EPA on February 16, 1994 (U.S. EPA 1994). The final section offers some reflections on the process.

OVERVIEW OF ISSUES AND RULEMAKING

During the spring and into the summer of 1991 the regulatory negotiation committee met every two to three weeks, attempting to reach an agreement within the timetable of the Clean Air Act. Concurrently, EPA began the process of regulatory development, initiating a contract for refinery modeling and making contingency plans for the RIA and subsequent rule in case the negotiated rulemaking failed.

First Notice of Proposed Rulemaking

Concern about the statutory deadline of November 15, 1991, for issuing a final Phase I rule led EPA to publish on July 9, 1991, the first *notice of proposed rulemaking* (NPRM) on RFG (56 *Federal Register* 31176). At that early date, a consensus had not been established within the regulatory negotiation committee. Normally, an agency in a negotiated rulemaking would wait to issue a rule until the committee had reached agreement.

EPA's NPRM was "inclusive" in the sense that any option advocated by an interest group was included within the range of alternatives for which the agency solicited comments. This approach was necessary to keep the regulatory negotiations alive. While EPA believed that consensus of the regulatory negotiation committee was possible, publishing an NPRM before that consensus had evolved opened EPA to possible legal challenges since the inclusive nature of the NPRM would be unlikely to withstand challenge as sufficient notice of the agency's final rule to follow. In many ways, the NPRM was more like an *advance notice of proposed rulemaking*, with EPA struggling to meet the Clean Air Act's November 1991 deadline. Despite the obvious risks, EPA succeeded with the NPRM in that parties to the regulatory negotiation remained at the table and reported a consensus on August 16, 1991.

The Regulatory Negotiation Agreement

By August 1991, the regulatory negotiation committee produced an agreement in principle outlining a two-step process for RFG. The first step, to take effect in 1995, would use a "simple model" for certifying that gasoline met the emission reduction standards. The simple model bases certification on a fuel's RVP as well as its content of benzene, aromatics, and oxygen. In the second step, EPA would propose a "complex model" to replace the simple model for emission certification. The complex model would be used for certification four years after it was promulgated. EPA also acknowledged that it would issue in a timely fashion the more stringent Phase II emission performance standards.

Limits on benzene and minimum oxygen content were not controversial since they were part of the Clean Air Act. The need to reduce aromatics to control toxic emissions also was not controversial since the relationship between aromatics and toxic emissions was reasonably clear and the use of oxygenates would reduce aromatics in any case. With the regulatory negotiation committee, the debate concerned RVP limits, which provided most of the mandated 15% VOC reduction, and a maximum oxygen content, which was intended to prevent an increase in NO_x emissions.

The debate over RVP was lengthy and contentious. While small reductions in RVP were easy for refiners to achieve, the petroleum industry argued that larger reductions required new investments and would be much more costly. Normally, EPA's MOBILE model (a computer-based model that predicts fleetwide motor vehicle emissions) would provide a precise estimate of the RVP necessary to achieve a 15% reduction in fleet VOC emissions. However, concurrent with the RFG regulatory negotiation, EPA was revising the then-current version MOBILE 4.0. Based on new information on in-use exhaust VOC emissions from two state inspection and maintenance programs, EPA was considering increasing its projections of exhaust VOC emissions dramatically and reducing the predicted efficacy of RVP reductions. From week to week, the modeled effects of RVP on VOC emissions would change, lending uncertainty to the regulatory negotiation committee's work and prolonging negotiations.

Eventually, EPA selected a draft version of MOBILE 4.1 for use in RFG certification that supported the sufficiency of the RVP levels of the simple model. In the following months, however, the final MOBILE 4.1 and a newer MOBILE 5.0 predicted much higher exhaust VOC emissions, indicating that the simple model would produce less than the required 15% VOC reduction. EPA chose to retain the simple model RVP limits negotiated in August 1991 and use the newer information in the complex model for certifying Phase II RFG in 2000 and later.

The most controversial aspect of the simple model concerned the maximum oxygen content. As noted, Section 211(k) required that RFG not increase NO_x emissions. Testing of older vehicles showed that adding oxygen to gasoline, which Section 211(k) required to be at least 2% by weight in RFG, usually increased NO_x emissions. Testing of newer vehicles, which was required to measure RFG performance, was less conclusive. Ether blends with 2.0% to 2.7% oxygen by weight tended to show NO_x emissions steady or declining, while ethanol blends at higher oxygen content often increased NO_x. Many RFG areas were also in nonattainment status for CO, meaning they were required to use gasoline with a minimum of 2.7% oxygen by weight during the winter. Consequently, the simple model specified a minimum of 2.1% oxygen by weight and a maximum 2.7% oxygen by weight, with the provision that a state could waive the 2.7% limit in the winter if it had no winter ozone violations.

On April 16, 1992, EPA issued a *supplemental* notice of proposed rulemaking (SNPRM) reflecting the agreement reached in the regulatory negotiation (U.S. EPA 1993a). The SNPRM described standards and enforcement systems for RFG and conventional gasoline and included proposals to be implemented for gasoline certification and enforcement. EPA also proposed the complex model and the Phase II VOC and toxics standards. Going beyond the requirements in Section 211(k), EPA proposed a 6.8% NO_x reduction standard using its general authority to regulate fuels in Section 211(c).

Special Treatment for Ethanol

Standards developed in the regulatory negotiation process and proposed in the SNPRM were designed to be fuel neutral. Ethanol producers saw their fuel additive at a disadvantage, however, since the addition of ethanol to meet summer oxygenate requirements would raise the RVP of the resulting blend by about 1 psi and make it more difficult to meet the RVP and VOC performance standards compared to blends using MTBE, which did not affect RVP. To use ethanol as an oxygenate would require blendstocks with RVP low enough to offset ethanol's impact, and ethanol producers complained that obtaining such blendstocks would be difficult and expensive and effectively eliminate ethanol from the RFG program. This result, they argued, was contrary to the intent of the oxygen content requirement of Section 211(k)(2), which they saw as motivated in large part by congressional intent to expand markets for ethanol.

Ethanol industry representatives argued that the benefits of using ethanol fully justified its inclusion in the program, despite the higher vapor pressure of ethanol blends. They noted that because ethanol is domestically produced, it would improve U.S. energy security. Further, most ethanol is produced from corn, a commodity whose producers benefit from several government support programs. Greater demand for corn, they argued, would reduce the need for price supports. Because of these side benefits of ethanol use, ethanol representatives lobbied EPA and the White House to obtain more favorable treatment in the RFG program. One suggestion they offered was a one-pound (psi) RVP waiver. If the waiver were not available, an alternative would be to have the VOC reduction requirement account for the differing ozone-forming potential of individual constituents of VOC. While ethanol demonstrably raised the RVP of gasoline blends, they argued that the ozone-forming potential of the resulting VOCs did not increase correspondingly.

Shortly before the 1992 elections, President Bush announced an initiative designed to permit ethanol to compete in RFG. In lieu of a one-pound RVP waiver or special treatment of VOC reactivity, his initiative was grounded in provisions of Section 211(k)(1) that directed the administrator of EPA to consider cost, energy requirements, and other factors in establishing RFG performance standards. The initiative called for EPA to set rules for RFG that have the effect of granting a one-pound RVP waiver for the first 30% market share of ethanol blends, while maintaining the environmental benefits that would have otherwise occurred (that is, the volatility of the entire gasoline pool must not increase). In effect, the volatility of gasoline-ether blends would have to be reduced by an additional 0.3 psi in northern ozone nonattainment areas and 0.2 psi in southern ozone nonattainment areas to accommodate the ethanol mandate. Despite misgivings over its legality and energy security benefits, and despite questions about its environmentally neutral consequences, EPA proposed rules implementing President Bush's initiative as part of the RFG rulemaking.

Intervention by the White House was viewed by many of the participants in the regulatory negotiation process as invalidating the regulatory negotiation agreement, since the administration was one of the signatories to the agreement. While the ethanol industry was pleased that it would approximately double its share of the gasoline market, MTBE and methanol producers stood to lose. Environmentalists expressed concern that NO_x would increase, and the petroleum industry complained that it would bear additional costs to lower the RVP of gasoline.

Final Rule on RFG and Conventional Gasoline

During 1993, new EPA emissions analysis predicted an increase in VOC emissions in summer months from ethanol blends. A report from the Department of Energy (DOE) showed that using ethanol to meet the oxygenate requirement would not help U.S. energy security (DOE analysis showed ethanol production actually increased petroleum imports relative to MTBE use as an oxygenate). This evidence led EPA to separate the ethanol initiative from the RFG final rule (U.S. EPA 1994). The final rule provides no special treatment for ethanol but argues that ethanol blends will find a ready market during the winter months, especially so in the Midwest where ethanol production is subsidized. On February 16, 1994, EPA proposed a separate rule that 30% of RFG must satisfy the oxygen requirement through the use of "renewable oxygenates," a proposal that would expand the use of ethanol substantially.

By July 1994, EPA promulgated final rules for the so-called ROXY (renewable oxygenates) program. The program avoided the emission impacts of ethanol use by defining ethanol to be a renewable oxygenate only in the winter. Ethyl tertiary butyl ether (ETBE), which does not adversely affect RVP, qualified as a renewable oxygenate year-round. Later, the petroleum industry successfully challenged the ROXY rule; an appeals court vacated the program on the grounds that EPA did not have authority to promulgate such a mandate.

THE RIA FOR REFORMULATED GASOLINE

The regulatory impact analysis for reformulated gasoline, which was not completed until late in 1993, is unusual in several respects. It does not follow the standard format of the OMB RIA guidelines, beginning with a statement of need for the regulation, a review of alternatives, and consideration of the benefits and costs of the alternatives. Rather, the RIA follows the format of the final rule, sharing a great many similarities but offering more detail in its exposition, in essence providing the technical justification for the rule. The RIA begins with a lengthy discussion of the treatment of ethanol, noting that it was "one of the most controversial issues in the reformulated gasoline rulemaking." There is no statement of the need for the regulation or of the statutory background. The second major section describes EPA's work on the Phase I standards, including establishment of the emissions baseline and why EPA set caps on three gasoline parameters. Section three elaborates on the then-ongoing process of setting fuels and emissions baselines for the complex model. The fourth section of the RIA describes the complex models EPA developed to predict exhaust and nonexhaust emissions using readily measurable fuel properties. Section five covers Phase I RFG performance standards for the simple model, including an analysis of cost-effectiveness. The following section contains the supporting analysis for Phase II RFG performance standards, as well as NO_x standards that EPA promulgated concurrently. Thereafter come three more sections: antidumping requirements for conventional gasoline, compliance with the Regulatory Flexibility Act, and state opt-in provisions. Economic arguments of cost, emission effects, and cost-effectiveness dominate the analysis in the two sections supporting the Phase I and Phase II RFG performance standards.

Treatment of Ethanol

This section provides support for EPA's decision in the final RFG rule not to provide regulatory advantages for ethanol. After reviewing the many comments EPA received in support of an RVP waiver, for special consideration of VOC emission reactivity, and for a defined market share for renewable oxygenates, EPA observed that several considerations argued against preferential treatment for ethanol-blended RFG. EPA reviewed the legislative history of the 1990 CAAA, concluding that Congress did not intend for ethanol to receive a one-pound RVP waiver. EPA's general counsel opined that the 1psi waiver for conventional fuels in Section 211(h) does not apply to RFG requirements in Section 211(k).

Next, EPA observed that granting a 1psi waiver for ethanol in RFG would have "significant adverse environmental impacts." A 1psi waiver would result in emissions from RFG ethanol blends approximately 20% higher than baseline gasoline, potentially having a large impact on the VOC emission inventory in the areas where RFG was required. EPA drew a sharp distinction between a 1psi waiver for ethanol in RFG and a 1psi waiver it granted to ethanol blends in its 1987 rulemaking on gasoline volatility. In the 1987 rulemaking, EPA was concerned that if it did not grant the waiver to ethanol blends, the ethanol industry would face financial ruin. Further, EPA asserted that the 1987 waiver did not have a significant impact on fleetwide VOC emissions. The blends comprised only about 8% of the market and the market share of ethanol blends was unlikely to increase as a consequence of the waiver, so a 1psi waiver would result in a loss of only about 0.08 psi across the entire gasoline market, or about 3% of the volatility control anticipated from the rule. Further, the use of ethanol reduced exhaust VOC emissions relative to nonoxygenated gasoline, mitigating the evaporative increase. The ethanol industry should not face financial ruin as a consequence of EPA's failure to grant a 1psi waiver for RFG; indeed, EPA suggested the ethanol market would grow due to wintertime oxygenate requirements for CO nonattainment areas and the possibility of converting ethanol into ETBE, an oxygenate that did not increase gasoline volatility.

EPA dismissed calls for adding or excluding emission products from those it labeled reactive VOC. Under the authority of Section 211(k), EPA has this authority. Previously, it had deleted methane and ethane from the list of ozone-forming VOCs on the basis of their low reactivity (methane had long been excluded from VOCs; EPA excluded ethane in a separate rulemaking in 57 *Federal Register* 3941). All reactive VOCs eventually form ozone. The reactivity of VOC species measures how fast they react. Those that are relatively slow to react may be transported out of a nonattainment area before reacting to form ozone. Such species are less of a concern than those with high reactivity.

The ethanol industry suggested that CO be included as a VOC. EPA noted that the ozone reactivity of CO is roughly one-fiftieth of the average gasoline constituent. CO is a large constituent of vehicle emissions, about five grams per mile in 1990 vehicles, whereas VOC is about one gram per mile. If CO were included in the measure of VOC, the VOC control program would turn into a CO control program and eliminate most of its ozone control benefits. EPA also received suggestions that it exclude ethanol from the list of reactive VOC. The agency noted, however, that while ethanol is less reactive than some hydrocarbon species in gasoline, it is five times as reactive as ethane, the most reactive hydrocarbon species deleted from the list of VOC, and twice as reactive as MTBE. EPA concluded that excluding ethanol would be inappropriate, since the reduced reactivity did not offset the impact of the RVP increase.

Simple Model

Section 211(k) of the Clean Air Act required EPA to promulgate standards for RFG relative to emissions from baseline vehicles using baseline fuels. Section 211(k) called for control of VOC during the "high ozone season," a term EPA interpreted to be from May 1 to September 15. The Clean Air Act specified the formulation of baseline high ozone season fuel (see Table 1), but not that for winter baseline fuel. In its July 9, 1991, NPRM, EPA offered a description of the data and methods used to determine winter baseline fuel, later modifying this slightly in the 1992 SNPRM. The base-

Table 1. Baseline Fuel Composition.

Parameter	*Summer*	*Winter*
Sulfur (ppm)	339	338
Benzene (vol. %)	1.53	1.64
RVP (psi)	8.7	11.5
Octane (R+M/2)	87.3	88.2
T10 (degrees F)	128	112
T50 (degrees F)	218	200
T90 (degrees F)	330	333
Aromatics (vol. %)	32.0	26.4
Olefins (vol. %)	9.2	11.9
Saturates (vol. %)	58.8	61.7

Note: vol % = percentage by volume.

Source: U.S. EPA 1993b, p. 50.

line fuel composition assumes no lead or oxygenate. One minor change in composition appeared in the final rule: saturates in winter fuel were set at 61.9% by volume. Most of the parameters are self-explanatory; T10, T50, and T90 refer to the temperature at which 10%, 50%, and 90% of the fuel has evaporated.

Section 211 specified that RFG must result in reduced emissions relative to emissions from "representative" model year 1990 vehicles using baseline gasoline. EPA interpreted "representative vehicles" to mean vehicles using emission control technology comparable to that in 1990 vehicles. Thus, EPA included as baseline vehicles certain vehicles manufactured as early as the 1986 model year and as recently as then-current (1994) model year vehicles if they used adaptive learning, a technique used by nearly all 1990 model year vehicles.

EPA modeled in-use emissions of a fleet consisting entirely of 1990 vehicles with a draft version of MOBILE 4.1. The MOBILE model is a set of equations used to predict fleet emissions (both tailpipe and nonexhaust) at any point in time based on the fuel, temperature, and other parameters, such as inspection and maintenance (I/M), vapor recovery systems on gasoline pumps (termed *Stage II controls*), and the California Low Emission Vehicle (LEV) program. It assumes a 25-year life for a vehicle, with all vehicles older than 25 years lumped into the 25-year-old group. In general, the emissions of each model year's vehicles increase as the vehicles age. The MOBILE model calculates fleet organic emissions in any one of five forms: total hydrocarbons, nonmethane hydrocarbons, volatile organic compounds, total organic gases, and nonmethane organic gases, as well as CO and NO_x emissions.

Prior to running the model, EPA had to specify a number of parameters. EPA used national weather statistics to estimate daily low and high temperatures for southern and northern RFG cities on high ozone days.

EPA assumed that Stage II refueling controls and a basic I/M program would be in place in all RFG areas. EPA made other necessary assumptions for calculating baseline toxic emissions (benzene, 1,3-butadiene, formaldehyde, acetaldehyde, and polycyclic organic material, which are not calculated in the MOBILE model). Table 2 portrays EPA's calculated simple model baseline emissions.

Section 211(k)(3) requires that Phase I RFG meet the more stringent of two VOC and toxic emission requirements: a 15% reduction from the baseline performance standard or the compositional requirements in Section 211(k)(3)(A), which specified a fuel with no more than 1% benzene by volume, no more than 25% aromatics, no less than 2% oxygen by weight, and that met other Section 211 requirements for detergent additives and lead content. EPA determined that VOC emissions from formula fuels possible under the compositional requirements did not meet the 15% reduction requirement. Therefore, EPA based the VOC emission requirements on the 15% reduction minimum performance standard. With respect to toxic emissions, EPA found that the type of oxygenate used in the formula fuel determined whether fuels meeting the compositional requirement would satisfy the 15% minimum toxic emission reduction requirement ETBE and ethanol did not, while MTBE and tertiary amyl methyl ether did. Consequently, EPA determined that the 15% minimum emission reduction for toxic emissions was more stringent than the compositional requirement.

EPA established an optional averaging program for toxic emissions, since some oxygenates would more than meet the 15% reduction requirement while others would fall short. Arguing that averaging made it both cheaper and easier to reduce toxic emissions, EPA set an average toxic

Table 2. Simple Model Baseline Emissions.

	Summer, northern region	*Summer, southern region*	*Winter*
Exhaust VOC (g/mi)	0.444	0.444	0.656
Nonexhaust VOC (g/mi)	0.856	0.766	0
Total VOC (g/mi)	1.30	1.21	0.656
Exhaust benzene (mg/mi)	30.1	30.1	40.9
Evap benzene	4.3	3.8	0.0
Running loss benzene	4.9	4.5	0.0
Refueling benzene	0.4	0.4	0.0
1,3-Butadiene	2.5	2.5	3.6
Formaldehyde	5.6	5.6	5.6
Acetaldehyde	4.0	4.0	4.0
POM	1.4	1.4	1.4
Total toxics (mg/mi)	53.2	52.1	55.5

Source: U.S. EPA 1993b, p. 57.

emission reduction of 16.5% from baseline levels, 1.5% more stringent than what the Clean Air Act required on a per-gallon basis. This extra 1.5% reduction also provided assurance that each RFG city would experience at least the required 15% reduction.

The simple model, which refiners will use as a guideline for compliance with Phase I RFG requirements from January 1, 1995, until January 1998, shows refiners how to achieve a calculated 15% reduction in VOC and toxic emissions from baseline levels on a per-gallon basis (16.5% reductions with averaging). The simple model is based on data relating four fuel parameters to emissions: oxygen content, Reid vapor pressure, aromatics, and benzene. It includes maximum RVP limits of 7.2 and 8.1 psi for southern and northern RFG cities (7.1 and 8.0 psi if the averaging option is selected), respectively. It also includes the minimum and maximum oxygen requirements described earlier. Other parameters such as sulfur, T90, and olefins were capped at refiners' 1990 baseline levels to prevent them from undermining the effectiveness of controls on the three parameters of the simple model. High levels of these parameters were known generally to increase emissions, but not with sufficient accuracy to include them in the simple model. The complex model incorporates several additional parameters whose impact on emissions was not well established at the time the simple model was developed.

Complex Model

EPA developed the complex model to predict emissions as a function of a broader set of fuel properties for use in certifying RFG in 1988 and beyond. Baseline fuels were the same as those for the simple model (see Table 1). EPA's complex model includes the following parameters: oxygen, sulfur, RVP, E200, E300, aromatics, olefins, and benzene. Rather than having to measure E200 and E300 directly, the complex model allows the user to convert levels of T50 and T90 to estimate the percentage of fuel evaporated at temperatures of 200° F and 300° F (E200 and E300).

$$E200 = 147.91 - (0.49)(T50)$$

$$E300 = 155.47 - (0.22)(T90)$$

As was the case for the simple model, the complex model develops baseline emissions from the MOBILE model, in this case MOBILE 5.0. MOBILE 5.0 was initially designed to supplement MOBILE 4.1 by adding post-1990 emission controls. During the development of MOBILE 5.0, EPA also revised its estimate upward of in-use emissions from pre-1991 model vehicles. To generate the baseline summer and winter emissions

Table 3. Summer Baseline Emissions.

	Phase I Class B	*Phase I Class C*	*Phase II Class B*	*Phase II Class C*
Exhaust VOC (gm/mi)	0.446	0.446	0.907	0.907
Nonexhaust VOC (gm/mi)	0.860	0.769	0.559	0.492
NO_x (gm/mi)	0.660	0.660	1.340	1.340

Note: Class B and Class C areas refer to ASTM designations and are based on temperature. Class B are also referred to as southern and Class C as northern areas.

Source: U.S. EPA 1993b, p. 69.

(from which the performance of RFG would be calculated), EPA again considered only 1990-technology vehicles. EPA did this by running MOBILE 5.0 for the year 2015, with all post-1990 vehicle emission control programs assumed inoperative but with Stage II and enhanced I/M. The resulting summer baseline emissions for Phase I and Phase II RFG are reported in Table 3.

EPA also developed baseline estimates for toxic emissions. Since the MOBILE model does not predict toxic emissions, EPA ran the complex model with the baseline fuel parameters to calculate baseline toxic emissions of benzene, 1,3-butadiene, acetaldehyde, and formaldehyde. EPA developed the complex model from a database that included EPA's emission factors studies, the Auto/Oil Test Program results, the RVP/oxygenate study sponsored by General Motors, the California Air Resources Board and the Western States Petroleum Council, the RVP/oxygenate and aromatic study sponsored by the American Petroleum Institute, ARCO's EC study, Chevron's distillation study, and UNOCAL's RFG study. EPA estimated nonexhaust benzene emissions using a General Motors vapor equilibrium model that estimates the fraction of benzene in the vapors in a tank of gasoline as a function of fuel composition.

With the more than 16,000 usable observations in this database, and input from both the vehicle and petroleum industries, EPA developed a set of equations that predicted VOCs, NO_x, and toxic emissions reductions relative to the baseline fuel parameters. This was termed the *complex model*. With the parameters of interest for a batch of gasoline, a refiner could predict VOC, NO_x, and toxic emissions and determine whether the required percentage reductions from baseline in VOCs and toxics had been achieved.

Environmental and Economic Impact of Phase I RFG

EPA contracted with Bonner and Moore to develop estimates of the cost to produce Phase I RFG. Modeling showed that costs were sensitive to the cost of oxygenates and also varied somewhat by region. From the range of estimates provided by Bonner and Moore and after taking into account cost differentials across refining regions and uncertainties in future oxygenate costs, EPA put the average cost of producing Phase I RFG across all regions at 3.9 cents per gallon, with a range of 2.3 cents to 6.5 cents, including a fuel economy penalty of 1.6 cents per gallon. Table 4 compares the baseline fuel parameters with those of Phase I RFG which underlie the Bonner and Moore cost calculations (Bonner and Moore 1993).

Using MOBILE 5.0 and the complex model, EPA estimated that VOC emissions would be reduced to 1.917 grams per mile in Class B areas and 2.059 grams per mile in Class C areas for vehicles fueled with RFG and subject to basic I/M requirements. With enhanced I/M, the VOC emission reductions would fall to 1.33 grams per mile in Class B areas and 1.421 grams per mile in Class C areas. To estimate the total tons of VOC reduction, EPA took a projected 1998 fuel consumption of 113.84 billion gallons, projected fleet fuel economy of 20.58 miles per gallon, and the assumption that RFG would be sold in the nine mandated areas plus those areas that had opted in by mid-summer 1993 (12.8% of national fuel consumption in Class B areas and 18.2% of national fuel consumption in Class C areas). With these assumptions, plus the treatment of the RFG program as if it were in place all year, EPA forecast a total emission reduction of 501,000 tons of VOC from using Phase I RFG. EPA treated RFG as if it were in use all year "since the RFG summer VOC emission benefits provide the same in-use ozone reduction as a year-round program which reduces summer emissions to the same degree but also reduces emissions in the winter" (U.S. EPA 1993b). Lareau (1994) offers an alternative approach for calculating the cost-effectiveness of programs with seasonal effectiveness.

Without providing details on how the numbers were developed, the RIA estimated the number of cancer cases avoided due to the reduction of

Table 4. Baseline and Phase I RFG Composition.

	Baseline fuel	*Phase I RFG*
Oxygen (wt.%)	0.0	2.1
Sulfur (ppm)	339	309
RVP Region 1 (psi)	7.8	7.1
RVP Region 2 (psi)	8.7	8.0
E200 (vol.%)	41.0	46.7
E300 (vol.%)	83.0	84.9
Aromatics (vol.%)	32.0	25.5
Olefins (vol.%)	9.2	13.1
Benzene (vol.%)	1.53	0.95

Note: vol.% = percentage by volume; wt.% = percentage by weight.

Source: U.S. EPA 1993b.

toxic emissions: twenty-four cases if only basic I/M were in place and sixteen cases with enhanced I/M.

While EPA did not compute the cost-effectiveness of Phase I RFG, it may be estimated as the cost (3.9 cents per gallon for 5.5 months and 3.5 cents per gallon for 6.5 months times 31% of 113.8 billion gallons or $1.37 billion) divided by the emission reduction (501,000 tons) or $2,750 per ton of VOC if no credit is given for the avoided cancers and about $2,500 per ton of VOC if the cancer cases are valued at $7 million each.

Phase II Standards

Section 211(k) of the Clean Air Act requires that by the year 2000 and beyond RFG achieve a 25% reduction in VOC and toxic emissions relative to emissions from baseline gasoline. EPA may adjust the 25% requirement to provide greater or lesser reductions (but in no case less than a 20% reduction) based on considerations of the cost of achieving the emission reductions, health and environmental impacts, and energy requirements. The act also requires that emissions of NO_x not increase (for both Phase I and Phase II RFG). EPA, however, elected to go beyond the Clean Air Act minimums and require reductions in emissions of NO_x because of its role in the formation of ozone. The issue explored in the RIA was how stringent the VOC, NO_x, and toxics specifications should be.

Since the cost of meeting a Phase II RFG standard was stated explicitly in the act as a criterion, the RIA focused on the cost-effectiveness of alternative RFG performance standards in achieving VOC, NO_x, and toxic emission reductions. Under contract to EPA, Bonner and Moore conducted extensive refinery modeling that provided the foundation for EPA estimates of the cost of producing RFG to various specifications. In developing refining cost estimates, EPA also considered refinery modeling conducted by Oak Ridge National Laboratories for DOE and by Turner and Mason, an engineering consulting firm, for the petroleum industry (a series of studies done for the American Petroleum Institute, the Western States Petroleum Association, the Auto/Oil Air Quality Research Program, and the National Petroleum Council). While the Bonner and Moore study provides the backbone for EPA's RIA, the American Petroleum Institute now believes that the Bonner and Moore results are flawed because of certain "overoptimization" assumptions (Deal 1996).

Table 5 summarizes EPA cost projections developed for these studies showing the estimated incremental costs of producing RFG as parameters of the formula are changed. The table reports these costs by region, as a national average cost and as adjusted for fuel economy (the principal effects come from lowering RVP, which increases energy content, and adding oxygenates, which lowers the energy content). The table con-

Table 5. Incremental RFG Costs as Parameters Change.

Parameter and control level	*Region 1 cost*	*Region 2 cost*	*Region 3 cost*	*National average cost*	*Fuel economy adjusted cost*	*Units (in cents/ gallon)*
RVP to 7.3 psi	0.492	0.477	0.352	0.41	0.19	c/gal/psi
RVP 7.3 to 7.1 psi	0.461	0.533	0.347	0.42	0.20	c/gal/psi
RVP 7.1 to 6.5 psi	0.839	0.529	0.223	0.43	0.21	c/gal/psi
Sulfur to 250	0.00106	—	0.0024	0.0020	0.0021	c/gal/ppm
Sulfur 250 to 160	0.00736	—	0.0058	0.0063	0.0063	c/gal/ppm
Sulfur 160 to 100	0.0135	—	0.0122	0.0125	0.0125	c/gal/ppm
Sulfur 100 to 50	0.0241	—	0.0154	0.0178	0.0178	c/gal/ppm
Olefins to 8.0 (vol.%)	—	—	—	0.138	0.138	c/gal
Olefins 8.0 to 5.0 (vol.%)	—	—	—	0.736	0.736	c/gal
Oxygen 2.1 to 2.7 (vol.%)	1.44	1.02	1.08	1.35	1.99	c/gal
Aromatics to 28 (vol.%)	0.169	—	0.0268	0.0664	0.152	c/gal
Aromatics 28 to 24 (vol.%)	0.324	—	0.298	0.305	0.391	c/gal
Aromatics 24 to 20 (vol.%)	0.370	—	0.365	0.376	0.452	c/gal
E300 to 88%	—	—	0.251	0.285	0.311	c/gal
E300 88% to 91%	—	—	0.568	0.646	0.671	c/gal
E300 91% to 94%	—	—	0.559	0.635	0.660	c/gal
E200 to 58%	0.384	0.102	0.02	0.125	0.125	c/gal
E200 58% to 61%	0.0867	0.151	0.315	0.439	0.439	c/gal
E200 61% to 64%	2.50	0.400	0.466	0.989	0.989	c/gal

Source: U.S. EPA 1993b.

forms to expectations of increasing marginal cost as RVP and sulfur, aromatic, and olefin contents are lowered. Also the table shows a considerable variation across regions in the cost of achieving incremental reductions in many of the parameters, suggesting that there could be differences among refiners in the Phase II blends they produce to meet Phase II standards and/or the costs they experience in producing RFG.

Using a methodology similar to what it used to estimate costs, EPA derived a series of fuel parameter interrelationships, showing whether (and if so, how) refineries would adjust other fuel parameters in response to a change in a single parameter. For example, would refiners reduce the olefin content if they had to lower sulfur levels? EPA incorporated the resulting interrelationship parameters into its cost-effectiveness calculations if they represented a significant and consistent relationship.

Tables 6 and 7 summarize the cost-effectiveness analysis performed by EPA. Both tables show the incremental cost and incremental cost-effectiveness of changes in RFG specifications. Table 6 deals with VOC reductions and Table 7 with NO_x reductions. While EPA did not analyze the combined cost-effectiveness of VOC plus NO_x, it would have led the agency to consider modestly stricter requirements. In theory, if one is

Table 6. VOC Cost-Effectiveness.

Fuel parameter	*Incremental cost (c/gal)*	*Cumulative reduction (%)*	*Incremental cost effectiveness ($/ton)*	*Cumulative cost effectiveness relative to Phase I*
RVP to 7.0 psi	0.18	22.9	400	400
RVP to 6.7 psi	0.08	25.5	600	400
Sulfur to 250 ppm	0.12	26.1	3,700	600
Sulfur to 160 ppm	0.56	27.1	11,000	1,300
Sulfur to 100 ppm	0.75	27.7	22,000	2,300
Sulfur to 50 ppm	0.89	28.3	32,000	3,300
Olefins to 8.0 vol.%	0.73	26.8	—	4,700
Aromatics to 20 vol.%	1.77	28.2	24,000	6,600
Oxygen to 2.7 vol.%	1.20	28.7	57,000	7,900
Olefins to 5.0 vol.%	2.77	27.7	—	12,000
E300 to 88 vol.%	0.35	27.7	49,000	12,000
E300 to 91 vol.%	2.01	27.8	199,000	15,000
E200 to 44 vol.%	0.38	28.1	37,000	15,000
E200 to 47 vol.%	1.32	28.8	36,000	16,000
E200 to 50 vol.%	2.97	29.4	96,000	19,000

Note: All costs are allocated to VOC.

Source: U.S. EPA 1993b.

Table 7. NO_x Cost-Effectiveness.

	Incremental cost (c/gal)	*Cumulative reduction (%)*	*Incremental cost-effectiveness ($/ton)*	*Cumulative cost-effectiveness relative to Phase I ($/ton)*
RVP to 7.0 psi	0.18	0.2	9,000	8,900
RVP to 6.7 psi	0.08	0.4	8,500	3,200
Sulfur to 250 ppm	0.12	2.4	1,300	3,500
Sulfur to 160 ppm	0.56	5.8	3,700	4,200
Sulfur to 100 ppm	0.75	8.6	5,800	5,000
Sulfur to 50 ppm	0.89	11.2	7,700	5,500
Olefins to 8.0 vol.%	0.73	13.1	8,500	7,800
Aromatics to 20 vol.%	1.77	14.1	40,000	9,400
Oxygen to 2.7 vol.%	1.20	14.6	54,000	12,000
Olefins to 5.0 vol.%	2.77	15.8	51,000	13,000
E300 to 88 vol.%	0.35	15.8	—	16,000
E300 to 91 vol.%	2.01	15.8	790,000	16,000
E200 to 44 vol.%	0.38	15.6	—	19,000
E200 to 47 vol.%	1.32	15.4	—	23,000
E200 to 50 vol.%	2.97	15.2	—	—

Note: All costs are allocated to NO_x.

Source: U.S. EPA 1993b.

willing to spend $5,000 per ton for either VOC or NO_x, one should be willing to spend $5,000 per ton for the sum of the two.

VOC-controlled RFG is marketed for 5.5 months per year, whereas nearly all other VOC control programs to which EPA would compare its cost-effectiveness are year-round. Nearly all of these other programs reduce VOC emissions year-round. To provide comparability, two types of adjustment are possible: count only summer VOC emissions in the measure of effectiveness for all programs or treat RFG as if it were in effect year-round. For EPA, the second alternative was simpler than recalculating the cost-effectiveness of the year-round programs to which RFG would be compared.

From Table 6, EPA noted that RVP reductions to 6.7 psi (and perhaps below) and sulfur reductions to 250 parts per million (ppm) have incremental cost-effectiveness relative to baseline gasoline well below the $5,000 to $10,000 per ton range for VOC suggested in the February 1992 notice of proposed rulemaking. The incremental cost-effectiveness of other parameter changes is worse.

EPA's analysis in the RIA shows sulfur reductions are the most cost-effective means of lowering NO_x emissions. EPA interpolated between the data points in Table 7 to identify the point at which sulfur reductions had an incremental cost-effectiveness of $5,000 per ton, its chosen cost-effectiveness cutoff, establishing 134 ppm sulfur as the upper limit in Phase II RFG. It is interesting to note that adding oxygenates to 2.7 percentage by volume (vol.%) has a cost-effectiveness far above $5,000 per ton for VOC and NO_x, for each pollutant individually as well as combined. EPA notes elsewhere in the RIA that the cost-effectiveness of the first 2 vol.% of oxygenate is also not particularly cost-effective but does not consider that in its analysis, probably because the increment to 2 wt.% was required by the Clean Air Act. Recent analyses have shown that significant RVP reductions, such as those required in RFG, but without the other fuel requirements of RFG, can decrease E200 and E300 to the point that exhaust VOC emissions increase substantially. The addition of oxygenates counteracts this effect. Thus, some portion of the benefit of RVP control should be attributed to the addition of oxygenates, implying that the cost-effectiveness of oxygenates is better than what was reported in the RIA. The RIA also did not try to quantify any energy-related benefits of oxygenates.

For toxic emission reductions, EPA conducted a similar cost-effectiveness analysis, except that for effectiveness EPA used cancer cases avoided. If all costs are allocated to toxic emission control, the most cost-effective measure was sulfur reduction to 250 ppm, with a cost per cancer case avoided of $40 million. Next was olefin reduction to 8 vol.% at $50 million per case, followed by sulfur to 160 ppm at $90 million per case.

EPA compared the estimated VOC and NO_x cost-effectiveness of limiting various RFG parameters with other stationary and mobile source controls. VOC can be reduced for $1,000 to $4,000 per ton at the painting operations in automobile and light-duty truck plants. VOC emissions produced during the manufacture of pneumatic tires can be reduced for $150 to $18,800 per ton. According to the RIA, control of VOC emissions from floating roof tanks used to store crude and refined petroleum can cost up to $3,700 per ton, though other estimates that take product savings into account suggest this measure is free or nearly so (such as OTA's). In the 1994 RIA for the petroleum refinery NESHAP rule (Pechan 1994) estimated the costs of additional VOC controls for refineries located in ozone nonattainment areas at $800 per ton. The joint AMOCO-EPA study of the Yorktown refinery found many VOC measures that could be implemented at a cost of $500 per ton or less. Low NO_x burners for utility boilers, under consideration by EPA at the time it prepared the RIA, would cost up to $1,000 per ton of NO_x. In many areas, EPA asserted that selective catalytic reduction may be needed, at a cost of $3,000 to $10,000 per ton of NO_x, though it provided no supporting evidence.

Mobile source programs are generally less cost-effective, with basic I/M estimated by EPA to cost $5,000 to $6,500 per combined ton of VOC and NO_x. Enhanced I/M was estimated by EPA to cost $900 to $1,700 per ton of VOC. Other independent estimates, however (for instance, Anderson and Lareau 1992; McConnell and Harrington 1992), are much higher, on the order of $30,000 to $50,000 per ton of VOC and NO_x for basic I/M and $6,000 to $15,000 per ton for enhanced I/M. Tier I emission standards for light-duty vehicles, required beginning with the 1994 models, cost about $6,000 per ton of VOC and $2,000 to $6,000 per ton of NO_x using EPA cost estimates.

Based on the cost-effectiveness analysis, with a cost cutoff of $5,000 per ton of VOC and NO_x, and other considerations such as energy requirements and technological feasibility, EPA set the final VOC standards at 29.4% reduction in Region 1 and 27.7% in Region 2, and selected a NO_x decrease of 6.8% in both regions, levels that can be achieved through RVP reductions to 6.7 psi and by controlling sulfur to 134 ppm. EPA viewed the cost-effectiveness of toxic emission controls beyond a 20% reduction in emissions as questionable (indeed, one should question whether any toxic emission control is warranted based on EPA's analysis of the cost per cancer case avoided). The analysis presented in the RIA shows that average toxic emissions will be reduced by 26% through compliance with the VOC and NO_x requirements. EPA suggested that some refiners would not be able to meet a 25% reduction standard automatically when controlling VOC and NO_x emissions. Consequently, EPA selected a 20% reduction as the toxic emission standard for Phase II RFG.

Antidumping Requirements for Conventional Gasoline

Section 211(k)(8) of the Clean Air Act requires that EPA issue regulations to ensure that conventional gasoline marketed outside the RFG areas not be more polluting (in terms of VOC, CO, NO_x, and toxic emissions) than 1990 baseline gasoline. While this appears to be a fairly straightforward task, the RIA notes a number of difficult judgment calls for the agency in establishing individual refinery baselines. Among these calls are how to allow for increases in the capacity of a refinery, how to allow for work in progress at a refinery, how to deal with the fact that some refiners marketed oxygenated gasoline in 1990 and others did not, and how to deal with product imports. Finally, EPA had to deal with the sensitive issue of whether to publish each refiner's 1990 baseline gasoline specifications (it decided not to after being sued by the petroleum industry on grounds the information had strategic value to competitors).

EPA set antidumping requirements on NO_x and toxics (from tailpipes only) to ensure that conventional gasoline did not increase these emissions, but did not do so for VOC and CO since these emissions would likely be lower than 1990 levels in attainment areas with or without the antidumping requirements due to the Phase II RVP and oxygenated fuel programs implemented in 1992.

Compliance With Regulatory Flexibility Act

The Regulatory Flexibility Act of 1980 requires that agencies evaluate whether a regulation would have adverse effects on a substantial number of small entities. EPA determined that the RFG rule had this potential. One option available to EPA would have been to promulgate less stringent regulations for small entities; however, as EPA noted, this would detract from enforcement due to the fungible nature of gasoline supplies and make it more difficult to meet the statutory percentage reductions in emissions. EPA indicated that it made accommodations where "possible and appropriate," such as allowing refiners to establish more than one baseline by treating their refineries singly or in groups for the purposes of setting baseline compliance requirements.

State Opt-In Provisions

Under state opt-in provisions of Section 211(k)(6), which allows nonattainment areas to join the RFG program, EPA decided not to permit states a partial opt-in to Phase I gasoline only. EPA's rationale was that this otherwise attractive alternative would burden refiners with the need to manufacture gasoline to three distinct specifications beginning in 2000. EPA reasoned that refiners would be challenged to meet Phase I and Phase II RFG requirements and did not need additional complexities forced on them.

Certification

In order to determine whether the fuel content specifications have been satisfied, Section 211(k)(4) directs EPA to establish certification procedures. The procedures set forth in the RFG rulemaking proved contentious; however, the issue of how fuels would be certified was not addressed in the RIA. Some of the issues in certification, such as averaging and trading to meet RFG requirements, and whether modeling or testing should be required, involve both costs and benefits and could have been addressed in the RIA.

OBSERVATIONS

Section 211(k) of the Clean Air Act Amendments of 1990 gives EPA relatively little room for constructive analysis. It sets a minimum requirement for oxygen content, a maximum limit for benzene content, and emission reduction requirements for VOC, toxics, and NO_x. Within these constraints, EPA conducted its regulatory impact analysis to determine what other fuel parameters could be adjusted cost-effectively. Interestingly, EPA's analysis suggests that several of the fuel specifications and performance requirements required by Section 211(k) probably did not meet the agency's cost-effectiveness threshold (such as minimum oxygenate requirement, benzene, and toxic limits) set for Phase II standards.

To the extent Section 211(k) gave the agency flexibility to set standards for RFG, the regulatory impact analysis did influence the final rule. EPA set fuel specifications and emission performance requirements based on extensive (if somewhat uncertain) refinery modeling and cost-effectiveness analysis.

The RFG RIA differs substantially in content from the OMB RIA guidelines, a feature shared by other RIAs prepared by EPA's Office of Mobile Sources. The RFG RIA is long on engineering analysis (of a technical nature as well as cost-effectiveness) and short when it comes to what might be termed economic or policy analysis. The RIA is virtually devoid of analysis of many items called for in the OMB guidelines: benefits, market-based alternatives, impacts on competitiveness and trade, and impacts on segments of the industry that might be adversely affected.

Economic analysis in the RIA centered on determining the cost of incremental changes in the composition of gasoline and the effectiveness of these changes in reducing emissions of VOC, NO_x, and toxic emissions. Other than for toxic emissions, where the RIA determined that the cost per cancer case avoided was far higher than rough limits the agency had set using willingness-to-pay criteria, the RIA did not attempt to quantify benefits or compare benefits to costs. Such an analysis would have required detailed information on the manner in which VOC and NO_x interact to form ozone, the impact of ozone on human health and the environment, and willingness to pay to avoid the various types of adverse effects. Quantifying benefits, if possible, would have enormously complicated the RIA. Benefit-cost analysis would have had the advantage, however, of providing an overall assessment of the social desirability of the RFG program and other ozone and toxic emission control efforts, something that is lacking when programs such as RFG are mandated by Congress.

In directing EPA to establish rules for reformulated gasoline, the Clean Air Act gave EPA the discretion to set alternative percentage reduction requirements for Phase II RFG. Cost-effectiveness analysis is an appropriate analytic technique for choosing cutoff points, in that it can ensure that at the margin air quality improvement investments are equally effective and that environmental goals are met for the least cost.

In selecting $5,000 per ton as the NO_x cost-effectiveness cutoff, EPA may be on weak ground. As partial justification, the agency offered the opinion that selective catalytic reduction at utility boilers is likely to be required to control NO_x in at least some ozone nonattainment areas. The agency offered no analysis to support this opinion and had not yet begun a rulemaking that would require such controls. EPA also pointed to congressionally mandated Tier I standards and enhanced I/M that have relatively high NO_x control costs.

The role of NO_x in ozone formation is still a matter of active debate and research, with NO_x controls sometimes helping control ozone and sometimes making matters worse. Indeed, on December 8, 1995, the American Petroleum Institute petitioned the agency to reconsider the NO_x requirement for Phase II RFG on the grounds that it was not cost-effective, that other control strategies for stationary sources were readily available and cost no more than $500 to $2,000 per ton.

There is no single cost for producing RFG. The cost of making RFG is likely to vary among refineries for a number of reasons. One reason is that as the percentage of RFG in total gasoline output at a refinery increases, the cost per gallon of RFG increases. The first few gallons of RFG cost a refiner little extra to produce but successive increments become more costly. Other factors affecting relative cost include the sulfur and benzene content of the crude being processed, and the capacity and efficiency of refinery processing units. The key cost drivers that all refineries face in producing RFG include adding oxygenates, reducing sulfur, removing butane to meet the RVP limits, and satisfying the benzene limit. Individual refineries will differ in the ease with which they can make these changes.

The petroleum industry had Phase I reformulated fuels available by the January 1, 1995, start date for the program. Despite concerns that spot shortages of RFG could result in prices far above those for conventional gasoline, the opposite occurred. Some areas that had opted into the RFG program dropped out as motorists voiced concerns over cost and the potential health effects of the oxygenate MTBE. Anticipating a larger market than actually developed, refiners produced more RFG than could be absorbed readily in the markets where it was required and marketed some RFG as conventional gasoline. During 1995, at least, the price premium for RFG was very close to that forecast by EPA (based on the Bonner and Moore study) in the RIA. To support this point, Table 8 compares the average differentials in price for spot RFG and contract (rack) RFG versus conventional gasoline for the first eight months of 1995 with production cost differentials as estimated by the Department of Energy using the Oak Ridge National Laboratories refinery model (Zyren, Dale, and Riner 1996). Table 8 offers support for the RIA cost analysis: EPA's predicted average 2.3-cent-per-gallon increase in refinery cost (a fuel economy penalty of 1.6 cents per gallon brings the net cost to the motorist to 3.9 cents per gallon) is quite close to current market cost differentials, the difference likely explained by temporary shortages of methanol during the early part of 1995 (spot methanol prices reached $1.65 per gallon, up from an anticipated price range of $0.50 to $0.70 per gallon) and the consequent somewhat higher than expected prices for MTBE (Ragsdale 1994). Methanol prices quickly receded to previously anticipated levels by the fall of 1995, leading one to expect a very close match from late-1995 on between actual RFG costs and those forecast in the RIA.

The RIA did not consider the possibility that areas that opted into the RFG program might later decide to opt out. Indeed, this possibility is not discussed in the Clean Air Act and was not treated in the final rule. Opting out, while nearly costless for a nonattainment area, does impose costs on refiners, costs that may not be fully recoverable in the market. An RIA conducted in the full spirit of the OMB guidelines might have anticipated this problem, presented an analysis of the potential impacts of such an

Table 8. RFG/Conventional Gasoline Price Differential and Refiners' Costs. (in cents per gallon)

	Spot price differential	*Average refinery contract price differential*	*DOE estimate of average cost difference*	*EPA estimate of average cost difference*
New York region	2.9	3.9	5.7 to 6.1	2.3
Gulf Coast region	4.0	2.6	4.0 to 4.5	2.3

Note: The production cost differentials are those estimated by the U.S. Department of Energy using the Oak Ridge National Laboratories refinery model.

Source: Zyren, Dale, and Riner 1996.

alternative, and led to opt-out provisions in the final rule (such as preventing areas from opting out without providing advanced notice). In this same vein, the RIA might have analyzed the impacts of variation in the number of opt-in areas, rather than using a single number as it did.

Following EPA's promulgation of the final RFG rule, Venezuela asked for permission to ship gasoline to the New York area that did not fully conform to the RFG requirements. EPA denied this request. While EPA's RFG rule would allow Venezuelan producers to use their 1990 gasoline as the baseline from which to calculate emission reductions, Venezuelan refiners did not keep such records for all relevant fuel parameters. Absent such records, EPA's rules required that Venezuela ship gasoline that met the average quality standard for the United States, not a more lenient standard that they otherwise might have satisfied based on their 1990 baseline. Domestic refiners also did not maintain sufficient records of 1990 fuel parameters, but were allowed to make estimates based on later data.

Venezuela and Brazil sued EPA before the World Trade Organization (WTO), charging that the RFG rule placed foreign refiners at a competitive disadvantage since they had to meet a higher standard than some of their North American competitors. In essence, Venezuela and other foreign producers would have to comply with tighter environmental standards than the domestic refiners. The RIA did not explore this issue. While the OMB guidelines call for a review of significant impacts on trade and competitiveness, and EPA could have anticipated the problems that foreign refiners would face, foreign refiners were not invited to the regulatory negotiation process. Consequently, the agency was not pressured to include this issue in the RIA. Not having analyzed impacts of the RFG rule on foreign refiners seems unwise in retrospect following the April 29, 1996, WTO decision finding the EPA rules were a "disguised restriction on international trade." The decision means the EPA will have to rewrite some parts of the RFG rule that deal with environmental standards for imported gasoline.

ACKNOWLEDGMENTS

The authors wish to thank David Deal, managing attorney, American Petroleum Institute, for his helpful comments and criticism.

REFERENCES

Anderson, Robert C. and Thomas J. Lareau. 1992. *The Cost-Effectiveness of Vehicle Inspection and Maintenance Programs*. Research Paper 067. Washington, D.C.: American Petroleum Institute.

Bonner and Moore. 1993. *Study of the Effects of Fuel Parameter Changes on the Cost of Producing Reformulated Gasoline.* Houston: Bonner and Moore Management Science

Deal, David T. 1992. Mobile Source Fuels and Fuel Additives. In Timothy A. Vanderver Jr. (ed.) *Clean Air Law and Regulation* Washington, D.C.: Bureau of National Affairs.

———. 1996. Personal communication with the authors, February 16, 1996. (David T. Deal is managing attorney with the American Petroleum Institute.)

Gushee, David. 1996. *Alternative Transportation Fuels and Clean Gasoline: Background and Regulatory Issues.* (IB91008). Washington, D.C.: Congressional Research Service.

Lareau, Thomas J. 1994. *Improving Cost-Effectiveness Estimation: A Reassessment of Control Options to Reduce Ozone Precursor Emissions.* Research Study 075. Washington, D.C.: American Petroleum Institute.

McConnell, Virginia D. and Winston Harrington. 1992. *Cost-Effectiveness of Enhanced Motor Vehicle Inspection and Maintenance Programs.* Discussion Paper QE92-18. Washington, D.C.: Resources for the Future.

Pechan (E.H. Pechan & Associates and Mathtech, Inc.). 1994. *Regulatory Impact Analysis for the Petroleum Industry NESHAP, Revised Draft.* Prepared for the Office of Air Quality Planning and Standards. Washington, D.C.: U.S. EPA.

Ragsdale, Ralph. 1994. U.S. Refiners Choosing Variety of Routes to Produce Clean Fuels. *Oil and Gas Journal.* March 21, pp. 51–7.

Sierra Research. 1995. *Institutional Support Programs for Alternative Fuels and Alternative Fuel Vehicles in California.* Prepared for Western States Petroleum Association. Sacremento: Sierra Research.

U.S. Congress. Office of Technology Assessment. 1989. *Catching Our Breath: Next Steps for Reducing Urban Ozone.* OTA-O-412. Washington, D.C.: U.S. GPO.

U.S. EPA (Environmental Protection Agency). 1991. *Regulation of Fuels and Fuel Additives; Standards for Reformulated Gasoline; Proposed Rule.* 56 *Federal Register* 31176.

———. 1992. *Regulation of Fuels and Fuel Additives. Standards for Reformulated and Conventional Gasoline; Proposed Rule.* 57 *Federal Register* 13416.

———. 1993a. *Regulation of Fuels and Fuel Additives; Standards for Reformulated Gasoline; Proposed Rule.* 58 *Federal Register* 11722.

———. 1993b. *Final Regulatory Impact Analysis for Reformulated Gasoline.* (December 13).

———. 1994. *Regulation of Fuels and Fuel Additives; Standards for Reformulated and Conventional Gasoline; Final Rule.* 59 *Federal Register* 7716, February 16.

Zyren, John, Charles Dale, and Charles Riner. 1996. 1995 Reformulated Gasoline Market Affected Refiners Differently. *Petroleum Marketing Monthly.* DOE/EIA-0380(96/01). Washington, D.C.: U.S. GPO.

15

Great Lakes Water Quality Guidance

Eloise Trabka Castillo, Mark L. Morris, and Robert S. Raucher

By the mid-1980s, fish tissue studies conducted in the Great Lakes Basin suggested that earlier declines in contaminant levels of bioaccumulative pollutants (such as PCBs, DDT, and dieldrin) had leveled off. Without new policy initiatives, many uses of the Great Lakes would be threatened. Alarmed about the effects of these pollutants on the ecosystem as well as anglers and other consumers of local fish, the Great Lakes governors agreed to take common action, thereby eliminating the incentive for industry to pressure individual states for leniency. In 1990, the Clean Water Act was amended to require that EPA promulgate the Great Lakes Water Quality Guidance for all eight states and thereby create a level playing field among the Great Lakes states. The benefits and costs of EPA's guidance were evaluated in a regulatory impact analysis (RIA), completed in early 1995.

The Great Lakes case, as the authors of this chapter show, highlights a number of issues critical to developing a quality economic analysis, particularly in a highly charged political environment. First, estimating the economic effects of a regulation that affects multiple industries is a particularly challenging task. Second, when a large portion of the pollution likely derives from nonpoint sources it is crucial that these sources be represented in the

MARK MORRIS was the director of the Great Lakes Task Force that coordinated the development of the final Guidance and benefit-cost analysis with U.S. EPA. ELOISE CASTILLO and ROBERT RAUCHER, with Hagler Bailly Consulting, Inc., consulted with EPA in developing the benefit-cost analysis.

analysis. Failure to do so may cast doubt on the credibility of the effort, including the efficiency of proposed solutions. Third, when the overwhelming benefits of a regulation turn out to be recreational as opposed to health related, contingent valuation studies become a necessary but controversial part of the analysis. Fourth, initiating an economic analysis late in the process obviously limits the opportunity to use the RIA to select economically efficient solutions. Finally, as opposed to a national analysis where accuracy is only required "on average," a great deal more precision is required to assess the benefits for a particular region.

INTRODUCTION

In 1986, the governors of the eight Great Lakes states signed the Great Lakes Toxic Substances Agreement. The governors recognized the Great Lakes to be an important economic and environmental resource and directed their environmental administrators to jointly develop an agreement for coordinating the control of toxic releases within the basin. This coordinated effort among the states contributed to the development of the Great Lakes Water Quality Initiative (or the Initiative) in 1989.

The Initiative was a voluntary effort intended to provide a forum for Great Lakes states and EPA to engage other stakeholders in the basin and jointly develop uniform water quality standards and permit programs in accordance with the Clean Water Act (CWA).[1] The participants in the Initiative were particularly concerned about pollutants exhibiting the potential to produce impacts throughout the basin. Based upon observed impairments to the Great Lakes, the states believed that these pollutants were one of the greatest threats to the ecosystem's health, and that further reductions in loadings of such pollutants from all sources should be pursued. Their intent was to prevent the beneficial uses of the ecosystem from becoming further impaired in the future. The Initiative's steering committee charged EPA and state staff to define those pollutants warranting more stringent controls and to draft approaches for controlling them.

The enactment of the Great Lakes Critical Programs Act (CPA) of 1990 (Public Law 101-596, November 16, 1990) codified the ongoing Initiative effort into the CWA (Section 118(c)(2)). The CPA required EPA to publish proposed and final water quality guidance for the Great Lakes Basin that conforms with the objectives and provisions of the Great Lakes Water Quality Agreement (under which the U.S. and Canadian governments established common water quality objectives for the Great Lakes Basin), and is no less restrictive than provisions of the CWA and national water quality criteria and guidance. The CPA required EPA's Great Lakes Water Quality Guidance (or the Guidance) to specify minimum requirements in three areas: water quality criteria, antidegradation policies, and implementation procedures.

Scope of the Rulemaking

The final Guidance establishes minimum water quality standards, antidegradation policies, and implementation procedures for the Great Lakes and their tributaries in the states of Illinois, Indiana, Michigan, Minnesota, New York, Pennsylvania, Ohio, and Wisconsin, including waters within the jurisdiction of Native American tribes. Specifically, the Guidance specifies numeric criteria for selected pollutants to protect aquatic life, wildlife, and human health within the basin and provides methodologies to derive numeric criteria for additional pollutants discharged to these waters. The Guidance also contains specific procedures for implementing these criteria (how criteria are translated into permit limits for the discharge of pollutants from point sources) and an antidegradation policy.[2] Under the CWA, Great Lakes states and tribes must adopt provisions that are consistent with the Guidance into their water quality standards and National Pollutant Discharge Elimination System (NPDES) permit programs by March 23, 1997, or EPA will promulgate the provisions for them.

Bioaccumulative Chemicals of Concern. The Guidance focuses on the chemicals identified by the Initiative's committees as having the greatest potential for impact in the basin. Special provisions were developed for these chemicals, known as *bioaccumulative chemicals of concern* (BCCs), and were incorporated into the final Guidance: more stringent antidegradation procedures, the phase out and elimination of mixing zones,[3] and the development of water quality criteria to protect wildlife that feed on aquatic prey. This approach was designed to prevent toxic-related problems from emerging in the future. Experience with BCCs such as DDT and polychlorinated biphenyls (PCBs) has indicated that it takes many decades to overcome the ecosystem damage caused by even short-term discharges, and that prevention would have been dramatically less costly than cleanup. Thus, the Guidance was intended to provide a coordinated ecosystem approach for addressing possible pollutant problems before they produce adverse and long-lasting impacts basinwide, rather than waiting to see what the future impacts of the pollutants might be before acting to control them.

Scope of the RIA

The proposed and final Guidance were accompanied by detailed cost and benefit studies, and regulatory impact analyses (RIAs). The analyses encompassed the four-year period from 1992 through 1995, and included substantial involvement of EPA staff and contractors. Much of the contractor effort was directed toward the analysis of costs and pollutant load reductions resulting from Guidance implementation; also, a greater proportion of the total effort occurred between proposal and promulgation of the final rule and RIA, rather than before proposal.

The RIA for the proposed rule included an analysis of costs for the entire Great Lakes Basin, as well as preliminary benefit-cost analyses for three river basins within the Great Lakes Basin, as case study sites. An extensive comment period of 150 days was allowed for the proposal, after which EPA published four documents identifying corrections and requesting comments on additional related materials. EPA received over 26,500 pages of comments, data, and information from over 6,000 people in response to these documents, and from meetings with members of the public. Subsequent analysis and revision of both the Guidance and the RIA were directed in response to the numerous comments.

KEY POLICY ISSUES IN THE DEVELOPMENT OF THE FINAL GUIDANCE

The RIA that accompanied the proposed rule, which showed benefits and costs in the same general range, played an important role in ensuring that the rulemaking made it through interagency review, thus moving the regulatory process forward. However, issues raised during the public comment period had a significant impact on subsequent development of the rule. Several key areas of policy interest emerged that were explored in depth by EPA as part of the development of the final RIA. These issues—those involving significant cost impacts, environmental justice, and the interrelated issues of nonpoint sources and appropriate attribution of benefits—are described below.

Significant Cost Impacts (Cost Drivers)

The shape of the final rulemaking was greatly influenced by comments addressing the high cost of the proposal. The agency's concern subsequent to the publication of the proposed Guidance centered on those provisions in the regulation that resulted in high costs, but that couldn't be justified on the basis of producing high benefits in terms of reduced pollutant loadings. Examples of the types of issues identified as having a significant impact on costs ("cost drivers") include the impact of technological improvements enabling detection of pollutants at lower levels than currently possible, granting dischargers credit for pollutants that already exist in their water supply (intake credits), and allowing a mixing zone at the end of a pipe where standards for BCCs would not have to be met until after the discharge mixes with the receiving water.

In evaluating the regulatory options associated with the cost drivers, EPA performed an unprecedented number of sensitivity analyses of the costs and benefits (in terms of pollutant-loading reductions) for both more lenient and more stringent alternatives. In total, the agency evaluated forty regulatory provisions considered for incorporation into the final Guidance. Twenty-seven of these provisions and their regulatory options were evaluated qualitatively because many of these requirements were procedural and difficult to quantify in terms of costs and reduced pollutant loadings. However, the remaining thirteen provisions were technical in nature and more responsive to quantitative evaluation. These provisions (suspected of having significant implementation costs) and their regulatory alternatives were analyzed for cost and pollutant load reduction.

As an example, one analysis indicated that the provisions of the Guidance for regulating discharges to different bodies of water had a relatively small impact on costs, however, eliminating intake credits for discharges to the same body of water resulted in a 600% increase in costs. As a result, dischargers were granted intake credits in the final Guidance.

Environmental Justice

A second area, the issue of environmental justice, as set forth in Executive Order 12898, was also prominent in the development process. The issue of environmental justice—of concern for development of the Guidance—encompassed the protection of minorities consuming high quantities of fish, specifically, the assumed rate of fish consumption that would be used to calculate water quality criteria for the protection of human health. EPA evaluated the impacts of using fish consumption rates of 6.5 grams per day (gpd) (the current rate used by EPA to develop national water quality criteria), 15 gpd, and 45 gpd. Increasing the current rate to 15 gpd or 45 gpd had a negligible impact on the estimated compliance cost and expected pollutant load reductions because the resulting difference in water quality criteria was not significant enough to change the control options selected at a particular facility. This was particularly the case for most BCCs, for which criteria developed using consumption rates of 6.5, 15, or 45 gpd remained below the analytical detection level for the pollutants. Because current analytical detection levels "masked" the fish consumption rate analysis, EPA's risk policy decision was to use a consumption rate of 15 gpd for the final Guidance and require that Great Lakes states and tribes modify human health criteria based on site-specific data to provide additional protection appropriate for highly exposed subpopulations. Thus, where a state or tribe finds that a population of high-end consumers would not be adequately protected by criteria derived using the 15-gpd assumption, the state or tribe would be required to modify the criteria to provide appropriate additional protection.

Nonpoint Sources and Benefits Attribution

A key set of interrelated issues also arose in the context of having the costs of the Guidance estimated as if totally borne by point-source discharges and of attributing an appropriate share of potential water quality benefits to these cost-bearing sources. This process involved accounting for the contribution to water quality problems posed by current point-source dischargers relative to pollutant loads coming from nonpoint sources.

The problem arises because the Guidance implies ambient water quality standards, which the states would strive to meet by controlling any or all of the sources of the pollutants of concern. These pollutant sources include not just ongoing discharges from point sources in the watershed, but also nonpoint sources that, in this instance, are broadly defined as including historical loads (discharges from past periods) that are bound in the sediment, plus ongoing pollutant input from the deposition of air emissions, as well as runoff from nonpoint sources throughout the watershed. Thus, in actual implementation, the Guidance might impose costs on point source dischargers as well as a number of other sources. The benefits of the Guidance would then also be attributable to the shares of the problem addressed by the various sources controlled.

Because of the analytic difficulty of predicting how states would implement the ambient standards in terms of imposing a mix of more stringent controls on point and nonpoint sources, and the difficulty of predicting how much it would cost to control the nonpoint sources, the RIA is based on the presumption that all costs are borne by point sources. Thus, even though all of the problem cannot be attributed to point sources, and some of the costs may ultimately be borne by nonpoint rather than point sources, the analysis makes it appear as if all costs are imposed exclusively on point sources. Since all costs in the analysis are placed on point sources, the benefit-cost analysis needed to focus on the benefits that might arise from controlling only point sources. Thus, the issue became one of attributing the share of potential water quality benefits that are due to the control of point sources.

Compounding the issue of attribution is the nature of how water quality improvement benefits can be realized in water bodies. In general, benefits arise from fairly large steps in water quality improvements (for instance, from boatable to fishable, or from supporting a sport fishery that is toxic impaired to supporting a "toxic-free" fishery in which no consumption advisories are needed). Any single regulatory initiative is unlikely, in and of itself, to generate such a discreet step of water quality improvement; instead, significant improvements in water quality (and, hence, the realization of discernible types of benefits) are typically the joint product of many efforts to reduce pollutant loadings and sustain or improve habitat. Thus, the benefits of the Guidance are conceived as being a portion of the larger set of benefits that might be attained when the Great Lakes waters are no longer toxic impaired. Accordingly, the benefits of the Guidance are assumed to be a fraction of the total benefits that are predicted for moving from baseline to conditions where the waters and fisheries are no longer toxic impaired. The question of how many of these benefits to realistically attribute to point-source controls under the Guidance was a key issue for the RIA, and is addressed in greater detail in the critique of the RIA and the regulatory process.

ANALYSIS OF COSTS AND POLLUTANT-LOADING REDUCTIONS

A detailed analysis of the costs to comply with the Guidance in its proposed form, and the anticipated reductions in pollutant loadings to the Great Lakes, was developed to accompany the rule at proposal. This analysis received significant attention in the public comments, which resulted in significant revisions to both the costing methodology and the rule itself, via analysis of the sensitivity of costs and loadings to changes in the proposed rule. EPA concentrated its efforts in revising the cost analysis to more accurately predict the impact on the regulated community, and these efforts were exhaustive: even a highly critical study of the RIA commissioned by the Council of Great Lakes Governors was revised to endorse the final cost analysis (DRI/McGraw-Hill 1995).

This section provides a description of the methodology used to estimate costs and loadings for the proposed Guidance, the revisions employed for analysis of the final Guidance, and the final results. The Guidance is an extremely complex and technical rule, thus significant detail on the provisions of the rule and the costing methodology designed to reflect the rule are not presented here.[4] The material on this benefits analysis is based on the cost report for the final Guidance (U.S. EPA 1995).

The Cost Analysis for the Proposed Guidance

The general methodology for estimating the cost of compliance with the proposed Guidance was to develop detailed cost estimates for a randomly selected subset of point-source dischargers in the basin, and then extrapolate these costs to all dischargers in the basin. Estimates of baseline pollutant loadings, and the reduction in pollutant loadings anticipated to result from the Guidance, were also developed in this manner. In addition, the estimated costs and loadings for the sample facilities were also extrapolated to three case study areas in the basin, for which EPA also conducted monetized benefit analyses.

A sample of 50 facilities was selected to represent the estimated 588 major dischargers in the basin. Sample facilities were selected from each of the major categories of facilities, which included nine primary industrial groups and a category for municipal wastewater treatment facilities. Industrial sectors included mining, food and food products, pulp and paper, inorganic chemical manufacturing, organic chemical manufacturing and petroleum refining, metals manufacturing, electroplating/metal fabrication, steam electric power plants, and miscellaneous facilities (such as remedial cleanup discharges and tire manufacturers). Facilities were also selected to ensure representation across facility size (as measured by discharge flow volume) through stratification by flow within each category.

A sample of 9 minor facilities was randomly selected to represent the 3,207 minor dischargers.[5] Minimal compliance costs were anticipated for minor dischargers, and this limited sample was analyzed to verify that assumption. Also, because discharge flow data were limited for minor dischargers, it was not possible to adopt a flow-stratified sampling plan similar to that used for major dischargers.

The most current permit data and background information from the NPDES were collected for each sample facility to calculate the permit limits that would result from current regulatory requirements (if not already incorporated into the facility's current permit) and to develop additional permit requirements based on the proposed Guidance.

The Guidance includes a two-tiered approach, consisting of methodologies to develop water quality criteria (Tier I) when sufficient data are available, and methodologies to calculate water quality values (Tier II) when fewer data than the minimum required for a Tier I criterion calculation are available. The purpose of Tier II methodologies is to provide a uniform approach for evaluating and controlling pollutants when there are insufficient data to develop Tier I criteria.

Permit limits were developed for the thirty-two pollutants for which numeric Tier I criteria were proposed, but at a given facility, only those pollutants that were detected or expected to be present in a facility's discharge were evaluated. If a facility's existing effluent limits were not reflective of current state standards, a revised baseline permit limit was calculated for the facility to more accurately reflect the differences between newly revised state standards and the procedures in the Guidance. In this way, only the costs of complying with the Guidance, and not state standards that would be enforced when a facility's permit is renewed, were calculated.

Nonetheless, the estimated compliance costs for the proposed Guidance reflected the uncertainties associated with the highly complex nature of the rulemaking, and numerous data limitations. The site-specific background pollutant concentration data, effluent pollutant concentration data, and pollutant data for industrial and commercial facilities discharging to publicly owned treatment works (indirect dischargers) required to calculate the impact on the sample plants were limited. Technological limitations, such as the level of pollutants detectable in water and effluent, also added uncertainty to the results, especially when new standards are below the detection level.

Revisions to the Cost Analysis

In general, the basic methodology described above for the proposed Guidance was employed to estimate compliance costs and pollutant load reductions for the final Guidance. However, the approach was revised, based on comments received, to more accurately project the costs to the regulated community and to better account for pollutant load reductions. One of the most significant revisions was the updating of data and information on the sample facilities. The use of more recent data produced a shift in the baseline of permit requirements: the baseline was lowered due to more stringent NPDES permit requirements being applied by permitting authorities. The overall effect of lowering the permit baseline was that estimated compliance costs and pollutant load reductions were not as substantial as originally projected for the proposed Guidance.

The cost analysis was also revised to evaluate an expanded list of pollutants. The proposed Guidance, while generally applying to all pollutants, was structured to provide an initial focus on 138 pollutants identified as those known or suspected of being of primary concern in the Great Lakes Basin. Of those 138 pollutants, numeric criteria to protect aquatic life, human health, and wildlife were specified for, and the cost analysis for the proposal was based on, a subset of 32 pollutants. Because of concern that the thirty-two pollutants did not represent all the possible pollutants that may contribute to potential costs, additional pollutants were evaluated for inclusion in the cost analysis of the final Guidance. The additional pollutants were evaluated based on occurrence, loadings, and toxicity and, as a result, the total number of pollutants increased to sixty-nine in the final analysis. The remaining pollutants that were not included in the cost analysis were not detected in monitoring data, or, when detected but below quantitation levels, were not expected to be present in concentrations that would violate water quality standards.

Other changes included revisions to the Tier I criteria and Tier II values for the protection of aquatic life, human health, and wildlife (reflecting changes in the final Guidance for calculation of bioaccumulation factors, and updated toxicity data),[6] including the promulgation of criteria for metals in the dissolved form, as opposed to the total form, for aquatic life. The final Guidance expresses the criteria for metals as dissolved because the dissolved metal more closely approximates the bioavailable fraction of metal in the water column than does the total recoverable metal. In general, these revisions resulted in criteria that were less stringent than the criteria that were originally proposed. For the costs analysis, this resulted in lower estimated costs and pollutant-loading reductions for the final Guidance, compared with the proposed Guidance.

A revised compliance cost decision matrix was also reflected in the final cost analysis. The decision matrix, used to determine a facility's choice of control measures for complying with the Guidance, was revised to establish specific rules for selecting control options or regulatory relief in a consistent manner (see the appendix to this chapter for details on this matrix). The underlying assumption was that a facility would consider lower-cost alternatives prior to incurring the expense of end-of-pipe treatment. The assumed costs per toxic pound of pollutant removed that would "trigger" a facility to pursue regulatory relief correspond with the upper and lower cost estimates. These revisions also resulted in lowering the cost estimate for the final Guidance.

Finally, for the case study areas, all major and minor facilities located in each of the three areas were evaluated (and not just a sample). Costs and loadings were estimated for those pollutants with reasonable potential to exceed Guidance-based limits.

Estimated Costs and Loading Reductions

EPA estimated the total annualized costs of the final Guidance to range from $60 million to $376 million, as shown in Table 1. Costs were annualized at a 7% real interest rate, assuming a ten-year capital life. Direct dischargers bear the bulk of the total costs (67% and 98%, respectively) under both the low- and high-cost scenarios. Major municipal (direct) dischargers account for between 39% and 69% of costs for the low and high cost scenarios, respectively, while major industrial (direct) dischargers account for between 25% and 29% of total costs. Of the affected industries, the pulp and paper sector incurs the largest share of costs (14% and 23%), with per-plant costs estimated at between $151,000 and $1,583,000; miscellaneous facilities incur the next largest share of costs.

The estimated compliance cost reflects a downward revision from the proposed Guidance, which was estimated to cost between $53 million

Table 1. Annualized Compliance Costs of the Final Guidance.

Discharger category	*Number of facilities*	*Estimated costs (millions, first quarter $1994)*
Major direct industrial	272	$15–108
Major direct municipal	316	$24-260
Minor direct	3,207	$2
Indirect	3,528	$20–7
Total	7,323	$60–376

Note: Costs are annualized at a 7% real discount rate assuming a ten-year capital life. Ranges reflect low- to high-cost scenarios.

Source: U.S. EPA 1995.

and $452 million per year (with a most likely estimate of $113 million per year). Moreover, if the revised costing methodology had been used for the proposal, the estimated cost would have ranged from $302 million to $645 million per year. The downward revision in costs was attributed largely to the intake credit provisions and use of dissolved metals criteria for the final Guidance, which tended to eliminate metals from the cost and loadings analyses. The use of updated facility data for the cost analysis, which showed more stringent criteria were already in place, also had a large impact on estimated costs and pollutant-loading reductions. In addition, for the final analysis, EPA did not credit the Guidance for pollutant-loading reductions when a facility was assumed to pursue regulatory relief.

EPA estimated that the final Guidance would result in the reduction of between 5.8 million and 7.6 million toxic-weighted pounds of pollutant loadings to the Great Lakes Basin each year for the low- and high-cost scenarios, respectively (see Table 2). These reductions represent 16% and 22% reductions in baseline pollutant loadings, respectively. Dieldrin and lead account for over 50% of the pollutant load reduction under the low-cost scenario, while heptachlor, dieldrin, and lead account for about 70% of the load reduction under the high-cost scenario. Approximately 80% of the pollutant load reduction, under either scenario, is attributable to reducing BCCs. For several BCCs—PCBs, 2,3,7,8-TCDD (dioxin), mercury, toxaphene—however, little or no reduction from the baseline is estimated because of the method used to derive load reductions.[7]

ANALYSIS OF BENEFITS

The estimated reductions in pollutant loadings provided a basis for estimating water quality–based benefits from the rule. However, several factors had a significant impact on the analysis. For the proposal, EPA got a

Table 2. Estimated Basinwide Toxic-weighted Pollutant Loadings and Reductions Anticipated from the Guidance.

Pollutant	*Baseline loadings*[a]	*Reduction*[b] *(low)*	*Percent change*	*Reduction*[b] *(high)*	*Percent change*
Acrylonitrile					
Aldrin					
Aluminum	2,379,890	25,419	–1.1%	25,419	–1.1%
Antimony					
Arsenic (III)	21,975	21,556	–98.1%	21,556	–98.1%
Benzene	164	1	–0.6%	1	–0.6%
Benzidine					
Benzo(a)pyrene					
Beryllium					
Cadmium	344,827	0	0.0%	0	0.0%
Carbon tetrachloride	648	527	–81.3%	562	–86.7%
Chlordane	975,523	664,604	–68.1%	664,604	–68.1%
Chlorobenzene					
Chloroform	129	7	–5.4%	7	–5.4%
Chlorpyrifos					
Chromium (III)					
Chromium (VI)					
Chrysene					
Copper	2,401	744	–31.0%	744	–31.0%
Cyanide, free	95,940	10,623	–11.1%	10,623	–11.1%
Cyanide, total					
4,4-DDD	45	23	–51.1%	23	–51.1%
4,4-DDE	21	10	–47.6%	10	–47.6%
DDT	88,152	212	–0.2%	212	–0.2%
3,3-Dichlorobenzidine	110,466	68,619	–62.1%	90,465	–81.9%
1,1-Dichloroethane					
1,2-Dichloroethane	29	19	–65.5%	19	–65.5%
1,1-Dichloroethylene	62	0	0.0%	0	0.0%
1,2-trans-Dichloroethylene					
1,2-Dichloropropane	5	5	–100.0%	5	–100.0%
Dieldrin	3,190,719	2,092,368	–65.6%	2,092,368	–65.6%
2,4-Dimethylphenol					
2,4-Dinitrophenol					
Endosulfan					
alpha-Endosulfan					
beta-Endosulfan					
Endrin	189,557	183,778	–97.0%	183,778	–97.0%
Fluoranthene					
Fluoride	73,584	0	0.0%	0	0.0%
Heptachlor	2,324,390	434,659	–18.7%	2,201,441	–94.7%
Hexachlorobenzene	542,816	195,908	–36.1%	195,908	–36.1%
Hexachlorocyclohexane	34,675	33,172	–95.7%	33,172	–95.7%
alpha-Hexachlorocyclohexane	82,945	81,721	–98.5%	81,788	–98.6%
beta-Hexachlorocyclohexane	23,117	22,423	–97.0%	22,423	–97.0%
Hexachloroethane					

continued on next page

Table 2. Estimated Basinwide Toxic-weighted Pollutant Loadings and Reductions Anticipated from the Guidance—*Continued*

Pollutant	*Baseline loadings*[a]	*Reduction*[b] *(low)*	*Percent change*	*Reduction*[b] *(high)*	*Percent change*
Iron	17,732	0	0.0%	0	0.0%
Lead	1,794,813	1,080,141	–60.2%	1,080,141	–60.2%
Lindane	5,366	0	0.0%	5,289	–98.6%
Mercury	519,286	66,304	–12.8%	67,878	–13.1%
Methylene Chloride	5	2	–40.0%	2	–40.0%
Nickel	88	76	–86.4%	76	–86.4%
Parathion					
PCBs	454,908	0	0.0%	0	0.0%
Pentachlorobenzene	443,840	441,523	–99.5%	441,528	–99.5%
Pentachlorophenol	6,742	0	0.0%	5,891	–87.4%
Phenanthrene					
Phenol					
Selenium, total					
Silver	426,685	0	0.0%	0	0.0%
2,3,7,8-TCDD	3,989,245	0	0.0%	0	0.0%
1,2,4,5-Tetrachlorobenzene	388,895	376,802	–96.9%	386,536	–99.4%
Tetrachloroethylene					
Thallium					
Toluene	12	0	0.0%	2	–16.7%
Toxaphene	16,833,496	36,956	–0.2%	36,956	–0.2%
1,1,1-Trichloroethane					
Trichloroethylene	8	6	–75.0%	6	–75.0%
2,4,6-Trichlorophenol					
Zinc	1,735	76	–4.4%	76	–4.4%
Totals	35,364,937	5,838,284	–16.5%	7,649,509	–21.6%

[a]These loadings are weighted for toxicity relative to the toxicity of copper and are based on current permit limits, not actual loadings. (Pollutants with no entries in this table had no reasonable potential for exceeding the Great Lakes Water Quality Criteria.)

[b]These low reductions are based on the difference between current permit limits and the projected Guidance-based limits.

Source: U.S. EPA 1995.

late start to the analysis and was also faced with resource constraints. Preliminary analyses were limited both in funding and to a three-month time frame. In addition, because of the time schedule, the analysis of benefits and the analysis of costs were conducted simultaneously. That is, benefits analyses had to be under way prior to receiving final estimates of pollutant load reductions.

A methodology for the benefits analysis was thus designed to work within this framework. A case study approach was selected because of the site-specific nature of water quality benefits. Benefits transfer methods were used due to resource constraints; no new research on the value of reducing toxic pollutant loadings to the Great Lakes could be conducted. The timing problem was accommodated by employing a total benefits approach. That is, data and information were gathered to estimate total benefits of reducing toxic discharges. Then, when estimated reductions in pollutant loadings were made available, the information was used to apportion a percentage of total benefits to the rule at hand.

Although it was possible to derive monetized benefits estimates for several benefits categories at each case study site using this methodology, the estimates provide only rough measures that are not well suited for sensitivity analyses of small changes in the rule. Time constraints also precluded EPA from carrying the analyses of marginal benefits (in terms of pollutant-loading reductions described above) through to estimates of monetized benefits. The analysis of monetized benefits was intended to illustrate how the rule would generate benefits. However, a precise link between water quality changes and changes in values was not available or developed. In addition, methodology to monetize some benefits categories, such as human health benefits from reductions in noncarcinogens, was not available.

For the final Guidance, several new analyses were conducted to elaborate on the analysis of the proposed rule, and in response to comments from the public and the Office of Management and the Budget (OMB). However, the original benefits values transferred from studies conducted elsewhere in the basin and the United States remained a large part of the estimated benefits.

This section describes the analysis of benefits of the proposed Guidance, the major issues of controversy surrounding the analysis, and the revisions made for analysis of the final Guidance. Estimated benefits, and the comparison of benefits to costs, are also presented.

The Benefits Analysis of the Proposed Guidance

The benefits analysis for the proposed Guidance consisted of three case studies. The case study sites were selected on the basis of data availability and to provide geographic diversity, and included the Fox River and Lake Michigan's Green Bay in Wisconsin, the Saginaw River and Lake Huron's Saginaw Bay in Michigan, and the Black River that drains into Lake Erie in Ohio. Monetized benefits estimates were developed to the extent feasible and included recreational fishing, recreational boating and/or water contact activities, commercial fishing, nonconsumptive wildlife recreation, waterfowl and other hunting, and nonuse values.

Benefits Methodology. The general methodology entailed evaluating baseline resource values and water quality impairments at each site, and the impact of improved water quality (with respect to reduced levels of toxic contaminants such as PCBs) on baseline values. EPA employed benefits transfer techniques using peer-reviewed recreation studies for this purpose. Finally, a portion of the estimated benefits was attributed to the Guidance.

This methodology was employed because water quality improvements often involve thresholds, such as action levels for fish consumption advisories. However, water quality regulations often contribute only a portion of the improvement needed to surpass a threshold. Although individuals may (or may not) have a willingness to pay for incremental steps toward crossing a threshold, when the threshold is surpassed (for instance, fish consumption advisories are lifted), every action that contributed to the effort should be allocated a portion of the benefits. This was accomplished by allocating a portion of the total toxic-free benefits (proportional to the reduction in loadings) to the final Guidance.

The alternative approach is to allocate all of benefits to the particular regulation that made the final improvement required to surpass the threshold, and zero benefits to all regulations that brought water quality conditions close enough to make this possible, which seems inaccurate. In addition, if the timing of different regulations were altered, then the particular action that results in surpassing the threshold would no longer do so, and a different regulation would be attributed with the benefits. Therefore, total benefits are apportioned to each regulatory action based on their respective contribution toward meeting the water quality goal.

The methodology is illustrated for the Fox River/Green Bay case study below.

Fox River/Green Bay Case Study. The benefits considered in this case study involved recreational fishing, commercial fishing, and nonconsumptive wildlife-related recreation. Also considered were nonuse/ecological benefits.

Recreational Fishing Benefits. The recreational fishery in the Fox River/Green Bay area is affected by fish consumption advisories for PCBs, and numerous other contaminants are found in low levels in fish. EPA used two approaches to estimate potential benefits to the recreational fishery from the Guidance.

In the first approach, EPA derived the value of water quality improvements to Wisconsin's Great Lakes trout and salmon fishery and Green Bay's yellow perch fishery based on two studies. The first study (Lyke 1992) used contingent valuation to assess the value of a "toxic-free" fishery to Great Lakes trout and salmon anglers. Lyke estimated an 11% to 31% increase in baseline value to trout and salmon anglers from eliminating contaminants that may threaten human health. EPA considered this scenario similar to the type of water quality improvement that would result from implementation of the Guidance, and adjusted the value estimated by Lyke to pertain just to Green Bay. (The percentage of Wisconsin's Great Lakes shoreline comprised by Green Bay was used as a rough proxy for allocating fishery benefits.) EPA then attributed a percentage of these benefits to the Guidance (for the proposal, EPA assumed that the Guidance could take half the credit for moving from baseline conditions toward a contaminant-free scenario).

The second study (Milliman, Bishop , and Johnson 1987) estimated the value of a plan to rehabilitate Green Bay's yellow perch fishery, and accompanying benefits in terms of increased consumer surplus for the sport fishery. EPA proposed that the annual benefit of the rehabilitation plan could be enhanced to the extent that reduced point-source pollutant loadings accelerate the population recovery (in terms of number and size of fish to be recreationally harvested). EPA then attributed a percentage of these benefits to the Guidance. (For the proposal, EPA assumed that the Guidance could contribute from as little as 20% of the rehabilitation value to as much as 100%, the entire rehabilitation effort.) EPA estimated recreational fishing benefits of the Guidance to be the combined benefits to the yellow perch and trout and salmon fisheries.

The second approach to estimating recreational fishing benefits calculated the baseline value of Green Bay's total sport fishery. The Wisconsin Department of Natural Resources possessed data on the number of recreational angling hours spent in Green Bay and its drainage. These hours were valued by applying a range of average consumer surplus values from the literature (estimated specifically for Green Bay's yellow perch sport fishery by Milliman and others (1991) and for anadromous coldwater fishing outings as derived by Walsh and others (1988, 1990). EPA then used the Lyke (1992) study described above to estimate the change in baseline value that would result from the water quality improvements (an 11% to 31% increase in consumer surplus) and attributed a portion of these benefits to the Guidance. For the proposal, EPA assumed that the Guidance could be credited with realizing one-half of such a projected change in value. The RIA presented the range of recreational fishing benefits estimated using both approaches.

Commercial Fisheries. Commercial fisheries in the case study area have also been affected by high levels of PCBs. (For instance, carp was closed to commercial fishing in 1984 due to PCB levels.) EPA estimated potential benefits of the Guidance in terms of contributing to the reopening of commercial fisheries using a rough rule of thumb developed in the literature (see Crutchfield and others 1982; Huppert 1990): the sum of consumer and producer surplus associated with commercial fisheries amounts to between 50% and 90% of the gross value of ex-vessel landings. Using this approach, EPA estimated the maximum benefits of reopening the commercial fishery based on catch while the fishery was open (lower benefits would be expected if the Guidance alone was not sufficient to reopen the fishery). EPA's estimate of commercial fishery benefits was minimal due to the size of recent past fishing operations.

Nonconsumptive Wildlife-Related Recreation. The toxic contaminants addressed by the Guidance have potential population impacts on many avian and mammalian species sought and valued by wildlife observers and photographers in the case study area. EPA estimated the potential nonconsumptive wildlife-related recreation benefits associated with the Guidance by calculating the baseline value of the resource and by estimating the increase in baseline value that could be attributable to the Guidance. EPA used state data on activity days at a wildlife sanctuary on Green Bay and the estimated value of these days from the literature (Loomis 1988; Walsh and others 1990). For the proposal, EPA assumed that the Guidance could be associated with a 5% increase in the baseline value of these activities.

Nonuse/Ecological Benefits. The Guidance has potential implications for reducing the direct mortality risk and improving the reproductive success of many aquatic, avian, and mammalian species of concern, including threatened and endangered species such as the bald eagle. EPA estimated the potential nonuse/ecologic (that is, individuals' values for the resource that are not associated with their direct or indirect use of the resource) using two approaches. The first approach used evidence in the literature (Fisher and Raucher 1984) that nonuse values have been estimated to be at least half as great as recreational fishing values. This study concluded that if nonuse values were potentially applicable to a policy action, using a 50% approximation was rough but, with proper caveating, preferred to omitting nonuse values from a benefit-cost analysis. EPA estimated nonuse benefits from the Guidance to be half of the estimated benefits to recreational fisheries.

The second approach was based on a nationwide contingent valuation survey of household willingness to pay for "fishable" waters (Mitchell and Carson 1984). As described above, both commercial and sport fisheries in the case study area are affected by toxins. EPA applied the value from this study that could be attributed to respondents' desire to have in-state waters made fishable to the number of households in the case study area. EPA then estimated the portion of these benefits that could be attributed to the Guidance, assuming that water quality had already improved to a halfway point toward meeting the fishable use goal since the year of the

Mitchell and Carson survey (1984). For the proposal, EPA assumed that the Guidance could be attributed with half the credit for attaining the fishery use given the applicable baseline (the halfway point). The RIA presented the range of nonuse benefits using both approaches.

Nonmonetized Benefits. In addition to the benefit categories described above, EPA also expected positive benefits in the recreational boating and swimming, subsistence fishing, waterfowl hunting, and human health categories. However, there were insufficient data available to monetize these benefits for the case study area.

Comments on the Benefits Analysis

The benefits analysis for the proposed rule generated comments related to several major issues, including:

- attribution of benefits to the Guidance;
- lack of measurable improvements in the Great Lakes from the Guidance;
- benefits methodologies (that is, the use of contingent valuation);
- risk assessment methodologies; and
- use of case studies that represent "hot spots" of contamination in the basin.

As a result, EPA conducted numerous new analyses in attempt to bolster the benefits analysis; these analyses are described below. EPA also exhausted a search for direct benefits of the rule that might have been overlooked and also considered the secondary economic benefits associated with development of a pollution control industry in the basin. However, no new primary benefits valuation studies were undertaken and the originally transferred values remained the basis for the final benefits estimates. Aside from changes in the cost and loading analysis, the analyses that resulted in the most substantial revisions to the estimated benefits of the final Guidance were related to the attribution issue.

The Attribution Issue. The case study benefits are based on a two-step approach:

1. Estimating the total value of a large-scale water quality improvement (such as moving from current conditions to a fishery free of consumption advisories) and
2. Attributing a portion of these water quality benefits to the Guidance.

Although the lack of data was not refuted, public comments focused on the uncertainty surrounding the attribution of benefits to the Guidance and the use of potentially "optimistic" attribution assumptions. For the proposal, EPA had evaluated the proportion of benefits that could reasonably be attributed to the Guidance given that there was very little known about the relevant water quality baseline for the Guidance (that is, the water quality levels that would result from past and ongoing regulatory efforts),[8] or the water quality level that would be achieved by implementation of the Guidance. As shown above, EPA assumed for illustrative purposes that the Guidance could be attributed with a portion of total benefits that varied depending on site-specific conditions (the percentage ranged from 1% to 100% over all of the case studies). These assumptions seemed reasonable given the substantial reductions in pollutants of concern for the case study sites (for instance, PCBs) estimated for the proposed rule. However, the issue of attributing benefits to the Guidance resulted in substantial controversy, more so than the valuation of water quality benefits themselves that were based on actual data and applied research.

OMB also commented that EPA should try to model the expected reductions in fish tissue contaminant concentrations (and changes in fish consumption advisories) resulting from the Guidance. In addition, many of the public comments suggested that EPA should be addressing nonpoint-source controls in addition to or instead of more stringent point-source controls. Thus, subsequent efforts were directed toward better quantification of the potential impact of the Guidance in bringing about future toxic-oriented benefits. These efforts are described briefly below.

Available Studies. EPA analyzed available data on total and point-source loadings of contaminants to the Great Lakes watershed for implications of the relative role of point sources in the basin's toxic-related problems. Estimates were available for only a few contaminants (including mercury, lead, and PCBs), and were often highly qualified. However, the estimates indicated that the contribution varied by lake. EPA used the studies to develop lake-specific attribution assumptions for use in the case study analyses. For two of the case study areas, the revised assumptions were a reduction from those used for the proposal.

For example, for Lake Michigan, EPA estimated that the share of total loadings attributable to point sources was between 5% and 10%, and this range was used in the attribution of benefits for the Fox River/Green Bay case study. In comparison, for the proposal, as much as 100% of total benefits were attributed to the Guidance at this site. The new analyses resulted in an attribution assumption of 5% to 10% for Lake Huron compared to the original 5% and 50% assumed for the case study of Lake Huron's Saginaw Bay. EPA revised the attribution assumption for Lake Erie used in the Black River case study to 10% to 15%, compared to the original assumption of 1% to 5%.

Modeling and Other Efforts. Additional analyses related to the attribution issue developed to supplement the RIA of the final Guidance included modeling the relative contribution of point-source loadings to fish tissue concentrations via a generalized Great Lakes exposure and bioaccumulation model, analysis of the cost-effectiveness of point-source controls versus sediment remediation in reducing toxic contamination in the Great Lakes, and a screening analysis to evaluate the contributions of air emissions as a major source of pollutants with significant Guidance-related reductions. In general, these analyses supported the position that point sources could play a role in addressing the basin's toxic-related problems, which also served to respond to concerns over the lack of measurable benefits from the Guidance.

Contingent Valuation. Contrary to comments voicing opposition to the use of contingent valuation (CV), EPA felt justified in the use of this method. Using CV was preferred to leaving important benefits categories unmonetized. (CV is currently the only method accepted by the U.S. Department of the Interior to estimate nonuse values. Also, the National Oceanic and Atmospheric Administration's Blue Ribbon Panel found CV an appropriate methodology for measuring nonuse values, and the method has withstood federal court review for its use in litigation contexts.) In general, as with all methods and research efforts, the merits of specific studies to be used must be evaluated. However, the application of high-quality CV studies in a benefits transfer context is appropriate given the time and resource constraints often faced by EPA.

Additionally, much of the criticism of CV is conceptual rather than based on empirical research, and where CV can be compared with other research techniques (for example, use values estimated by the travel cost methodology or the hedonic price method), CV is shown to yield similar values (see Brookshire and others 1982). A chapter of the RIA for the proposed Guidance was devoted to a review of issues surrounding the accuracy of the CV approach for measuring the use and nonuse values associated with changes in environmental commodities (such as improved water quality), and the effect of using CV-derived research results in the benefits analysis for the Guidance. Benefits estimates based on CV studies were not modified for the final RIA.

Risk Assessment. A preliminary assessment of human health risks to sport anglers and potential risk reductions resulting from the Guidance was included in the RIA of the proposal. For the final RIA, in accordance with the Executive Order on Environmental Justice, EPA collected additional data and information on the consumption of Great Lakes Basin fish to refine the analysis to reflect minority and low-income exposures and to conduct a separate assessment for Native Americans engaged in subsistence fishing in the basin. As described above, the issue of environmental justice was of special concern for EPA in revision of the Guidance. In response to comments, EPA used lake-specific fish tissue concentrations in revising the analysis.

The risk assessment indicated high baseline risk levels for some angler groups (such as a risk level of 10^{-2} for low-income minorities for Lake Michigan). However, at the basinwide level, estimated reductions in risk due to the Guidance were minimal (a potential reduction of 0.4 to 0.7 excess cancer cases per year) because baseline risks were primarily attributable to PCBs, and no reductions in PCB loadings were estimated as a result of the Guidance. This result pointed to the limitations of the sampling methodology used to analyze basinwide costs and loading reductions. That is, no reductions in PCBs loadings were found for the sample of facilities used to represent the basin. However, for two of the case study sites (where *all* facilities were evaluated), reductions in PCBs and thus human health risks were estimated.

"Hot Spots". The public comments asserted that the benefits analysis was not representative of basinwide benefits because it was based on case studies of three "hot spots." EPA's original selection of case study sites was based on data availability, as well as other considerations. There was no reason to anticipate that the selected sites were atypical of the region. However, in response to the comments, EPA investigated the representativeness of the case study sites to the extent possible. Representativeness was assessed by comparing the percentage of total benefits estimated to accrue in the case study areas to the percentage of basinwide costs they will incur. Benefits-related measures (such as population, recreational angling days, and nonconsumptive recreation days) were used in place of total benefits for this analysis because there was no estimate of benefits for the entire Great Lakes Basin. The case studies were found to represent reasonably proportionate shares of costs and benefits. So while loadings (and expected benefits) at these hot spots may be unrepresentatively high, costs are also relatively high. Less-contaminated sites may show lower benefits but also lower costs.

Benefit-Cost Comparison of the Final RIA

For the RIA developed for the final rulemaking, the case study benefits estimates reflected revisions in the attribution assumptions, and included human health benefits based on the updated risk assessment. They were also tied directly to the estimated reduction in toxic-weighted pollutant loadings, assuming a linear relationship between water quality improvements and benefits. Given the downward revision to the estimated pollutant-loading reductions, and the more conservative attribution assumptions used for two of the sites, the case study benefits estimates of the final Guidance were also generally revised downward. Only the Black River case study showed a slight upward revision to the benefits estimates compared to the estimated benefits for the proposed rule.

The estimated annualized benefits for each site are shown in Table 3 and compared to annualized costs. Also shown are the streams of discounted benefits and costs (discounted at 3%). For this comparison, benefits are phased in over ten- and twenty-year periods to account for a time lag in the realization of benefits. As can be seen from Table 3, benefits and costs are commensurate in both the direct annualized comparison and the present value comparison of discounted benefits and costs.

CRITIQUE OF THE RIA AND REGULATORY PROCESS

The final Guidance was signed by the EPA administrator on March 13, 1995, and published in the *Federal Register* on March 23, 1995. Although analysis of regulatory impacts ceased at that point, legal proceedings and

Table 3. Potential Benefits and Costs of the Final Guidance for the Case Study Areas. (in millions of $1994)

	Benefits range[1]	*Midpoint of benefits range*	*Costs*
Fox River/Green Bay case study			
Direct annualized comparison	$0.3–8.5	$4.5	$3.6
Discounted benefits and costs			
10-year phase in of benefits	$5.4–133.9	$69.7	$71.8
20-year phase in of benefits	$4.1–101.4	$52.7	$71.8
Saginaw River/Saginaw Bay case study			
Direct annualized comparison	$0.2–7.7	$4.0	$2.6
Discounted benefits and costs			
10-year phase in of benefits	$2.6–120.9	$61.7	$53.0
20-year phase in of benefits	$2.0–91.5	$46.8	$53.0
Black River case study			
Direct annualized comparison	$0.4–1.5	$0.9	$2.1
Discounted benefits and costs			
10-year phase in of benefits	$6.4–22.7	$14.5	$42.7
20-year phase in of benefits	$4.8–17.2	$11.0	$42.7

Note: All direct annualized cost comparisons are based on annualized costs assuming a ten-year capital life and reflecting a 7% real interest rate on capital. All discounted benefits and costs are are based on present ($1994) values discounted over thirty years, assuming a ten-year capital life and a 7% opportunity cost of capital. (Benefits and costs are both discounted at a 3% real discount rate.)

defense of the rulemaking and RIA continued. EPA is faced with both challenges directed at the Guidance and critique from industry aimed at the benefits analysis. However, a tremendous amount of outside review supported the Guidance and the validity of the agency's analysis.

Many Great Lakes state environmental commissioners conducted their own evaluations of compliance costs and concluded that EPA's estimates were on target, or even a little high. In addition, the study commissioned by the Great Lakes Council of Governors to estimate the economic impact of the Guidance on the Great Lakes region supported EPA's cost range (DRI/McGraw-Hill 1995). In light of these reviews, EPA considered the RIA for the final rule an important success. Interestingly, these reviews did not address the analysis of monetized benefits for the RIA, but focused solely on the economic impact of these costs.

Notwithstanding the rule's supporters and detractors, the RIA provides insight into the state of current methodologies for estimating costs and benefits. These insights are discussed below.

Emphasis on Costs and Loading Reductions

The RIA for the Guidance reflects the agency's emphasis on the analysis of costs and pollutant-loading reductions, toward which a substantially greater amount of time and money was allocated for both the proposed and final Guidance relative to the analysis of benefits. The selection of regulatory options was directed primarily by the pollutant, the loading, and the cost to remove the loading. Then, this cost-effectiveness analysis was used to select the options that were further analyzed for assessing monetized benefits. The emphasis on costs and loadings is partly a result of the limited time frame within which regulatory decisions are made at EPA. It also reflects the fact that methodologies for estimating monetized benefits of water quality improvements are not well suited for analyzing national regulations (because water quality benefits are very site specific) or for assessing water body-specific benefits from small changes in water quality that often result from a single regulation.

The emphasis on costs served EPA well in many respects for the Guidance, however. The agency exhausted efforts to get a handle on compliance costs to the point-source dischargers in the basin, and in particular, on potential upper-bound costs. Extensive effort was directed toward analysis of the sensitivity of the costs to particular provisions in the rule. In addition, EPA used numerous assumptions for the final cost analysis that produced higher cost estimates while, in some cases, not crediting the rule with reductions in loadings (benefits). Although this conservative approach to estimating costs may not have stopped criticism of the Guidance, it helped EPA defend its analysis during discussions with outside parties.

Limitations of Benefits Methodology

Development of the RIA revealed other limitations of benefits methodology. For example, EPA surmises in retrospect that, if it could redo the analysis of benefits for the Guidance, it would conduct original benefits research (such as an angler survey) within the Great Lakes Basin. This consideration reflects on the effectiveness of benefits transfer techniques, which were used for the RIA of the Guidance, in producing convincing benefits estimates. However, given the time and resource constraints faced by the agency, benefits transfer often remains the only viable option. Judging by the emphasis in the public comments, this issue was of less concern for the benefits analysis of the Guidance than the more scientific issues, such as the role of point sources in the problem.

Another consideration for the Guidance is that traditionally, EPA has paid much greater attention to human health effects in terms of cancer cases. However, the Guidance was not a rule for which a large reduction in excess cancer cases was expected. In part, this was a function of limitations in the methodology employed for estimating pollutant loadings and reductions, and by analytical detection levels. In addition, reductions in PCBs, which drove the health benefits estimates, were not found in the modeled basinwide analysis although they were estimated from the more detailed assessment conducted for the case studies.

The limited amount of information on noncancer (systemic) risks, as well as a lack of methodology for translating these risks into potential cases, may have been of consequence for the Guidance, which addresses a large number of toxic pollutants. However, no monetization of benefits from reductions in noncancer health effects was possible for the Guidance. EPA is beginning to assess noncancer human health effects with greater interest, and these impacts may be better addressed in future regulations.

The emphasis on reductions in cancer cases also overshadows potential benefits from overall improvements in ecosystem health. The agency's Science Advisory Board is currently investigating ways to improve evaluation of "ecological" benefits. There is also a perception problem for toxic pollutants that are below the analytical detection level at the chronic effects level. That is, there is a tendency to forget that the pollutants are there, and may be a health threat to humans and wildlife, when the pollutants are below the analytical detection level. Thus, the controls provided by the Guidance are not as readily supported as those for conventional pollutants, or for detectable/acute levels of toxic pollutants.

The Benefit-Cost Analysis and Focus of the Regulation

As described above, the benefit-cost analysis for the Guidance was greatly influenced by the volume of comments on the proposal related to the economic efficiency requirements of RIAs. In revision of its cost analysis, EPA systematically evaluated each issue raised in the comments, conducted sensitivity analyses of the cost-effectiveness (cost per loading reduction) of various options, and significantly modified the rule to ensure cost-effective implementation. In its analysis, EPA paid particular attention to estimating an uppermost bound on potential costs of the rule.

Estimates of several categories of monetized benefits were made for the final rule and compared to costs, but alternative benefit-cost findings for a range of potential other options were not developed because the benefits methodology feasible under the agency's constraints does not lend itself to analyzing small changes in a regulation. No meaningful comparison of benefits to costs was possible at the basinwide level, since water quality benefits are very site specific. However, three case study analyses indicated that expected benefits were roughly equivalent to estimated costs. While issues were raised regarding the benefits methodologies (for instance, the appropriateness of CV), the most contentious issue was determining what portion of the estimated water quality improvement benefits should be attributed to the Guidance in general (and to point-source controls in specific).

EPA focused research efforts on quantifying the role of point sources in contributing to current conditions. Modeling, based on fish tissue contaminant concentrations, showed that point sources could play a cost-effective role in removing fish consumption advisories. Available data from mass loadings studies of the pollution sources loadings to the basin were used to apportion the estimated benefits of cleanup to the point-source controls of the final Guidance. However, data gaps remain to create some uncertainty in the analysis of benefits and costs of the Guidance's point-source controls.[9]

SUMMARY AND CONCLUSIONS

The RIA for the proposed Guidance was initiated late in the regulatory process, and thus it had a limited role in the process of regulatory option selection. However, the volume of public comments focused attention on the cost-effectiveness of the rule. As a result of the agency's efforts in addressing the comments through development of the final RIA, the RIA at promulgation played more of the intended role of pointing out optimum regulatory solutions.

The development of the RIA for the final Guidance was a major undertaking for EPA. The costing methodology addressed difficult technical issues, such as pollutant levels that are below analytical detection levels. Benefits methodologies addressed the difficult problem of estimating monetized benefits of water quality improvements from available literature and of working with time frames that required parallel development of estimates of pollutant-loading reductions and monetized benefits. Data gaps and uncertainties remain in the analyses of both costs and benefits, particularly with respect to the roles of point and nonpoint sources in the toxic-related problems in the basin. However, the agency has been able to defend itself in current litigation due to the completeness of its RIA and the consistency of the promulgated Guidance with the benefit-cost principles embodied in the RIA.

APPENDIX: COMPLIANCE COST DECISION MATRIX

To model the decision process used by actual facilities, a cost decision matrix was developed to determine appropriate, cost-effective control measures that could be used by a facility to meet final Guidance water quality based effluent limits (WQBELs; that is, permit limits). Specific rules were established in the matrix to provide reviewers with guidance in selecting options in a consistent manner. The matrix is presented in Figure 1. Under the decision matrix, costs for minor treatment plant operation and facility changes were considered first. Modification or adjustment of existing treatment was determined to be feasible where literature indicated that the existing treatment process could achieve the revised WQBEL and where the additional pollutant reduction was relatively small (for example, 10% to 25% of current discharge levels).

Where it was not technically feasible to simply adjust existing operations, the next most attractive control strategy was determined to be waste minimization/pollution prevention controls. These controls, however, were costed only when they were considered feasible based on the reviewing engineer's understanding of the processes at a facility. The practicality of techniques was determined based on several "rules of thumb" established in the decision matrix. Decision considerations included the level of pollutant reduction achievable through waste minimization/pollution prevention techniques, appropriateness of waste minimization/pollution prevention for the specific pollutant, and knowledge of the manufacturing processes generating the pollutant of concern. In general, detailed treatment and manufacturing process information was not available in NPDES permit files; therefore, the assessment of feasibility was based primarily on best professional judgment using general knowledge of industrial and municipal operations.

If waste minimization/pollution prevention alone was deemed not feasible to reduce pollutant levels to those needed to comply with the final Guidance criteria, a combination of waste minimization/pollution prevention and simple treatment was considered. If these relatively low-

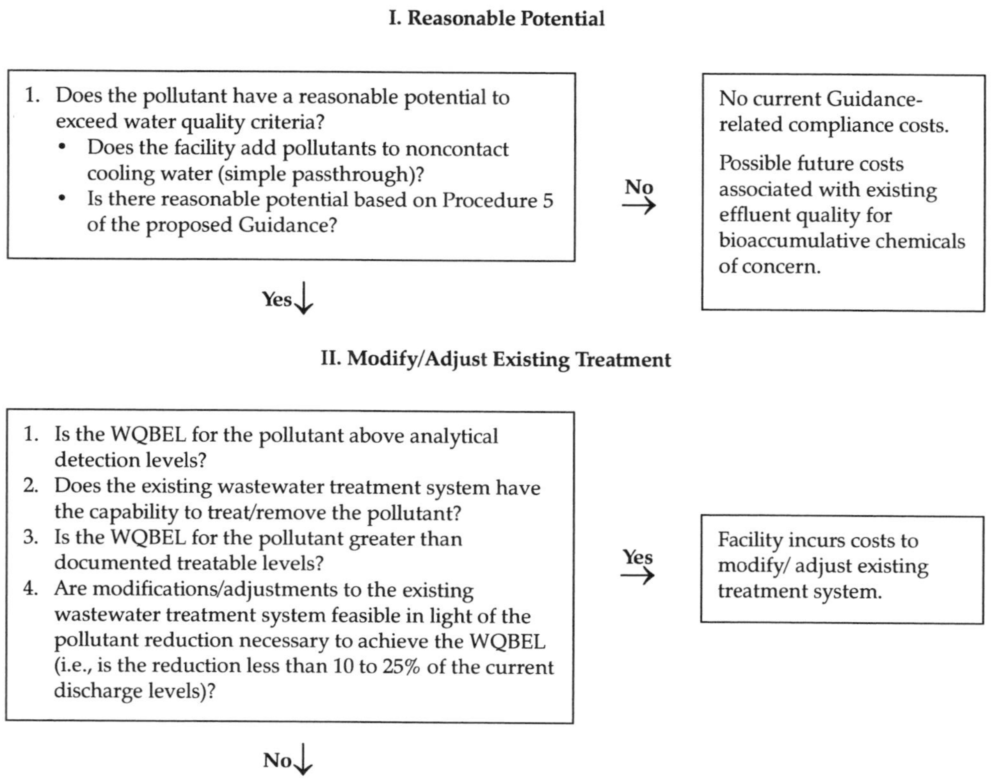

I. Reasonable Potential

1. Does the pollutant have a reasonable potential to exceed water quality criteria?
   - Does the facility add pollutants to noncontact cooling water (simple passthrough)?
   - Is there reasonable potential based on Procedure 5 of the proposed Guidance?

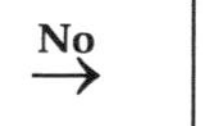

No →

No current Guidance-related compliance costs.

Possible future costs associated with existing effluent quality for bioaccumulative chemicals of concern.

Yes ↓

II. Modify/Adjust Existing Treatment

1. Is the WQBEL for the pollutant above analytical detection levels?
2. Does the existing wastewater treatment system have the capability to treat/remove the pollutant?
3. Is the WQBEL for the pollutant greater than documented treatable levels?
4. Are modifications/adjustments to the existing wastewater treatment system feasible in light of the pollutant reduction necessary to achieve the WQBEL (i.e., is the reduction less than 10 to 25% of the current discharge levels)?

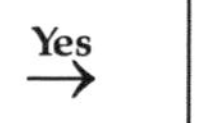

Yes →

Facility incurs costs to modify/ adjust existing treatment system.

No ↓

Figure 1. Compliance Cost Decision Matrix.

(continued on next page)

III. Waste Minimization/Pollution Prevention

1. Is the production process or source generating the pollutant amenable to waste minimization/pollution prevention techniques?

Industrials Only

2. Is the WQBEL for the pollutant above analytical detection levels?
   If no, is the production process or pollutant source amenable to control techniques expected to reduce pollutant to below analytical detection levels (e.g., product substitution)?
3. Is the level of pollutant reduction required to meet the WQBEL insignificant (i.e., less than 10 to 25% of current discharge levels)?

POTWs Only

2. Is the WQBEL for the pollutant above analytical detection levels?
3. Is the level of pollutant reduction required to meet the WQBEL insignificant (i.e., less than 10 to 25% of current discharge levels)?
4. Are increased industrial user/source controls feasible?

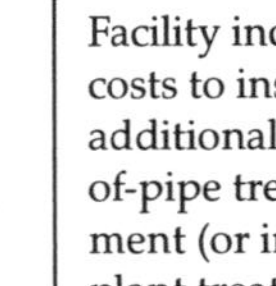

Yes → Facility incurs costs to implement waste minimization/pollution prevention.

No ↓

IV. New/Additional Treatment System

1. Is the WQBEL for the pollutant above analytical detection levels?
2. Is the WQBEL for the pollutant above documented treatable levels?
3. Is the new/additional treatment technology feasible in light of the:
   - existing treatment process(es)?
   - production process(es)?
   - pollutant source(s)?
   - level of pollutant reductions required to achieve the WQBEL (i.e., is it greater than 10 to 25% of the current discharge levels)?
   - cost to add the necessary treatment?

(*Note*: Under the low-end scenario, the cost trigger is $200/toxic lb-equivalent for a specific facility for a pollutant. Under the high-end scenario, the cost trigger is $500/toxic lb-equivalent for the industrial category.)

Yes → Facility incurs costs to install additional end-of-pipe treatment (or in-plant treatment).

No ↓

V. Other Controls

1. Is the pollutant concentration above treatable levels and is the WQBEL below analytical detection levels?
2. Is a combination of end-of-pipe treatment and waste minimization/pollution prevention feasible in light of the:
   - existing treatment process(es)?
   - production process(es)?
   - pollutant source(s)?
   - level of pollutant reductions required to achieve the WQBEL (i.e., is it greater
   - than 10 to 25% of the current discharge levels)?
   - cost to add the necessary treatment?

(*Note*: Under the low-end scenario, the cost trigger is $200/toxic lb-equivalent for a specific facility for a pollutant. Under the high-end scenario, the cost trigger is $500/toxic lb-equivalent for the industrial category.)

Yes → Facility incurs costs for other controls.

Figure 1. Compliance Cost Decision Matrix—*Continued*

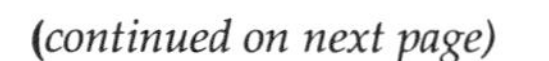
(continued on next page)

↓ **VIa. Phased TMDL**	↓ **VIb. Variances from water quality standards**	↓ **VIc. Site-specific criteria**	↓ **VId. Change designated use**	↓ **VIe. Alternative mixing zone**
1. Is the discharge to a nonattainment receiving water? 2. Are the other sources of pollutants to the receiving water known?	1. Is the pollutant naturally occurring? 2. Are there natural, ephemeral, intermittent, or low-flow conditions? 3. Are there human-caused conditions or sources? 4. Are dams, diversions, or other types of hydrologic modifications present? 5. Do the physical conditions related to the natural features of the water body contribute? 6. Would the controls result in substantial and widespread economic and social impact? If yes, will the discharge comply with anti-degradation requirements and cause no increased risk to human health and the environment?	1. Are local environmental conditions not reflected in criteria? 2. Are bioaccumulation factors appropriate?	1. Is the pollutant naturally occurring? 2. Are there natural, ephemeral, intermittent, or low-flow conditions? 3. Are there human-caused conditions or sources? 4. Are dams, diversions, or other types of hydrologic modifications present? 5. Do the physical conditions related to the natural features of the water body contribute? 6. Would the controls result in substantial and widespread economic and social impact?	1. If a tributary, does the discharge dilution factor need to be greater than 0.75 to provide relief?

VIa. Phased TMDL	**VIb. Variances from water quality standards**	**VIc. Site-specific criteria**	**VId. Change designated use**	**VIe. Alternative mixing zone**
Facility incurs future cost to comply with TMDL. If existing effluent concentration is greater than treatable levels, then go to IV, and facility incurs costs to reduce the effluent to treatable levels. If existing effluent concentration is less than treatable levels, then go to III, and facility incurs costs to implement waste minimization.	Facility incurs costs for preparing variance request and future compliance costs when variance expires. If existing effluent concentration is greater than treatable levels, then go to IV, and facility incurs costs to reduce the effluent to treatable levels. If existing effluent concentration is less than treatable levels, then go to III, and facility incurs costs to implement waste minimization.	Facility incurs costs for preparing request for site-specific criteria. If existing effluent concentration is greater than treatable levels, then go to IV, and facility incurs costs to reduce the effluent to treatable levels. If existing effluent concentration is less than treatable levels, then go to III, and facility incurs costs to implement waste minimization.	Facility incurs costs associated with preparing a use attainability analysis. If existing effluent concentration is greater than treatable levels, then go to IV, and facility incurs costs to reduce the effluent to treatable levels. If existing effluent concentration is less than treatable levels, then go to III, and facility incurs costs to implement waste minimization.	Facility incurs costs to prepare demonstration. If existing effluent concentration is greater than treatable levels, then go to IV, and facility incurs costs to reduce the effluent to treatable levels. If existing effluent concentration is less than treatable levels, then go to III, and facility incurs costs to implement waste minimization.

Figure 1. Compliance Cost Decision Matrix—*Continued*

cost controls could not achieve the Guidance-based WQBELs, then end-of-pipe treatment was considered.

Development of end-of-pipe treatment cost estimates began with a review of the existing treatment systems at each facility. Decisions to add new treatment systems or to supplement existing treatment systems were based on this initial evaluation. For determining the need for additional or supplemental treatment, sources of performance information included EPA development documents for effluent guidelines and standards for the facility's industrial classification, and the EPA Office of Research and Development, the Risk Reduction Engineering Laboratory (RREL) Treatability Database (Version 4.0). The pollutant removal capabilities of the existing treatment systems and/or any proposed additional or supplemental systems were evaluated based on the following criteria:

- the effluent levels that were being achieved currently at the facility;
- the levels that were achieved at similar facilities with similar treatment systems documented in the effluent guideline development documents; and
- the levels that are documented in the EPA RREL Treatability Database.

If this analysis showed that additional treatment was needed, unit processes that would achieve compliance with the Guidance-based effluent limitations were then chosen using the same documentation.

Following the calculation of end-of-pipe treatment costs, the relationship between the cost of adding the treatment and other types of remedies or controls was considered. Specifically, if the estimated annualized cost for removal of a pollutant exceeded $200 per toxic pounds-equivalent, then the decision matrix indicated that dischargers would explore the use of other remedies or controls. This cost trigger was based on the upper end of the range of the costs to comply with promulgated effluent guideline limitations and standards for direct discharger industrial categories. When it was assumed that facilities would pursue alternative relief, no treatment cost was estimated for a facility; however, a nominal cost for some efforts to reduce the pollutant until the relief is granted was included. In addition, pollutant load reductions were not calculated or credited for any pollutant for which alternative relief was assumed.

Finally, based on discussions with EPA regional and state permitting agencies and outside experts, the typical cost to facilities pursuing relief from Guidance-based WQBELs was estimated. These costs will be in the form of additional monitoring, performing special studies, and so forth, to support facilities' requests for relief from the Guidance-based WQBEL. The costs estimated by the regions and states for the relief mechanisms ranged from $1,000,000 per pollutant (for phased total maximum daily loads) to $20,000 (for criteria modifications). Table 4 summarizes the cost

Table 4. Cost Estimates for Pursuing Regulatory Relief.

	Regulatory relief mechanism (state costs/ facility costs)				
Source	*Phased total maximum daily load*	*Variance*	*Site-specific criteria*	*Change designated use*	*Alternative mixing zone*
Michigan DNR	NA	NA	NA	NA	NA
Minnesota DNR	NA	$2,000/NA	NA/$50,000	$2,000/NA	$5,000/NA
New York DEC	$400/NA	NA/ $100,000	NA/ $200,000	NA/ $75,000	NA/ $150,000
Indiana DEM	NA	NA	NA	NA	NA
Ohio EPA	NA	NA/ $150,000	NA	NA/ $40,000	NA/ $150,000
Pennsylvania DER	NA	NA	NA	NA	NA
Wisconsin DNR	NA/ $1,000,000	$10,000/ $75,000	$70,000/ $50,000	NA/ $1,000,000	NA/ $100,000

Note: NA: Not available. State agency acronyms—DNR = Department of Natural Resources; DER = Department of Environmental Resources; DEC = Department of Environment and Conservation; EPA = Environmental Protection Agency.

Source: State environmental agencies, U.S. EPA regional offices.

estimates provided by states and EPA regions. For purposes of estimating compliance costs, a midrange cost value of $200,000 per pollutant was used each time a relief mechanism was assumed necessary.

In developing and using the cost decision matrix, it is acknowledged that granting relief from WQBELs is dependent upon the specific circumstances at a facility, as well as the judgment and implementation procedures of the permitting authority. It is also acknowledged that opportunities for waste minimization are dependent upon the specific circumstances at a facility. The use of a $200 per toxic pounds-equivalent trigger for a facility assumes that the regulatory flexibility in the final Guidance would be available and granted to all facilities that exceed the cost trigger.

Acknowledging that the use of regulatory relief may be limited depending on the particular circumstances for a facility, costs were also estimated under a higher cost scenario that assumes regulatory relief would be granted only when the cost for the particular category of dischargers exceeds a cost trigger. Particularly, if the estimated annualized cost for a category of dischargers exceeded $500 per toxic pounds-equivalent, then it was assumed that dischargers within the category would be granted regulatory relief. This cost trigger was based on the highest costs to comply with promulgated effluent guideline limitations and standards for direct discharger industrial categories, which ranged from $1 to $500 per toxic pounds-equivalent per industrial category.

ACKNOWLEDGMENTS

The authors gratefully acknowledge the contribution of Jim Parker of Science Applications International Corporation, whose work on the cost analysis for the Great Lakes Water Quality Guidance is reflected in this study.

ENDNOTES

[1]Under the Clean Water Act, industrial and municipal dischargers (point-source discharges) are regulated through National Pollutant Discharge Elimination System (NPDES) permits. Since 1972, the NPDES program has grown from regulating conventional pollutants through technology-based treatment controls to regulating toxic pollutants through water quality–based controls. All point-source discharges to waters of the United States are required to have in place, at a minimum, technology-based treatment. If the level of pollutants in the effluent still have the potential to adversely affect the water body after installing technology-based treatment, the discharger must add additional treatment (that is, water quality–based controls) until the pollutant level is below numeric or narrative water quality standards set forth in the Clean Water Act. Recently, EPA has undertaken several initiatives that have ensured the proper regulation of all point-source dischargers. For example, the National Toxics Rule (*Federal Register*, December 22, 1992) ensures that water quality criteria are promulgated for all toxic pollutants in all states. The recent storm water regulations promulgated November 16, 1990, and the subsequent finalization of the baseline permit in September 1992, and the combined sewer overflow control policy (*Federal Register*, April 19, 1994) should result in NPDES permit controls for most wet-weather discharges. Finally, with the advances in analytical technology and a better understanding of ecosystem effects, EPA is currently developing additional environmental protection criteria (for example, sediment criteria, wildlife criteria, biocriteria) for use in establishing controls for point-source discharges. The Guidance is a watershed/ecosystem approach that will ensure that the Great Lakes states have consistent human health and environmental protection across the whole Great Lakes Basin.

[2]The Guidance sets forth minimum procedures for bioaccumulative pollutants to ensure that water quality is not further degraded. The state and discharger are required to follow the procedures when considering actions that may lower water quality in an area where water quality standards are already met by increasing the amount of a bioaccumulative pollutant released into that area.

[3]Mixing zones are areas in the immediate vicinity of industrial outfalls where pollutant concentrations are allowed to exceed water quality criteria until they are lowered as a result of mixing with a greater volume of water.

[4]Readers interested in these details are referred to *Regulatory Impact Analysis of the Final Great Lakes Water Quality Guidance* and *Assessment of Compliance Costs*

Resulting from Implementation of the Final Great Lakes Water Quality Guidance, both published in March 1995. These documents are available for a fee upon written request or telephone call to the National Technical Information Center, U.S. Department of Commerce, 5285 Port Royal Road, Springfield, VA 22161 (telephone 800-553-6847 or 703-487-4650).The cost report is NTIS #PB95187340; the RIA is NTIS #PB95187357.

[5]In the case of minor dischargers, discussions with state permit writers indicated that minor direct dischargers in the Great Lakes Basin were not discharging bioaccumulative pollutants or other toxic pollutants regulated under the Guidance in levels that were expected to violate water quality standards. Evaluation of the permit files for these 9 minor facilities, and the 184 minor facilities located in the three case study areas, verified this assertion.

[6]The calculation of bioaccumulation factors for the final Guidance uses a different model to determine food-chain multipliers for organic chemicals than was used for the proposed Guidance. Tier I criteria and Tier II values were differentiated in the proposal based on the quantity and quality of toxicological data only. There was no differentiation based on the quantity or quality of bioaccumulation factors. After reconsideration, EPA decided to differentiate the Tier I criteria and Tier II values based on both the quantity and quality of toxicological and bioaccumulation data.

[7]When existing permit limits or the water quality–based effluent limit (WQBEL) calculated to comply with the Guidance are below the analytical detection level, one-half of the method detection level is used for each. The result of this approach is that no pollutant reduction is estimated, regardless of whether the Guidance-based WQBEL is further below detection levels than the existing state permit limit.

[8]Although the cost analysis calculates baseline permit discharge levels and the changes in permits anticipated to result from implementation of the Guidance, the corresponding water quality levels were not modeled.

[9]EPA does include nonpoint sources in the rule by allowing for load allocations to be made among all sources to a water body. However, the costs and benefits of nonpoint-source controls were not explicitly evaluated. Instead, the RIA assumes all costs and benefits accrue to point sources. In actual implementation, nonpoint-source controls may be used to reduce the total costs placed on point sources, yet generate the same total water quality benefits as in the RIA.

REFERENCES

Brookshire, D.S., A. Randall, and J.R. Stoll. 1982. Valuing Increments and Decrements in Natural Resource Service Flows. *American Journal of Agricultural Economics* 62(3): 478–88.

Crutchfield, J.A., S. Langdon, O.A. Mathisen, and P.H. Poe. 1982. *The Biological, Economic, and Social Values of a Sockeye Salmon Stream in Bristol Bay, Alaska: A*

Case Study of Taimina River. Fisheries Research Institute Circular #82-2. Seattle: University of Washington.

DRI/McGraw-Hill. 1995. *The Great Lakes Water Quality Initiative: Cost Effectiveness Update.* Prepared for the Council of Great Lakes Governors. Chicago: DRI/McGraw-Hill..

Fisher, A. and R. Raucher. 1984. Intrinsic Benefits of Improved Water Quality: Conceptual and Empirical Perspectives. In V.K. Smith (ed.) *Advances in Applied Microeconomics,* Vol. 3. Greenwich, Connecticut: JAI Press.

Huppert, D.D. 1990. *Economic Benefits from Commercial Fishing.* Draft Report to National Marine Fisheries Service, NOAA.

Loomis, J.B. 1988. *Valuing Nonconsumptive Use and Values of Game and Nongame Wildlife in California: Results of Surveys on Deer, Birds, and Mono Lake.* Unpublished study. University of California–Davis.

Lyke, A.J. 1992. *Multiple Site Trip Generation and Allocation: A Travel Cost Model for Wisconsin Great Lakes Sport Fishing.* Draft Ph.D. thesis. Madison: University of Wisconsin.

Milliman, S.R., R.C. Bishop, and B.L. Johnson. 1987. Economic Analysis of Fihery Rehabilitation Under Biological Uncertainty: A Conceptual Framework and Application. *Canadian Journal of Fish and Aquatic Science* 44: 289–97.

Milliman, S.R., B.L. Johnson, R.C. Bishop, and K.J. Boyle. 1991. The Bioeconomics of Resource Rehabilitation; A Commercial-Sport Analysis for a Great Lakes Fishery. *Land Economics* July.

Mitchell, R.C. and R.T. Carson. 1984. *An Experiment in Determining Willingness to Pay for National Water Quality Improvements.* Washington, D.C.: Resources for the Future for U.S. EPA.

U.S. EPA (Environmental Protection Agency). 1995. *Revised Assessment of Compliance Costs Resulting from Implementation of the Final Great Lakes Water Quality Guidance.* EPA-820-95-010. U.S. EPA. Washington, D.C.: U.S EPA. Office of Water.

Walsh, R.G., D.M. Johnson, and J.R. McKean. 1988. *Review of Outdoor Receration Economic Demand Studies with Nonmarket Benefit Estimates.* Colorado Water Resources Research Institute Technical Report No. 54. Fort Collins: Colorado State University.

———. 1990. Nonmarket Values from Two Decades of Research on Recreation Demand. In V.K. Smith and A.N Link (eds.) *Advances in Applied Microeconomics,* Vol. 5. Greenwich, Connecticut: JAI Press.

16

Economic Analysis: Benefits, Costs, Implications

Richard D. Morgenstern and Marc K. Landy

Given the diverse array of issues considered in this volume and the uneven application of economic analysis, it is difficult to draw any general conclusions. Despite these limits, a number of cross-cutting questions can usefully be explored:

- Did the economic analyses help increase the benefits and/or lower the costs of the rules?
- Did the value of the resulting efficiency improvements exceed the costs of conducting the analyses?
- What overall conclusions can be drawn about the practice of economic analysis at EPA?
- How might one improve both the conduct and use of economic analysis at EPA?

OVERVIEW

It is of obvious importance to question whether the estimated benefits of a regulation parallel the estimated costs. As with any such sweeping inquiry, the devil is in the details. Which regulations are selected for study, what methods are used for estimating benefits and costs, what time periods are examined—all affect the ultimate calculus. Studies by U.S. EPA (1987) and by Hahn (1996) have tallied the benefits and costs of selected groups of rules in an attempt to draw inferences about whether Americans are getting their money's worth from environmental investments. Using EPA data, both the EPA and the Hahn studies concluded that in the aggregate

MARC K. LANDY is Professor of Political Science at Boston College and a senior fellow at the Gordon Public Policy Center at Brandeis University.

the benefits exceeded the costs of the rules studied, although many of the individual rules did not demonstrate net benefits.

If one applied such an approach to the rules studied in this volume, the aggregate benefits of the rules studied would clearly outweigh, by a considerable margin, the aggregate costs. In this sample, the benefits of three rules alone—lead in gasoline, lead in drinking water and CFCs—far exceed the costs of all twelve rules combined. Yet this approach is incomplete, for it cannot answer many questions. For example, how representative is the sample of rules studied? What are the implications for the rules for which it was not possible to generate credible benefit estimates in excess of cost? What do we know about the nonquantifiable benefits? From an economic efficiency standpoint would we have been better off scaling back or even eliminating some of these rules?

Even without addressing these questions, the case studies do permit some generalizations. In all the cases examined, the economic analyses did, in fact, contribute to improving the rules: the value of such improvements likely dwarfs the one-time cost of conducting the analyses. Despite this finding, in many instances the analyses did not prove terribly useful to decisionmakers. It is certainly fair to say that significant opportunities for rule improvements were missed in many cases. Numerous factors, including politics, could explain this situation.

In some instances the economic analyses served little purpose, as the underlying risk assessments on which they were based were weak. In other instances, the policy options explored were so narrow that the resulting economic analyses were of limited relevance to the policy process. While there was considerable range in the quality as such of the economic analyses, none was fundamentally flawed and a number were truly outstanding.

Reforming the regulatory process in order to improve both the conduct and use of economic analysis is a challenging task. A number of recent congressional proposals would require an elaborate and time-consuming peer review process and, more importantly, would superimpose economic decisional criteria onto existing statutes. Further, they could subject the entire process to judicial review, thereby making the courts the final arbiter of both the quality and the appropriate use of economic analysis.

Although none of these proposals has yet been enacted into law, they have stimulated a great deal of debate about the appropriate role for economic analysis in environmental decisionmaking. One of the findings from this volume is that *significant opportunities exist to improve the conduct and use of economic analyses within the exisiting legal framework.* Specifically, a more open process for rule development, earlier consultation with stakeholders for purposes of both designing and conducting the analyses, and consideration of a broader set of policy options to be studied might improve the overall quality of the final regulations. A model for such reforms is found in the "scoping" process for conducting environmental impact statements under the National Environmental Policy Act.

CONTRIBUTION OF ANALYSES TO THE RULES STUDIED

In reviewing individual rulemakings, it is often difficult to determine whether the economic analysis or other influences are responsible for particular aspects of a rule. Sometimes the role of the analysis is clear. As argued by case study authors Ronnie Levin and Albert Nichols, economic analyses played pivotal roles in determining the level of stringency and the implementation schedules in both the lead in drinking water and lead in gasoline rules. At other times, support for the specific parts of a rule can be found in the economic analysis as well as in the public comments or staff studies not part of the formal economic analysis. Often, multiple influences are at work and even the most informed "insider" is unable to make a clear determination about cause and effect.

Notwithstanding this somewhat murky process of determining the source of rule improvements, Table 1 contains examples cited by the case study authors of specific improvements in the rules associated with the twelve economic analyses examined in this volume. In most cases the case study authors do not argue that the changes occurred solely because of the economic analyses. Nor do they argue that the same or similar changes might not have occurred for other reasons. But the case study authors who, in virtually all cases, were themselves directly involved with the rule development, believe that the economic analyses played a significant role in stimulating these specific changes.

In all twelve cases, the economic analyses supported specific cost-saving rule improvements. A number of these changes are also cited for enhancing the administrative feasibility of the rules and increasing the likelihood of compliance. Typical rule changes that lead to cost reductions include marginal increases in emissions or discharge rates, diminished number of source categories or products under control, and substitution of monitoring of indicator pollutants for broad scale monitoring requirements. In the case of lead phase-down, the economic analysis is cited as the source of the innovative banking and trading system.

Five of the case study authors attribute specific benefit enhancing measures to the economic analysis: lead in gasoline, lead in drinking water regulations, reformulated gasoline (RFG), Navajo Generating Station, and the organic chemicals effluent guidelines. In these cases, typical rule changes included decreased emission or discharge rates, expanded

Table 1. Rule Improvements Associated with Economic Analyses.

Rule	*Improvements leading to increased benefits*	*Improvements leading to decreased costs*
Asbestos		Reduction in number of products subject to ban; tying of phase-down schedules to costs of substitutes
CFCs		Adoption of tradable CFC consumption permits (within the U.S.); permitting of production trades internationally (allowed by the Montreal Protocol); use of consumption taxes (within the U.S.) to further control CFC use
Farmworker protection		Tying of field re-entry to economic impact on crops
Great Lakes		Scaling back of numeric criteria; selection of cost-effective implementation procedures
Inspection and maintenance		Supported decision to lengthen interval between inspections from 1 to 2 years
Municipal landfill		Size, rainfall exemptions (later reversed in courts); relaxation of monitoring requirements
NGS/visibility	Adoption of more stringent controls (90% vs. 70%)	Relaxation of averaging time
Organic chemicals	Encouragement of VOC air emissions control	Scaling back of requirements for some heavily impacted segments
Lead in gasoline	Adoption of more stringent standard; earlier implementation	Banking and trading
Lead in drinking water	Adoption of more stringent standard	Phasing in of monitoring and implementation; extended timing for lead-pipe replacement
RFG	Greater VOC and NO_x control	Reduced costs for toxics (offsetting higher costs for controlling VOC)
Sludge		Scaling back of numeric criteria; reduction in number of pollutants regulated

coverage of source categories, and shortened time period for emitters to come into compliance. In the two lead rules, reformulated gasoline, and the Navajo Generating Station the economic analyses clearly contributed to more stringent standards. Although the limited benefit-cost analysis in the organic chemicals effluent guideline did not support particularly stringent water discharge rates, the analysis did help to identify a potential cross-media transfer of volatile organic compounds from water to air.

In turn, this stimulated efforts by the agency to address these air emissions in the organic chemicals effluent guideline and in subsequent regulations issued under the Clean Air Act.

In the cases of asbestos and CFCs, even if the economic analyses did not explicitly strengthen the rule, they did help support the rule in the face of OMB and public scrutiny. In the case of CFCs, the analysis demonstrated that if no action was taken, the projected growth in demand would cause significant increases in stratospheric ozone depletion, and that relatively low-cost alternatives did exist for many uses. The real impact of the analysis was on the government's initial decision to adopt the Montreal Protocol to the Vienna Convention. In the case of asbestos, the economic analysis identified many products that could be removed from the marketplace at low to moderate cost. Apparently for reasons of clarity and simplicity, as well as some skepticism about the cost analysis on the part of EPA management, the agency chose to delay some of the low-cost reductions and, instead, proceeded to ban some of the higher-valued uses in the early phases.[1] The fact that the rule was later successfully challenged in the courts, however, is not attributable to a failure of the economic analysis.[2]

Even a cursory examination of the rule modifications associated with the RIAs suggests that, overall, they are quite significant. While it is not feasible to develop systematic estimates of the dollar values of either these increases in benefits or decreases in costs, in a number of cases the improvements were quite sizable, amounting to a third or more of the benefits or costs of the rule. In both lead rules, for example, some might argue that most of the benefits are attributable to the economic analysis.

ASSESSING ASSESSMENT: THE BENEFITS AND COSTS OF ENVIRONMENTAL ECONOMIC ANALYSIS

One way to consider the value added of economic or regulatory impact analyses is to consider them in the same benefit-cost framework they embody. Accordingly, we begin with a discussion of the benefits and costs of conducting economic analyses of proposed environmental rules, with specific reference to the twelve case studies.

The Benefits

As discussed in Chapter 3, economic analyses of proposed environmental regulations provide a means for comparing alternative environmental goals as well as a basis for comparing alternative means of reaching the same goal. They also provide a framework for collecting and organizing information. The template character of an economic analysis enables the decisionmaker to determine the adequacy of the information collected and to identify what important information is missing. In that way, analysis is able to shed light on both our knowledge and our ignorance regarding elements of the proposed action. The theory, assumptions, methods, and procedures of a well-executed analysis will be clearly explained and linked to the results. This transparency can improve the public's understanding of the rationale for the proposed action and thus increase the accountability of government.

A 1987 report (U.S. EPA 1987) by EPA indicated five specific areas where RIAs have influenced the development of regulations:

- Guiding the regulation's development
- Adding new alternatives
- Eliminating non-cost-effective alternatives
- Adjusting alternatives to account for differences between industries or industry segments
- Supporting decisions

Economic analysis of proposed regulations can also lead to improved understanding of the implications of federal regulatory activity for officials of all three branches of government. As Paul Portney has argued, regulators and members of the legislative, judicial, and executive branches "would all know less about regulation than they know now were it not for the development of ... a tradition of scrutinizing regulatory proposals (Portney 1984)."

The broader educational value of economic analysis is particularly important when one recognizes the ongoing nature of most environmental decisions. Rules are written, implemented, enforced, often litigated, and sometimes revisited by Congress. As a result, there are many opportunities for economic analyses to enter the decision process. A prominent example can be found in the case of CFCs. The congressional decision to impose a tax on the production and importation of CFCs was made several years after the initial promulgation of the rules. That decision was based, in part, on the data and analysis contained in the RIA on the profitability of the relatively limited CFC production allowed under the regulation.

The Costs

Like the costs of regulation itself, the costs of regulatory analyses are both direct and indirect. The direct costs of regulatory analysis arise from the conduct of the benefit-cost analysis and related studies of both proposed and final regulations, and the review of those studies by EPA personnel regardless of whether the studies are conducted by EPA employees or by outside contractors. Direct costs should also include the subsequent review of the studies by OMB and other relevant agencies. For example, the Department of Agriculture spent considerable effort reviewing the potential benefits and costs of the farmworker rule.

The direct costs readily lend themselves to quantification. At EPA, typically a large proportion of the technical work on economic analyses has been performed by outside contractors, under the direction of agency staff. Interestingly, contractors who conduct these studies argue that a remarkably large portion of the costs are associated with gathering information rather than from conducting actual analysis. The costs of conducting the economic analyses vary enormously—sometimes by more than an order of magnitude. The costs depend on the nature and breadth of the benefits and cost assessments conducted, the state of the existing knowledge on physical effects, exposure, and other factors. Often it is feasible to simply transfer results from one particular setting to another, which typically involves a limited amount of original data gathering and analysis. In fact, most of the benefits analyses described in this volume involve some benefits transfer although, as discussed below, benefits transfer often can be a source of controversy.

The situation is similar on the cost side of the equation. For example, the existence of a previously developed linear programming model of petroleum refineries made it relatively inexpensive to develop detailed estimates of the costs of the lead phasedown and reformulated gasoline rules. Where previous work does not exist, the costs of an economic analysis will depend, in part, on the nature of the industry being regulated: costs may be more difficult to estimate in an industry with many heterogenous firms than in one with a small number of relatively similar facilities, or where multiple industries are involved, as with the Great Lakes Water Quality Guidance. Costs may also vary depending on the technological complexity of the policy options being considered.

Table 2 presents estimates of the costs of preparing an RIA, ranging from $731,000 (Portney 1984) to $1.035 million (U.S. EPA 1995b). All estimates include extramural and intramural costs and are expressed in 1995 dollars. Table 3 presents estimates for selected RIAs underlying the case studies in this volume. The average of these selected RIAs is about $2.5 million. Not including the asbestos RIA, the average dropped to $1.6 million. The least expensive RIA was for the farmworker rule that was performed largely on an intramural basis for a total cost of $79,000. The most expensive one—$8 million—was for the asbestos rule, which involved extensive extramural and intramural effort over a ten-year period.

The indirect costs of regulatory analyses take several forms. One includes the costs of any delay associated with their conduct. In cases where the rule truly generates net social benefits, the cost of these delays

Table 2. Estimated Costs of a Regulatory Impact Analysis. ($1995)

Source	*Extramural costs*	*Intramural costs*	*Total costs*
Portney (1984, 228–33)	$585,000	$164,000	$731,000
U.S. EPA (1987, 6–5)	$662,000	$325,000	$987,000
U.S. EPA (1995b)	$875,000	$160,000	$1,035,000

Table 3. Costs of RIA Preparation: Selected Cases. ($1995)

Rule	*Extramural costs*	*Intramural costs*	*Total costs*
Asbestos	$4 million	$4 million	$8 million
Farmworkers	$79,000	$100,000	$179,000
Lead in gasoline	$715,000	$800,000	$1,515,000
Municipal landfill	$2.3 million	$1 million	$3.3 million
Navajo Generating Station	$575,000	$200,000	$775,000
Reformulated gasoline	$1.25 million	$850,000	$2.1 million
Sludge	$1.25 million	$600,000	$1.85 million

Note: Intramural costs are assumed to be $100,000 per full-time equivalent employee.

could be considerable. Another is the potential increase in uncertainty concerning the outcome of a rulemaking created by the RIA. While industry often sees itself as benefitting from RIAs, there are a number of cases—five of the twelve examples in this volume—where industry ultimately faced more stringent regulation as a result of EPA's analysis of the benefits and costs.

Once an economic analysis is performed, it tends to have a life of its own. Even when an analysis makes a case for strong regulation, it can create headaches for EPA management by pointing out some contrary information or by making transparent certain costs that, even if clearly outweighed by benefits, are still seen as burdens to some group in the society. Such "on the one hand and on the other hand" sorts of statements provide ammunition for the policy's opponents to use in court or in the media, particularly when taken out of context.

If the economic analysis is not used to help make decisions but is ignored or simply used to justify decisions made on political grounds, public cynicism about the efficiency of government regulatory efforts may increase. However, the importance of this potential cynicism issue should not be overstated. Notwithstanding the fact that some statutory provisions may set the agency up for failure by promising unachievable goals, economic analyses have the potential to decrease public cynicism as well as to increase it. If a tradition of well-designed and well-executed economic analyses were established—as in the case of the Congressional Budget Office—the public would have less reason to assume that regulatory decisions are being made on arbitrary grounds and might become more trusting of the notion that decisions are made on the basis of sound analysis.

Net Benefits

The relatively small direct cost coupled with real doubts about claims of large indirect costs means that it would not take too much in the way of increased benefits or decreased costs of a rule to enable economic analyses to pay for themselves. Given the modest one time costs of these analyses compared to the annual costs of the rules—over $100 million—a reduction in costs or an increase in benefits in the final rule of well under one percent—probably closer to one-tenth of one percent—could easily justify the cost of an economic analysis. For the group of cases in this volume, the present value of the annual increase in benefits and/or decrease in costs of the rules attributable to the economic analyses clearly outweighs the one-time cost of carrying out the analyses.

CROSSCUTTING OBSERVATIONS

Should advocates of economic analysis take comfort in the finding that economic analysis did, in fact, measurably contribute to the improvement of environmental regulations studied in this volume? While some may debate whether the glass is truly half empty or half full, the view of the case study authors and reviewers involved in this project is that despite the documented contributions of the economic analyses, overall, and in the context of existing statutes, many opportunities were missed for significant rule improvements. It is undoubtedly so that even apart from some of the legal constraints on using economic analysis, implementation concerns along with bureaucratic and political factors all play a role. Nonetheless, it is worth trying to see to what extent inadequacies of the economic analyses effectively limited their usefulness in decisionmaking. Accordingly, this section addresses a number of possible shortcomings.

Inadequacies in the Risk Assessment

Quantitative risk assessment typically consists of four phases: hazard identification, dose-response analysis, exposure analysis, and risk characterization. The environmental sciences relevant to risk assessment include epidemiology, toxicology, ecology, and exposure assessment. Scientific uncertainties in these fields derive from both data limitations and the incomplete nature of the models themselves. Efforts to identify hazards can be impaired when meager experimental or human data exist to indicate the potential for hazard. Dose-response analyses can be flawed for lack of data or inappropriate modeling assumptions. Exposure analysis can be hampered by the absence of quality monitoring data or by the inability to model patterns of airflow, water transport, pharmacokinetics, metabolism, or human activity. Uncertainties in the first three phases—hazard identification, dose response, and exposure analyses—can be magnified in the final phase, risk characterization.

When the uncertainties about any of the four stages of risk assessment concern the magnitude of a quantity that can be measured or inferred from plausible assumptions, reasonable estimates of the uncertainties can often be developed. However, when the uncertainties stem from a lack of knowledge of basic cause and effect relationships, it is virtually impossible to develop quantitative estimates. In such situations one must rely on scientific theories rather than data driven or even model driven assessments. While theories can be quite useful in many situations, they generally provide limited guidance for developing quantitative estimates of risks and benefits

Economic analysis of the benefits of environmental policies typically depends on the existence of a well-done risk assessment. In general, economic analysis only adds value when scientists have laid the groundwork by developing credible quantitative estimates of the underlying physical relationships. Risk assessments based on biological evidence of human exposure can be especially compelling and can reinforce the results of an economic analyses. For example, the fact that the neurotoxic properties of lead had been so well researched and that human exposure was so clearly linked to a well-defined biomarker (blood-lead levels) certainly strengthened the findings of the subsequent economic analysis.

By the same token, when the underlying risk assessment is weak, then the economic analysis will also be weak. Even a cursory review indicates that in several of the cases chronicled in this volume the available scientific information was seriously inadequate to develop a comprehensive and robust benefit-cost assessment. One of the reviewers for this volume, in commenting on the case study on the Navajo Generating Station, asked, in effect, whether there is any scientific data so bad that one shouldn't base an economic analysis on it. In fact, a National Academy of Sciences study had cited Navajo as a "possible cause" of visibility impairment in the Grand Canyon. Notwithstanding the uncertainties, the economic analysis conducted by EPA assumed a strong link between the two. In contrast, the median value of the industry-sponsored study was zero, reflecting these scientific uncertainties. Our reviewer suggested the uncertainties were so great that analyzing the economic value of those improvements was not a useful exercise: no matter what the economic analysis concluded, it would always be subject to the criticism that the underlying science did not support the results.

Nor is the issue of inadequate science limited to the question of the underlying dose-response function. Critics have argued that omitting nonpoint sources, key components of water quality, was a critical flaw in the economic analysis underlying the Great Lakes Water Quality Guidance. That omission left unchallenged the widely held view that improvements in water quality associated with reductions in point source discharges would be overwhelmed by continued increases in nonpoint source discharges. It also left unchallenged the view that the marginal costs of controlling nonpoint sources were lower than those of point sources. Facing the criticism that focusing on industrial sources was like "looking under the lamp post for one's keys," along with the suspicion that nonpoint sources were, on the margin, less costly to control, the agency was forced to significantly scale back the final rule compared with the proposal.

A third example involving a seriously inadequate risk assessment was the farmworker rule. Because of limited data on baseline health conditions, exposure, and potential confounding risk factors, EPA was not able to demonstrate improvements in health status associated with extended field re-entry restrictions for farmworkers. Some might argue that the basis for stringent controls was obvious—especially for a disadvantaged group like farmworkers. Malnutrition, lack of health care, poverty, exposure to other pollutants, and chronic infections are all widely understood to be important determinants of health. Yet, as case study author Louis True argues, the inability to demonstrate significant improvements in health from the farmworker rule clearly weakened the final regulation.

In all these cases, one might argue that the case for stringent regulation was obvious, and weak science or inadequate risk assessment should not have affected the rule. That in itself is a policy judgment. The fact remains that in only one of these three cases—the Navajo Generating Station—did the rule proceed in relatively stringent form. However, as noted in that study, the Navajo regulation was bolstered by the confluence of favorable politics and the finding that higher control levels were more cost-effective. In both the Great Lakes and farmworker cases, deficiencies in the risk assessment clearly reduced the stringency of the final rules.

In all three of these rules, the best one can say is that the regulatory decisions represent compromises. We do not have great confidence that these rules rest on the strongest scientific footing, and certainly we do not have confidence that the economic analyses had a particularly significant impact on the outcomes. This suggests a simple observation: in cases where the underlying science or risk assessment is extremely uncertain, a full-scale economic analysis often serves little useful purpose. In such cases, the agency might be well served to abandon any pretext that it is possible to assess whether or not benefits exceed costs. Rather, it might be better to conduct a more limited cost-effectiveness analysis and focus on a frank discussion of the science, including the uncertainties. At least implicitly, one might assess the costs of inaction as well as the costs of action. Such an approach might enhance the agency's credibility in cases of seriously inadequate scientific information.

Limitations of the Economic Analysis

What can be said about the technical quality of the economic analyses conducted by EPA? Does a review of the twelve cases suggest that the agency has used the available economic science to conduct technically sound and robust economic analyses? What areas seem to be most in need of improvement?

It is appropriate to begin any discussion of technical quality with the observation that truly "perfect" studies do not exist: one can always conduct more and better analysis and one can always find fault with individual studies. It is also true that some of the older economic analyses chronicled in this volume do not reflect all the newest thinking laid out in the 1996 OMB Guidelines, such as the treatment of nonquantifiable benefits (and costs), distributional issues, and other issues.

In his review of all sixty-one RIAs conducted by EPA between 1990 and mid 1995, Hahn (1996) observed "...wide variation (in quality) from very poor to very good." For the rules reported in this volume, which were not chosen as a representative sample, the views of the authors and peer reviewers are that *per se* economic analyses are fundamentally sound. This does not deny room for improvement. Yet, the analyses reported here are not guilty of such gross errors as double counting, confusion of costs and benefits, equating of transfer payments with economic costs (or benefits), failure to discount, and so forth. If we were grading these economic analyses, most of them would get grades of "B" or "C." At least four of the twelve deserve "A's:" lead phase down, CFCs, lead in drinking water, and asbestos. None is guilty of the types of errors that would merit a "D" or an "F."

Having stipulated that the studies are generally credible, we focus on two key areas of concern: the development of baselines and the estimation of benefits, including the use of the results of one study in another application (so-called benefits transfer).

Baselines in Economic Analysis

The baseline refers to the status of environmental and economic conditions in the absence of a contemplated policy intervention. The determination of a baseline is central to the conduct of an economic analysis in that it poses the fundamental question: how does the proposed regulation compare to what would have happened anyway? Knowing what would have happened anyway involves predicting the future, an inherently difficult task. The lead in gasoline analysis was successful because it focused on a relatively short time period (less than a decade). Further, the analysis benefitted from the existence of a large knowledge base on the stock and turnover of the auto fleet, driving patterns, and other factors. Thus, it was possible to credibly forecast what lead emissions would be in the absence of new regulation. In contrast, the case of CFCs involved forecasting emissions produced by a product with rapidly changing uses, over a period of decades. Small changes in growth rate assumptions led to large changes in emissions fifty or seventy-five years later.

The municipal landfill case is another example that illustrates the difficulty of developing a credible baseline. Like lead phase down, this case involved relatively short time periods. Unlike the lead rule, however, current control levels were changing quickly as states adopted regulations on their own. When development of the federal regulation began, only a few states had issued their own rules. Three years later, when the rule was actually promulgated, many states had adopted rigorous requirements. Failure to recognize these new state requirements in the economic analysis led to an overstatement of the costs (and benefits) of the federal rule.

Certainly, baseline issues are not limited to the estimation of benefits. Cost issues can also pose major problems. In the case of lead phase down, EPA was dealing with a relatively mature product, supported by a wealth of industry and public sector analyses documenting the cost of both leaded and unleaded gasoline. In contrast, in the case of CFCs, the cost and availability of CFC substitutes were not well understood, at least for a number of the key product categories. In the case of landfills, at least the engineering costs of the technical upgrades, such as liners and groundwater monitoring, were well understood. However, since information was so limited on the extent of implementation of the evolving state regulations, total costs were highly uncertain.

As these examples suggest, determining the "way the world would look absent the regulation" is often a difficult task. What else is going on in the market other than responses to the proposed rule? What level of compliance with current regulations should be assumed? It may be especially difficult to decide what should be in the baseline when there are multiple regulatory actions under consideration, or when new rules are being promulgated before existing rules are fully implemented. In general, when more than one baseline appears reasonable or the baseline is highly uncertain, alternative baselines should be examined as a form of sensitivity analysis.

Benefits Estimation in Economic Analysis

As regards benefits estimation, some of the greatest problems arise in the area of benefits transfer. While it may seem straightforward, applying the results of one study to a wholly different set of circumstances can be quite difficult. In the case of Navajo Generating Station (NGS), for example, the principal studies on the valuation of enhanced visibility were performed in the summertime and applied—without modification—even though most of the visibility improvements from reducing the NGS SO_2 emissions were predicted to be in the wintertime, when relatively few people visit the Grand Canyon. In the landfill rule, property value studies involving leaking landfills were used as one method for estimating the benefits of stricter standards for new landfills. Yet, many of the original property value studies were not able to distinguish between the general nuisance aspects of a landfill and the fact that it may have leaked. Thus, the analysis was criticized as overstating the benefits of the stricter standards. The lesson is clear: the devil is in the details. Greater attention needs to be paid to the specifics of individual studies before attempting to transfer the results to other applications.

Developing monetary estimates across a broad range of effects is vital to the conduct of economic analysis. While the literature has expanded considerably over the past twenty years, it continues to evolve. In the air pollution field, key areas to target for more research include chronic health effects and material damages. For water issues, the valuation of ecological effects is paramount. For waste issues, the valuation of groundwater—the principal issue in the municipal landfill rule—is the single largest unknown. Especially large uncertainties surround the estimation of nonuse benefits in general.

One obvious way of evaluating the quality of an economic analysis is to compare the predicted values with actual (*ex poste*) data. Even on the cost side, however, where information is more readily obtainable, this can be a difficult task. There is undoubtedly some upward bias of *ex ante* cost estimates relative to actual because neither firms nor regulators can predict cost-saving innovations that will likely occur once a real effort is made to comply with the rules. Of course unforseen higher costs can also occur, but these are probably less common.

Three of the cases in this volume developed comparisons that can be used to judge the accuracy of the *ex ante* cost estimates. For CFCs, case study author James Hammitt shows that while early analysis considerably overstated marginal control costs, estimates developed for the final RIA proved to be quite accurate and may even have underestimated costs somewhat. For the lead phase down in gasoline, case study author Albert Nichols showed that the decline in sales of leaded gasoline proceeded much more quickly than anticipated by the RIA: actual 1990 sales of leaded gasoline were only one-twentieth the amount predicted in the RIA. He also found that banking of lead rights by refiners was even more popular than expected. Overall, he concluded, the costs were probably lower than predicted, based on the fact that refiners reduced their production of leaded gasoline so much more rapidly than anticipated. For reformulated gasoline, case study authors Robert Anderson and Richard Rykowski found that EPA's predicted cost differences were in line with those observed in the market in 1995.

Other Issues

Beyond the issues of baseline and benefits estimation, several other areas deserve mention. The RIAs on lead in gasoline, lead in drinking water, stratospheric ozone, asbestos, reformulated gasoline, and others were all based on detailed modeling frameworks that were able to address a broad array of policy issues. The existence of these modeling frameworks made it possible to conduct sensitivity analyses to address uncertainties, and to quickly incorporate new information or analyze new issues that arose during the policy deliberations. Thus, the availability of a modeling framework detailed enough to address new and sometimes subtle policy issues can enhance the likelihood that the analysis will play a significant role in decisionmaking.

The value of such detailed assessments is particularly great when dealing with site-specific (as opposed to national) issues. Yet, a review of the cases in this volume suggests that there are additional challenges to conducting economic assessments at the local or site-specific level. When a "real" as opposed to a "typical" site is involved, much more detailed knowledge is needed in order to develop credible estimates.

Both the Great Lakes water quality case and, to a lesser extent, the NGS case illustrate this point. In those instances, the rules concerned particular geographic areas. Questions arose over specific sources targeted for control. Did point-source discharges from large industrial facilities really represent the largest or most cost-effective control option in the Great Lakes? Similarly, was the Navajo Generating Station really a significant contributor to the summertime haze over the Grand Canyon? Certainly the claim that the bulk of the problem was associated with pollution transported from the Los Angeles Basin needed to be addressed. The point is straightforward: national analyses need only be right on average; when dealing with specific locales, greater detail is required.

Finally, one can look across the set of analyses presented here and see differences in approach among the different EPA program offices. For example, in the asbestos rule, despite its strong statutory basis to consider "unreasonable risks," the Office of Toxic Substances (now the Office of Toxic Substances and Pollution Prevention) was reluctant to conduct a full benefit-cost analysis. In contrast, benefit-cost analysis of major rules is more routinely conducted in the Air Office. Greater methodological standardization, as well as the use of a common set of assumptions across offices, seem like modest but logical improvements. Greater transparency in presenting results would also be desirable.

A number of the RIAs studied clearly benefited from technical support provided by more than one EPA office. For both lead rules, for example, the program office sought the specialized knowledge available in the agency's Policy and Research offices in conducting the RIA. In both cases, the result was clearly strengthened by this input. In fact, many possibilities exist for tapping specialized talents in both single media and cross-media offices. The quality of an RIA may be significantly enhanced by drawing on the broader EPA talent pool even if the skills lie outside of conventional bureaucratic boundaries.

Too Limited Policy Options

What can be said about the options selected for analysis by the agency? Did the options analyzed reflect a broad range of approaches to the environmental problem? Did they address the truly contentious issues? Did they focus on engineering-type solutions or did they examine what might be termed behaviorial or economic policy options, such as economic incentives?

Just as the use of average as opposed to marginal analysis can be misleading, so too the omission of relevant policy options from the analysis can obscure key differences among alternatives. Perhaps the clearest example of the failure to define and analyze a broad set of policy options is the 1992 inspection and maintenance (I/M) rule, which mandated procedures for states to follow in conducting periodic inspections of auto emissions. In that case the agency did not give serious attention to approaches other than the IM240 model, a testing procedure that required centralized testing centers and use of a complex apparatus to measure emissions under "load" as opposed to the current procedure that measures emissions only at "idle." That proved to be a serious mistake. As case study author Todd Ramsden argues, the RIA did not address many of the more contentious issues, including the effectiveness of decentralized (test-and-repair) networks versus the effectiveness of centralized (test-only) networks; the effectiveness of IM240 compared to the effectiveness of other loaded-mode, steady state tests; and the credit given to I/M programs that differed from the model program. Had EPA considered and analyzed other options in the original RIA, the need to revisit the issue in 1993, with its attendant political and economic costs, might have been avoided. As Ramsden notes, "Viewed perhaps cynically, this indicates that the policy choice selected the form of the RIA, not the other way around."

A second example of a failure to consider broad-scale options involves the case of reformulated gasoline (RFG). As case study authors Robert Anderson and Richard Rykowski argue, the RIA is long on engineering analysis and short when it comes to what might be termed economic or policy analysis. The RIA, they note, "is virtually devoid of analysis of many items called for in the OMB Guideline, (including) market-based alternatives, impacts on competitiveness and trade, and (other factors)."

Specifically, the RIA did not consider the possibility that areas that opted in to the RFG program might later decide to opt out. Anderson and Rykowski argue that an RIA conducted in the full spirit of the OMB Guidelines might have anticipated this problem, presented an analysis of the potential impacts of such an alternative, and led to opt-out provisions in the final rule (such as preventing areas from opting out without providing advance notice).

Similarly, the RIA did not address the problems associated with imported gasoline. As Anderson and Rykowski note, "EPA could have anticipated the problems that foreign refiners would face....(I)n retrospect...(as a later) World Trade Organization decision (found), the EPA rules were 'a disguised restriction on international trade.'"

An analogous situation arose in the analysis conducted for the sewage sludge rules. Here the issue was the failure of the agency to conduct a benefit-cost analysis at all, even when credible information was available to do so, rather than any inherent limits of the options considered. On the basis of the cost-effectiveness analysis, the agency endorsed beneficial use rather than disposal of sewage sludge. As case study authors Mahesh Podar, Susan Burris, and Robert Raucher argue:

> If the estimated benefits were monetized and decisions were made...(on the basis of) net...benefits, welfare-based policy conclusions would differ.... Although surface disposal is the costliest portion of the rule, it appears to have, by far, the smallest benefits....(Overall)...it appears that the rule is overly stringent for surface disposal, potentially shifting an undesirable quantity of sludges to other, riskier disposal or reuse options.

Other cases illustrate the same basic themes: if insufficient attention is paid to developing policy options and alternatives, the agency runs the risk of overlooking important alternatives. And if the analytical framework is not designed to incorporate the broad-scale implications of the policy options, these implications may be overlooked (or ignored) by decisionmakers.

CONCLUDING REMARKS AND RECOMMENDATIONS

In the previous sections we have explored three possible explanations for the positive but limited effectiveness of economic analysis in regulatory decisions:

- The underlying scientific and risk information was so uncertain that it provided an insufficient basis on which to conduct an economic analysis.
- The economic analysis itself was technically flawed in one or more critical ways.
- The economic analysis was not designed to address a sufficiently rich array of policy options and was thus rendered irrelevant to actual policy and regulatory decisions.

Our review suggests that all three explanations have currency, although the first and certainly the third seem to be the most important. That is, in some instances the economic analysis served little purpose as the underlying risk assessment on which it was based was so weak. In other instances, the policy options explored were so narrow that the resulting economic analyses were of limited relevance to the policy process.

Why are the regulatory options considered by the agency often so narrow? Legal restrictions do not necessarily inhibit consideration of broader options, as the I/M and RFG cases demonstrate. Certainly time and resource constraints are part of the explanation. Yet, in at least some of the examples cited here, resources do not seem to have been the critical factors. In one instance, namely reformulated gasoline, the case study authors argue that the process by which options were developed—namely, regulatory negotiation—itself served to exclude from consideration certain options. Specifically, they attribute the exclusion of the foreign refiners from the regulatory negotiation as the reason there was no "pressure" to consider foreign trade issues in the rulemaking. From a broader perspective, it may be that the current rule development process, which does not generally involve outside parties until the regulation is formally proposed and until a good deal of analytic work has already been conducted, discourages early identification of a full set of options.

A variety of institutional factors probably explain the failure to consider a sufficiently broad set of options. Apart from the brief discussion of legal and cultural issues in Chapter 2, it is beyond the scope of this volume to consider the full set of institutional and historical factors at play. Within the confines of existing statutes, however, probably the most practical way to address concerns about the breadth of the options and the depth of the analyses is to examine the *procedures* for developing and analyzing policy options. In this concluding section we focus on the process by which rules are developed as a significant and inherently remediable part of the problem. In the spirit of recent legislative actions that attempt to bring economic analysis into the mainstream of agency decisionmaking, such as the Safe Drinking Water Act, we propose three areas for improving the conduct and use of economic analysis at EPA.

Top down management support for generating adequate information to conduct valid risk and economic based decisions is essential. Former EPA Administrator William Ruckelshaus made "risk assessment" and "risk management" major themes of his second term. His leadership and support was a major factor in the promulgation of the rule phasing lead out of gasoline, which as case study author Albert Nichols argues, was driven by the analysis, not the statutes or the politics. It is clear that if EPA leaders really believe the oft uttered rhetoric about sound science and sound economics, they must be prepared to demonstrate those beliefs in making specific regulatory decisions.

Despite recent reforms, the current rulemaking process is still accurately characterized as a "bottom up" approach. Junior staff within program offices typically initiate the analysis, often without the benefit of significant involvement of senior management and often without the benefit of input from experts in industry or nongovernmental organizations.

As Victor Kimm, a twenty-five year EPA veteran who served in several program offices, has noted:

> My experiences...suggest that the ongoing participation by senior decisionmakers throughout the process in formulating options, deciding when available information is adequate to support policy choices, and in mobilizing the political leadership to bring closure on critical issues, offer the best chance for orderly progress in developing major regulations (Kimm 1996).

Economic analysis should be initiated at the beginning of the rulemaking process. Although the time it takes to issue an EPA regulation is highly variable—the asbestos rule took ten years to develop—on average it takes about three years from the date work begins until the time a rule is promulgated and appears in the Federal Register as a legally binding regulation (U.S. EPA 1995a; Kerwin and Furlong 1992). For major rules (that is, those with annual costs in excess of $100 million), it typically takes almost a year longer (Kerwin and Furlong 1992).

Although the practice is not uniform throughout the agency, in many offices economic analyses are only initiated after the rulemaking is well underway. Sometimes it is initiated after a preferred regulatory option has been developed, at least through the mid levels of agency management. In those situations it is difficult for economic analysis to make a major contribution. In fact, the economic analyst runs a risk of delivering "bad news" to his or her bosses if the analysis fails to support the preferred alternative. One obvious remedy to this problem is to initiate the economic analysis earlier in the rulemaking process.

Economic analysis should be conducted in a more open manner. Outside experts and stakeholders should be consulted on the design of the analysis to be conducted and the options to be considered. Although the practice is not uniform throughout the agency, most offices rely exclusively on internal experts to decide on the analysis to be conducted and the options to be considered. In most cases, it is not until a rule is formally proposed in the Federal Register that the underlying analysis can be examined by the public.[3]

While this practice may have administrative advantages to the agency, it fails to tap at the early stages the considerable expertise and knowledge base that exists in the private sector and the NGO community. Such expertise can be a source of innovative, "outside the box" type of thinking. While the agency might be concerned about increasing the complexity of the regulatory process, there are established, albeit rarely used, means of drawing on the expertise of outside communities.[4]

One model for opening up the process for obtaining outside advice on both the analytic design and the options considered in the rulemaking can be found in the 1978 revisions to the regulations implementing the National Environmental Policy Act. These revisions established an early "*scoping*" process, whereby an agency must act, at the very beginning of the planning process, to identify the important issues that require full analysis and to separate those issues from the less significant matters. Specifically:

> There shall be an early and open process for determining the scope of issues to be addressed and for identifying the significant issues related to a proposed action. This process shall be termed *scoping*....The lead agency shall invite the participation of affected...(government) agencies,...the proponent of the action (and others).... (The lead agency)... shall determine the scope and the significant issues to be analyzed in depth in the environmental impact statement....As part of the scoping process the lead Agency may...hold an early scoping meeting or meetings... (CEQ 1979).

Within EPA there are a limited number of examples of similar procedures.[5] Recently a panel of experts recommended that an analogous approach be adopted in the conduct of risk assessments. Specifically, they called for "...an appropriately diverse participation or representation of the spectrum of interested and affected parties, of decisionmakers, and of specialists in risk analysis, at each step (NRC 1996)."

Such an approach would maximize the likelihood that the design and initiation of the economic analysis occur at the beginning of the regulatory process and that its key methodological and analytic issues be vetted quickly and widely. The earlier the analytic template is laid out, the greater its claim to serve as the relatively impartial basis for subsequent policy discussion and debate. If it appears only after different parties have advanced their own analyses, it is less likely to guide the debate. The real lesson of the lead in gasoline and lead in drinking water stories is not that EPA should try to "pick winners." But rather that a timely, data rich and well executed RIA can indeed set the agenda for policy formulation.

ECONOMICS AND DEMOCRACY

These recommendations do not "solve" all the problems with conducting economic analyses but, within the existing statutory framework, they may help mitigate them. As long as some important benefits remain non-monetizable, tension will always exist between the effort to be inclusive and the effort to be analytically rigorous. However this tension can at least be made *transparent* if a full set of options is considered and plausible estimates of benefits and costs are available at an early stage of the regulatory formulation process. Under these circumstances, if agency officials choose to exclude a potential option from consideration they can be held accountable for doing so.

As noted earlier, economic analyses that demonstrate net benefits are likely to be more influential than those that do not. And it is certainly true that no amount of analysis can make politics go away. But timely and robust analysis may help innoculate decisionmakers from the claims of interested parties. Redesigning the RIA timetable cannot control the political clock. If politically powerful actors want to present last minute alternatives they will still be able to do so. But routines have a strong influence on public decisionmaking. If the norm is to analyze policy alternatives from the inception of a rulemaking process, exceptions will occur, but they will be just that, exceptions.

Political space and time will never jibe neatly with its analytic counterpart. But analysis is itself part of the political process. The more robust and clearly articulated it is, the more the various political actors need to take account of it. For all the political bias in favor of the status quo, the United States enjoys a very free and open political system. It provides so many different avenues for the exchange of ideas and information that credible, dispassionate analysis, served up in a timely fashion, is rarely suppressible. Even if it fails to have much impact on immediate agency decisionmaking, it can and will crop up again in one or another of the wide variety of forums that influence the long-term policy debate: congressional hearings, newspaper editorials, "think tank" reports, academic conferences, interest group position papers, and so forth.

In sum, there are two main virtues of formal economic analyses conducted by a government agency. First, they are relatively *dispassionate.* Bureaucrats are not saints, but they are typically more impartial than stakeholders. Second, the economic analysis they conduct at least has the potential to be *explicit and transparent.* A well-done economic analysis states its assumptions and explains the hows and whys of the various methodological presuppositions. Whether or not one agrees with those judgments and methods, one knows what they are and one is free to provide information and analysis that contradict them. Such transparency is the essence of democratic decisionmaking. It arms citizens to think and representatives to deliberate.

To paraphrase Winston Churchill, economic analysis may be the worst approach to environmental policymaking except for all the others that have been tried. Indeed it is inevitable. All the interested parties to a regulatory proceeding conduct their own regulatory impact analysis implicitly. How else could they decide what side to take in a particular policy dispute? By some means or other they are calculating the costs and the benefits of a proposed action as that action effects *them.* The best defense against biased studies is for government to engage in serious analytic efforts prior to taking action. In some circumstances, for example where the underlying science is very weak, there may be little value to conducting extensive economic analysis. But where the science is credible, careful economic study, initiated early, and conducted in an open environment where a full range of options can be explored, cannot help but improve the outcome.

ENDNOTES

[1]This point is based on remarks made by Victor J. Kimm, former Deputy Assistant Administrator in EPA's Office of Prevention, Pesticides and Toxics Substances at the authors and reviewers meeting held at Resources for the Future, Washington, D.C., June 13, 1996. Kimm noted that because auto manufacturers were replacing asbestos brake shoes with substitutes, apparently at lower cost than forecast by the RIA, EPA staff believed that cost estimates for other products might also be too high.

[2]In ruling against the government in the asbestos case the court found that all aspects of the rule could not be justified on a risk-cost or benefit-cost basis. This suggests that the regulatory decision was not fully consistent with the findings of the economic analysis.

[3]The principal exception is those cases when an Advanced Notice of Proposed Rulemaking (ANPRM) is issued before the Notice of Proprosed Rulemaking (NPRM). However, even when an ANPRM is issued, they are often lack specifics and thus fail to generate constructive comment.

[4]The Federal Advisory Committee Act has specific provisions regarding permissable consultations with outside experts or interest groups.

[5]One example is found in EPA's recent study on the Clean Air Act. (U.S. EPA 1996). As required by statute, an expert panel was convened at the outset of the six-year study to obtain advice on a wide range of issues. In this particular case the panel provided continuous information and advice to the agency. EPA staff involved in the process generally agree that the input of the expert panel led to a markedly improved product. Of course, this was a congressionally mandated study rather than a rulemaking, but the example may, nonetheless, be instructive.

REFERENCES

CEQ (Council on Environmental Quality). 1979. Regulations for Implementing the Procedural Provisions of NEPA. 40 CFR, Part 1501.7. Reprinted in *Environmental Quality: The Tenth Annual Report of the Council on Environmental Quality*. December. Washington, D.C.: Executive Office of the President.

Hahn, Robert W. 1996. Regulatory Reform: What Do the Government's Numbers Tell Us? In Robert W. Hahn (ed.) *Risks, Costs, and Lives Saved: Getting Better Results from Regulation*. Washington, D.C.: AEI Press.

Kerwin, Cornellius M. and Scott R. Furlong. 1992. Time and Rulemaking: An Empirical Test of Theory. *Journal of Public Administration Research and Theory* 2(2): 113–38.

Kimm, Victor J. 1996. Personal communication to the author, September 12.

NRC (National Research Council) 1996. *Understanding Risk: Informing Decisions in a Democratic Society*. (Paul C. Stern and Harvey V. Fineberg, eds.).Washington, D.C.: National Academy Press.

Portney, Paul. 1984. Will E.O. 12291 Improve Environmental Policy Making? In V. Kerry Smith (ed.) *The Benefits and Costs of Regulatory Analysis in Environmental Policy Under Reagan's Executive Order*. Chapel Hill: University of North Carolina Press.

U.S. EPA (Environmental Protection Agency). 1987. *EPA's Use of Benefit-Cost Analysis, 1981–1986*. (EPA-230-05-87-028; August). Washington, D.C.: U.S. EPA.

———. 1995a. Letter from U.S. EPA Administrator Carol Browner to Representative George Brown, January 31.

———. 1995b. Letter from David Gardiner (Assistant Administrator for Policy, Planning and Evaluation, U.S. EPA) to Representative Cardiss Collins, May 17. Additional information derived from unpublished supporting documents.

———. 1996. *The Benefits and Costs of the Clean Air Act, 1970 to 1990*. October (draft). Washington, D.C.: U.S. EPA.

APPENDIX

Peer Reviewers

The following individuals gave generously of their time to provide in-depth reviews of the chapters in this volume.

Frank Arnold
Environmental Law Institute

John Boland
The Johns Hopkins University

Robert Brenner
U.S. EPA

Maureen L. Cropper
World Bank

J. Clarence (Terry) Davies
Resources for the Future

Jim Democker
U.S. EPA

Roger Dower
World Resources Institute

Elizabeth J. Farber
Resources for the Future

Anne Forest
Environmental Law Institute

Arthur Fraas
U.S. OMB

A. Myrick Freeman
Bowdoin College

Teresa Gorman
LPI, Inc.

Winston Harrington
Resources for the Future

David Harrison
National Economic Research Associates

John Horowitz
University of Maryland

Michael Huguenin
Industrial Economics, Inc.

Victor J. Kimm
University of Southern California

Raymond J. Kopp
Resources for the Future

Alan J. Krupnick
Resources for the Future

Lester Lave
Carnegie Mellon University

Robert Lee
U.S. EPA

Lyn Luben
U.S. EPA

Randy Lutter
U.S. OMB

John McCarthy
American Crop Protection Association

Robert Mendelsohn
Yale University

Greg Michaels
Abt Associates

David Montgomery
Charles River Associates, Inc.

William O'Neil
U.S. EPA

George Parsons
University of Delaware

Neil Patel
U.S. EPA

Fred Pontius
American Water Works Association

Paul R. Portney
Resources for the Future

Katherine N. Probst
Resources for the Future

Lisa Robinson
Industrial Economics, Inc.

Kurt Schwabe
Duke University

Michael Shapiro
U.S. EPA

Ellen Silbergeld
University of Maryland

V. Kerry Smith
Duke University

Brett Snyder
U.S. EPA

Robert Stavins
Harvard University

Jonathan Wiener
Duke University

Tracey Woodruff
U.S. EPA